MEDICAL RADIOLOGY

Diagnostic Imaging

Editors:
A. L. Baert, Leuven
M. F. Reiser, München
H. Hricak, New York
M. Knauth, Göttingen

U. Bick · F. Diekmann (Eds.)

Digital Mammography

With Contributions by

C. Balleyguier · E. Bellon · U. Bick · H. Bosmans · T. Deprez · F. Diekmann · C. Dromain
J. Jacobs · N. Karssemeijer · E. A. Krupinski · J. M. Lewin · G. Marchal · R. M. Nishikawa
P. Skaane · P. R. Snoeren · R. Van Engen · C. Van Ongeval · A. Van Steen · M. J. Yaffe
K. C. Young · F. Zanca

Foreword by

A. L. Baert

With 132 Figures in 224 Separate Illustrations, 32 in Color and 16 Tables

Ulrich Bick, MD
Charité
Campus Mitte
Institut für Radiologie
Charitéplatz 1
10117 Berlin
Germany

Felix Diekmann, MD
Charité
Klinik für Strahlenheilkunde
Augustenburger Platz 1
13353 Berlin
Germany

ISSN 0942-5373

ISBN 978-3-642-17898-6 eISBN 978-3-540-78450-0

DOI 10.1007/978-3-540-78450-0

Springer Heidelberg Dordrecht London New York

Library of Congress Control Number: 2009933608

Cover design: Publishing Services Teichmann, 69256 Mauer, Germany

Printed on acid-free paper

9 8 7 6 5 4 3 2 1

Springer is part of Springer Science+Business Media (www.springer.com)

Foreword

Digital Radiography has been firmly established in diagnostic radiology during the last decade. Because of the special requirements of high contrast and spatial resolution needed for roentgen mammography, it took some more time to develop digital mammography as a routine radiological tool.

Recent technological progress in detector and screen design as well as increased experience with computer applications for image processing have now enabled *Digital Mammography* to become a mature modality that opens new perspectives for the diagnosis of breast diseases.

The editors of this timely new volume Prof. Dr. U. Bick and Dr. F. Diekmann, both well-known international leaders in breast imaging, have for many years been very active in the frontiers of theoretical and translational clinical research, needed to bring digital mammography finally into the sphere of daily clinical radiology.

I am very much indebted to the editors as well as to the other internationally recognized experts in the field for their outstanding state of the art contributions to this volume. It is indeed an excellent handbook that covers in depth all aspects of *Digital Mammography* and thus further enriches our book series *Medical Radiology*. The highly informative text as well as the numerous well-chosen superb illustrations will enable certified radiologists as well as radiologists in training to deepen their knowledge in modern breast imaging. Also gynecologists, oncologists, and breast surgeons will find it a very valuable instrument for better therapeutic management of their patients with breast disease.

I congratulate the editors with their fine contribution to the medical literature and I am convinced that their book will meet great success with our readership.

Albert L. Baert
Series Editor

Preface

The first article on X-ray examinations and breast cancer was published by A. Salomon, a surgeon at Charité university hospital in Berlin, nearly a hundred years ago. Although mammography has come a long way since then, there is an ongoing debate about its limitations. A major issue is that mammography alone does not allow reliable detection of breast cancer, especially in women with dense glandular tissue.

Full-field digital mammography – first introduced almost exactly 10 years ago – not only significantly facilitates different aspects of the workflow in mammography such as image viewing, communication, and storage, but also promises to improve image quality by enhancing contrast in dense areas of the breast.

A thorough understanding of the capabilities of this new technology is important for its users and their clinical partners alike. This book discusses the physical and technical concept of digital mammography along with clinical aspects such as differences between film-screen and digital mammography in visibility and appearance of mammographic findings. Digital mammography is a rapidly evolving imaging modality including novel and emerging mammographic techniques such as contrast-enhanced digital mammography and digital breast tomosynthesis.

Many renowned international experts in the field of digital mammography – both physicists and physicians – have contributed to this book. We believe that this mixture will provide a deeper insight into digital mammography and its current role.

We thank all authors and the staff of Springer for their patient and tireless support during the preparation of this book.

Berlin, Germany

Ulrich Bick
Felix Diekmann

Contents

Abbreviations

AAPM	American Association of Physicists In Medicine
ACR	American College of Radiology
ACRIN	American College of Radiology Investigational Network
AEC	Automatic exposure control
AGD	Average glandular dose
ANOVA	Analysis of variance
AUC	Area under the receiver operating characteristic curve
BIRADS	Breast Imaging Reporting and Data System
CAD	Computer-aided diagnosis; refers to the field as a whole and includes both CADe and CADx
CADe	Computer-aided detection; detection of lesions; used in screening mammography
CADx	Computer-aided diagnosis; classifying lesions as benign or malignant; used in diagnostic mammography
CC	Cranio-caudal
CCD	Charge-coupled device
CE-DBT	Contrast-enhanced digital breast tomosynthesis
CEDM	Contrast-enhanced digital mammography
CELBSS	Central and East London Breast Screening Service
CLAHE	Contrast limited adaptive histogram equalization
CNR	Contrast-to-noise ratio
CR	Computed radiography
d	Spot aperture
DBT	Digital breast tomosynthesis
DCIS	Ductal carcinoma in situ
Del	Detector element
DICOM	Digital imaging and communications in medicine
DMIST	Digital mammography imaging screening trial
DQE	Detective quantum efficiency
DR	Direct Radiography Systems
DSPP	Digital screening project preventicon
E	Energy
EC	European Commission
EMR	Electronic medical record
Eqn	Equation
EUREF	European Reference Organisation for Quality Assured Breast Screening
FDA	United States Food and Drug Administration
FFDM	Full-field digital mammography
FN	False negative
FPF	False-positive fraction
FROC	Free-response receiver operating characteristic
GSDF	Grayscale standard display function
η	Quantum detection efficiency
H&D	Hurter-Driffield curve
HIP	Health Insurance Program of New York Project
HIS	Hospital information system
HIW	Histogram-based intensity windowing

IAEA	International Atomic Energy Authority
IDC	Invasive ductal carcinoma
IHE	Integrated Healthcare Enterprise
IHE SUP	IHE mammo specific supplement
IHE TF	IHE technical framework
IHE WB	IHE mammo workbook
INBSP	Irish National Breast Screening Program
IP	Imaging plate
IPEM	Institute of Physics and Engineering in Medicine
JAFROC	Jack-knife free-response receiver operating characteristic
λ	Wavelength
LROC	Localization response operating characteristic
LUT	Lookup tables
MDM	Micro dose mammography
MIP	Maximum intensity projection
ML	Medio-lateral
ML-EM	Maximum likelihood expectation maximization
MLO	Medio-lateral oblique
MoniQA	Certain test pattern for quality control
MPPS	Modality performed procedure step
MQSA	Mammography Quality Standards Act
MTF	Modulation transfer function
Mx	Mammography
n(E)	Quantum interaction efficiency
n_a	X-rays interacting with the detector
NBCSP	Norwegian Breast Cancer Screening Program
n_d	Certain mean number of x rays
NEQ	Noise equivalent quanta
NHS	National Health System
NHSBSP	National Health Service Breast Screening Program
NLF	Non-lesion localization fraction
NPS	Noise power spectra
NPV	Negative predictive value
NS	Required number of signal steps
OD	Optical density
P	Centre-to-centre distance or pitch
PACS	Picture archiving and information systems
pAUC	Partial area under the receiver operating characteristic curve
PERFORMS 2	Personal Performance in Mammography Screening
PMMA	Polymethylmethacrylate
PPV	Positive-predictive value
PPV	Positive predictive value
PSP	Photostimulable Phosphor
QA	Quality Assurance
QC	Quality Control
RDOG	Radiology Diagnostic Oncology Group
RIS	Radiology Information System
ROC	Receiver operating characteristic
ROI	Region of interest
RTOG	Radiation Therapy Oncology Group
SART	Simultaneous algebraic reconstruction technique
SD	Standard deviation
SDNR	Signal difference-to-noise ratio
SFM	Screen-film mammography
SMF	Standard mammogram form
SNR	Signal-to-noise ratio

SWOG	Southwest Oncology Group
TDI	Time-delay integration
TFT	Thin film transistor
TG18	AAPM Topic Group 18
TN	True negative
TPF	True-positive fraction
VOI LUT	Values of interest lookup table
VPN	Virtual private network
w	Inverse amount of signal produced upon interaction for a particular X-ray energy
n_B	Number of x rays that are transmitted along path B
C_{rad}	Radiation contrast
σ^2_{tot}	Overall image noise
n_0	Absorbed radiation dose
n_{sq}	Number of secondary light quanta or electrons

Basic Physics of Digital Mammography

1

Martin J. Yaffe

CONTENTS

KEY POINTS

Digital mammography overcomes several technical limitations associated with screen-film mammography. An essential feature of digital mammography is that both the intensity and the spatial distribution of the X-ray transmission pattern are sampled to form the image. In the spatial domain, the interval between samples (pitch) and the response profile of the detector element (del) largely determine the spatial resolution of the imaging system. The dynamic range of the detector and the number of bits used to digitize the image determine the ability to image all parts of the breast with acceptable contrast and signal-to-noise ratio. Depending on the system design, it is possible to eliminate much of the structural or fixed-pattern noise associated with the detector and the X-ray beam to approach a quantum noise limited situation. In digital mammography systems, it is often possible to design detectors that allow efficient use of the incident X-rays without excessive loss of spatial resolution. This permits a substantial reduction in the radiation dose to the breast when compared with film mammography without sacrifice of image quality. Because of the differences in technology, the optimum exposure conditions may shift toward the use of higher energy spectra than would be used with film, particularly for dense or thick breasts.

1.1 Introduction

There are several key features of digital mammography that distinguish it from screen-film mammography and contribute to its potential advantages. Probably,

Martin J. Yaffe, PhD
Imaging Research Program, Sunnybrook Health Sciences Centre, University of Toronto, 2075 Bayview Avenue, Toronto, ON, Canada M4N 3M5

the most significant property of digital mammography is that it decouples the processes of image acquisition from the subsequent stages of archiving, retrieval, and image display. Unlike the situation in film mammography where these processes are inextricably linked, this facilitates optimization of each of the separate functions and great flexibility in the adjustment of image display characteristics. Because the image data are captured in numerical form, this implies that the image data are sampled both spatially and in the signal level. Sampling must be carried out appropriately because undersampling can cause potential limitations to image quality while oversampling can reduce the efficiency and increase the cost of operating the digital mammography system.

A well-optimized digital mammography system can provide the following benefits:

1. More efficient acquisition of the X-ray data for the mammogram because
   (a) The detector can be made thick enough to absorb a large fraction of the X-rays transmitted by the breast
   (b) Elimination of granularity noise
   (c) Reduction of radiation dose
2. Capture of the image data in numerical form
3. Control of display brightness and contrast that is independent from the amount or characteristics of the X-ray exposure
4. Image processing to adapt the image to match visual performance of the eye and overcome limitations of the display device
5. Ability to remove other structural (fixed-noise) patterns by flat-field correction
6. Quantitative imaging techniques, telemammography, CAD, tomosynthesis, contrast imaging

1.2 Characterizing Imaging Performance

To evaluate imaging systems or to compare the performance of a novel system to a conventional imaging device, it is necessary to have quantitative performance measures. Important imaging parameters to be considered are contrast, spatial resolution, noise characteristics, and dynamic range. While a detailed treatment of these is beyond the scope of this book and is available elsewhere (e.g., Barrett and Myers 2001), the essential descriptors of imaging performance such as radiation contrast, signal-to-noise ratio, MTF, DQE, and noise-equivalent quanta will be briefly introduced in the process of discussing the basic physics of digital mammography.

1.3 Basic Physics of Image Acquisition

It is instructive to analyze the key elements of the physics of X-ray image acquisition in mammography through a simple model of a breast containing a structure of interest (Fig. 1.1). This structure could be a tumor, a microcalcification, or some normal aspect of the breast anatomy.

For a monoenergetic X-ray beam, the mean number of X-rays transmitted along Path A through normal breast tissue and arriving at a hypothetical plane beyond the breast, referred to as the image plane, is:

$$n_A = n_0 e^{-\mu z} \tag{1.1}$$

where n_0 is the mean number of X-rays incident on the breast, z is its thickness, and μ is the X-ray attenuation coefficient of the tissue. Here, the divergence of X-rays from a point source has been ignored and the simplifying assumption has been made that no scattered radiation reaches the image plane. The number of X-rays that are transmitted along Path B passing

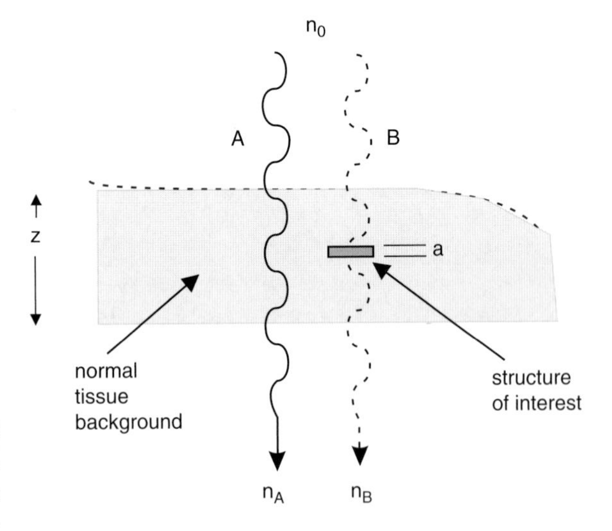

Fig. 1.1. Schematic diagram of the breast illustrating the basic imaging problem of detecting differences in X-ray transmission between Path A passing through normal tissue and Path B passing through a region containing a structure of interest such as a lesion in a breast of varying thickness (from Pisano et al. 2004. With permission)

through the structure of interest in the breast having X-ray linear attenuation coefficient, μ' is:

$$n_B = n_0 e^{-\mu(z-a)-\mu' a} \tag{1.2}$$

where a is the thickness of the structure in the direction of travel of the X-rays. The signal difference produced by the presence of the structure is

$$\mathrm{SD} = n_A - n_B \tag{1.3}$$

The resultant radiation contrast can be defined as:

$$C_{\mathrm{rad}} = \frac{n_A - n_B}{n_A + n_B} \tag{1.4}$$

Substituting (1.1) and (1.2) into (1.4), one obtains:

$$C_{\mathrm{rad}} = \frac{1 - e^{-(\mu-\mu')a}}{1 + \varepsilon^{-(\mu-\mu')\alpha}} \tag{1.5}$$

This expression demonstrates that the radiation contrast is determined by two factors, the difference in attenuation coefficient between the background breast tissue and the structure and the thickness of the structure. Note that in this simplified model, the contrast does not depend on n_0, z, or μ. In practice, where the X-ray spectrum is polyenergetic, and where some scattered radiation is recorded, C_{rad} will show some dependence on these variables.

The breast is composed primarily of fat and fibroglandular tissue. Measured attenuation coefficients of these materials as well as those of breast tumor specimens are plotted in Fig. 1.2. Note that both the attenuation coefficients of these materials and the difference in μ between any pair of materials decreases with increasing X-ray energy. Therefore, radiation contrast will decrease with increasing X-ray energy. The need to achieve an adequate radiation contrast is the reason why relatively low X-ray energies are employed in mammography. In Fig. 1.3, radiation contrast is plotted vs. energy for a modeled breast composed of 30% fibroglandular tissue and 70% fat and containing a 3-mm-thick tumor.

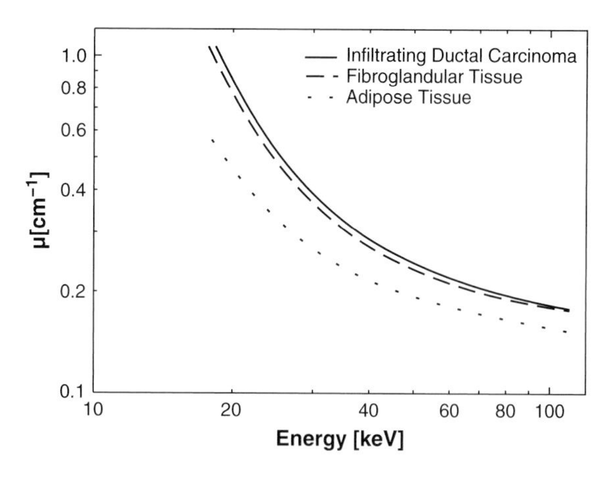

Fig. 1.2. Measured linear X-ray attenuation coefficients of fat, fibroglandular tissue, and tumor in the breast (from Johns and Yaffe (1987). With permission, IOP publications)

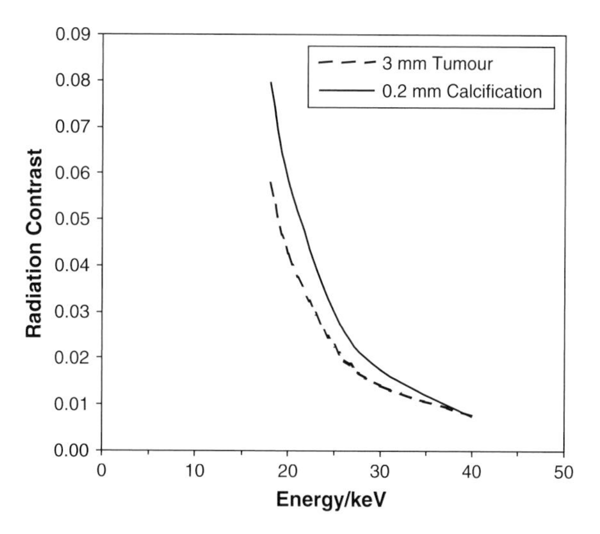

Fig. 1.3. Dependence on contrast of a breast mass and a calcification on X-ray energy. In this example, the breast is composed of 70% fat and 30% fibroglandular tissue. The tumor is modeled as being 3 mm thick, and the calcification is 0.2 mm thick

1.3.1 Detection of X-Rays

To form an image, the transmitted X-rays must be detected and their energy converted into a usable signal. Quantum detection efficiency, η, describes the fraction of the X-rays incident on the detector, that interact with it, producing at least some signal. The quantum detection efficiency is given by:

$$\eta(E) = 1 - e^{-\mu(E)d} \tag{1.6}$$

where μ(E) is the X-ray linear attenuation coefficient of the detector material, which depends on the X-ray energy, E, and d is the thickness of the active region of the detector, i.e., the region from which the signal is produced. The quantum detection efficiency increases with increasing d and μ. The value of μ depends on the density and atomic number of the absorber.

Properties of X-ray detectors for digital mammography will be discussed further in Chap. 2.

With reference to Fig. 1.1, the actual number of X-rays that will be detected for Paths A and B is $n_{dA} = \eta n_A$ and $n_{dB} = \eta n_B$, respectively.

1.3.2 Recording of the Image

There are fundamental differences between digital and film mammography in the way that the information carried by the detected X-rays is transferred and managed to form an image. In a film system, the intensifying screen produces an amount of light that is directly proportional to the amount of energy deposited by the X-rays. This exposes the film, which is subsequently chemically processed to produce a pattern of optical density, which comprises the image. The transfer characteristics (often referred to as the Hurter–Driffield curve) of a mammography screen–film combination are shown in Fig. 1.4a. In this plot of optical density (blackness of the processed film) versus the logarithm of relative X-ray exposure to the screen, it is seen that the response is highly nonlinear and it tends to flatten for exposures above and below a fairly restricted range. This limited range has important implications on image quality. The gradient or slope of this curve defines the amount by which the radiation contrast, C_{rad}, is either amplified or diminished in displaying the image. The value of optical density determines the brightness with which the image will appear on the viewbox. Where the curve becomes flat, the displayed contrast will be poor. The film characteristic imposes a compromise between the displayed contrast and the latitude or range of radiation over which the contrast is acceptable. Once a film image has been exposed, the display contrast characteristics are fixed and, if the image is of inadequate brightness or contrast, it must be re-exposed. For this reason, automatic exposure control (AEC) in mammography is very critical. The AEC attempts to terminate a film exposure at a point when the tissue above the sensor in the AEC has transmitted an appropriate number of X-rays to expose the film to a level where the gradient will be at or near its maximum value and the viewing brightness of the image is acceptable. Of necessity, the image corresponding to other areas of the breast may be sub-optimally exposed.

The characteristic curve of a typical digital mammography detector is shown in Fig. 1.4b. The detector inherently produces a signal that is linearly proportional to the intensity of X-rays transmitted by the breast. It has a very large dynamic range,[1] so that it is possible to produce a faithful representation of X-ray transmission for all parts of the breast. Furthermore, unlike film, the shape of the characteristic curve for these detectors is much less dependent on the level of radiation exposing the detector. Maintaining an adequate value of C_{rad} is still important; however, both the displayed image brightness and contrast as well as other viewing characteristics can be adjusted separately at a computer console during image viewing. Image display will be discussed in detail in Chap. 7.

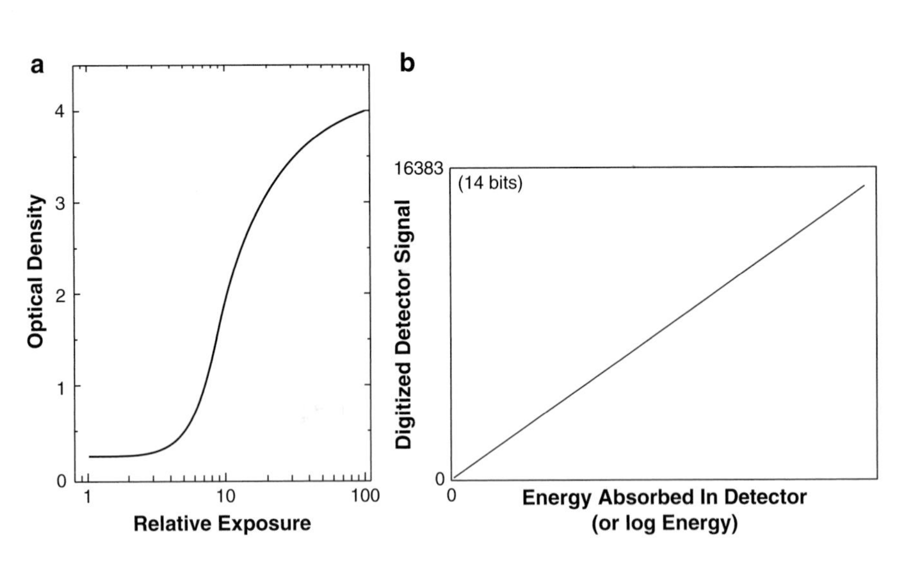

Fig. 1.4. (**a**) Characteristic (H and D) curve of a mammographic screen-film system. Optical density (OD) of the processed film is plotted versus the log of the relative X-ray exposure to the fluorescent intensifying screen. (**b**) Characteristic response of a detector designed for digital mammography (from Pisano et al. 2004. With permission)

[1] Dynamic range is loosely defined as the range of exposures over which the detector provides a reliably measurable signal. Some of the factors affecting dynamic range of the imaging system are discussed in the next section on sampling.

An AEC is important in determining the exposure level to the detector and to the breast, but its key roles are to assist in achieving a predetermined signal-to-noise ratio (see next section) and a reasonable radiation dose to the breast rather than determining the image brightness or contrast.

1.3.3 Sampling

A basic difference between the detectors used for screen-film mammography and for digital mammography is that in the former, the signal varies more or less continuously in both the spatial and intensity domains as in Fig. 1.5a, while in digital systems, the analog signal from the detector is sampled. Sampling occurs both spatially and in quantization of signal levels. Spatially, the digital image is represented by a matrix whose elements (pixels) are considered to be independent from one another. In an image, an analog-to-digital converter (Fig. 1.5b) is generally used to sample the analog signal from the detector such that it will be represented in each image pixel as an integer digital value (Fig. 1.5c). Owing to the sampling process, no meaningful spatial information in the image can be represented at a scale smaller than a pixel and no signal level can be subdivided more finely than the integers comprising the image data. Therefore, the pixel size and the interval between signal digitization levels can have an important influence on image spatial resolution and contrast sensitivity, respectively.

1.3.3.1 Some Spatial Sampling Concepts

As will be discussed in Chap. 2, different detectors use different approaches to achieve spatial sampling. Nevertheless, there are some important concepts that are common to all systems. Figure 1.6 illustrates part of a detector, divided into detector elements or dels. In image acquisition, each del provides one or (in the case of scanning systems) a series of discrete X-ray measurements to contribute to the image.

In the simplest case, the signal from one del will supply the information displayed in one pixel of the final image. The dels are arrayed with a given center-to-center distance or pitch, p. In addition, it is possible that only part of the del is actively sensitive to incoming X-rays, for example a square region of dimension d.

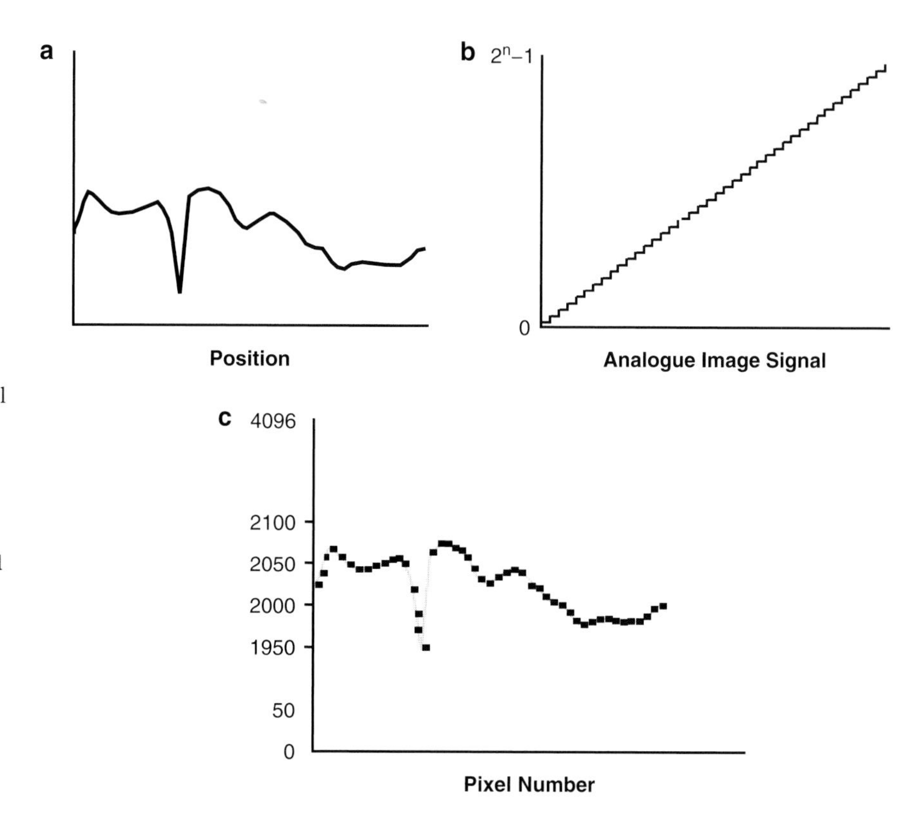

Fig. 1.5. Sampling of an X-ray pattern in the spatial and intensity domains to create a digital image. Here, only one of the two image dimensions is shown. Unlike the analog image (**a**), which is defined continuously in space and signal level, the digital image (**b**) is pixelated at discrete points and only a finite number of signal levels are recorded (from Pisano et al. 2004. With permission)

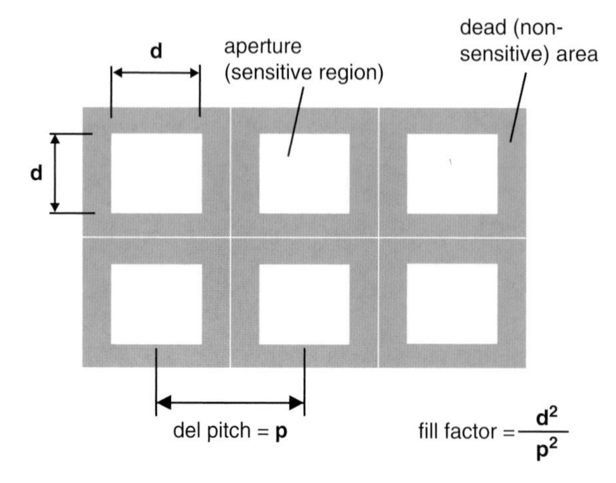

Fig. 1.6. Concept of the del and spatial sampling. A detector element (del) contains an active region with dimension d. Dels are spaced at a pitch p. Because of inactive detector material on the del, the fraction of the area that is sensitive to X-rays, d^2/p^2, also known as the "fill factor," can be less than 1 (reprinted from HAUS and YAFFE et al. (2000). With permission from Elsevier)

Here, d is referred to as the aperture size. If d is smaller than p, the loss of X-ray signal due to this geometric waste will cause the efficiency of the detector and its sensitivity to be reduced to d^2/p^2 of the possible value for d = p. Here, d^2/p^2 is known as the fill factor. In addition to the effect on efficiency and sensitivity, the values of d and p affect other aspects of imaging performance and these will be discussed later.

1.3.3.2 Sampling of Signal Level

Both the detector and the digitization of signal must be designed to cover the appropriate range of X-ray fluence to accommodate the most radiolucent areas (near the periphery of the breast) and the most radio-opaque areas of any breast that is to be imaged. In addition, to provide excellent contrast sensitivity (the ability to distinguish between structures providing only subtle changes in X-ray attenuation), the smallest digitized signal of interest that will occur in a location corresponding to the thickest, densest part of the breast, must be measured to the required precision. Therefore, the required dynamic range of analog signal from the detector must be adequate. Ideally, the characteristics of this response will be constant over that range, i.e., a detector designed to have linear response with fluence will maintain that response over the entire range and likewise with a system that provides logarithmic response.

The digitization process represents the X-ray fluence by a binary number between 0 and 2^n-1, where n is referred to as the *number of bits* of digitization. Including 0, this provides NS = 2^n signal steps. When the signal is subsequently digitized, the number of bits must be adequate to cover both the range of attenuation of X-rays by the breast, R, and also provide the needed precision at the bottom end of the range. For a linear system, the required number of signal steps is then:

$$\mathrm{NS} = 100R/q \tag{1.7}$$

where q is the required precision in percent. For example, if the maximum attenuation imposed by the breast is a factor of 60 and the required precision is 1%, then NS = 6,000. Digitization would require at least 13 bits as 12 bits provides only 4,096 steps. In practice, a 14-bit digitizer has been found to be more than adequate for digital mammography.

1.4 Noise

1.4.1 Quantum Noise

Both the production of X-rays and their interaction in a detector occur in a random manner whose statistics can be described by the Poisson distribution. That means that even for a part of the breast whose X-ray attenuation was absolutely constant, if the average number of X-rays recorded over a particular image area is $\langle n \rangle$ this number will fluctuate from location to location with a standard deviation of $\sigma = \sqrt{\langle n \rangle}$. This fluctuation occurs in the absence of variations in attenuation of the breast and is referred to as *quantum noise*. It is important to realize that the value of σ is determined, not by the number of X-rays incident on the detector, but by how many are used to form the image, i.e., by the number, n_d, interacting with the detector. The *signal-to-noise ratio (SNR)* can be defined as the ratio of $\langle n_d \rangle$ to σ and so is given by:

$$\mathrm{SNR} = \frac{\langle n_d \rangle}{\sqrt{\langle n_d \rangle}} = \sqrt{\langle n_d \rangle} \tag{1.8}$$

Therefore, if it is desired to reduce the apparent noisiness of the image (i.e., increase the SNR) to allow the perception of more subtle features, the radiation level absorbed by the detector, n_d, must be increased. This could be accomplished in two ways: by increasing the exposure factors (i.e., mAs) or by employing a detector with an increased value of η.

1.4.2 Structural Noise

In film-based mammography systems, another important noise source is the random fluctuation contributed by the granularity of the film itself. The film emulsion is composed of grains of silver halide and their random structure increases the total noise. The overall image noise can be estimated by treating noise sources as being statistically independent, giving:

$$\sigma^2_{tot} = \sigma^2_q + \sigma^2_{gr} + \cdots \quad (1.9)$$

where σ_q is the quantum noise and σ_{gr} the film grain noise. The presence of sources of noise other than quantum fluctuation will reduce the SNR and it is therefore desirable to minimize them. Other noise sources associated specifically with detectors will be discussed in Chap. 2.

In film imaging, each sheet of film has a different pattern of structural granularity and, therefore, there is no practical way to remove the effect of this noise from the mammogram. In most digital mammography systems, film granularity is eliminated; however, there is usually some structural variation across the image field, much of which is associated with spatial variations in detector sensitivity. Because these generally remain constant over time, they do not really represent noise in the traditional sense of being truly random in space and time, but are instead referred to as "fixed pattern noise" or "structural noise." The availability of data from the detector in digital form allows the effects of structural noise to be removed by image correction. An approach to such a correction will be described in Chap. 2.

1.4.3 Signal Difference-to-Noise Ratio

For characterizing the quality of the image in terms of the potential detectability of structures in the breast, it is useful to introduce the signal difference-to-noise ratio (SDNR). The SDNR is simply the ratio of the signal difference, SD, from (1.3) (but using the detected rather than incident number of X-rays) to the standard deviation, σ_{tot},

$$\mathrm{SDNR} = \frac{(n_{Ad} - n_{Bd})}{\sigma_{tot}} \quad (1.10)$$

The SDNR is related to the SNR and the radiation contrast by:

$$\mathrm{SDNR} = 2C_{rad}\mathrm{SNR} \quad (1.11)$$

Although both SNR and SDNR can be evaluated for single pixels in an image, it is generally more meaningful to consider these quantities over larger areas corresponding to a structure of interest or lesion and (in the case of SDNR) an equal area of adjacent background.

1.5 Radiation Dose

The absorbed radiation dose in the breast is proportional to, n_0, the number of X-ray quanta incident on the breast. If the required SNR is specified, this defines the number, n_d, of X-rays that must be detected (1.8), giving:

$$n_0 = \frac{n_d e^{\mu(E)z}}{\eta(E)} \quad (1.12)$$

The dose can then be determined by multiplying n_0 by the appropriate conversion factors (ICRU 2005).

1.6 Scattered Radiation

In mammography, some X-rays will pass through the breast without interaction, some will be absorbed, and some will scatter in the breast and escape. At mammographic energies, for an average breast, 40% or more of the X-rays directed toward the detector may have scattered one or more times in the breast (Barnes and Brezovich 1977, 1978). In projection radiography, scattered radiation is not considered to carry any useful information. There are several effects of scattered radiation on radiographic imaging; however, there are differences in the significance of these effects in digital and film mammography. Recording of scattered X-rays uses part of the dynamic range of the detector. In screen-film mammography where the dynamic range is already quite limited, this can be an important negative factor. In digital mammography, the detector generally has a sufficiently large dynamic range that this is not a concern.

Second, there is a fairly uniform haze that is imposed over the entire image. This reduces C_{rad}. Again, in film mammography, this effect can be significant while in digital mammography it is less so because image processing can be used to recover displayed contrast.

Finally, recording the scatter adds statistical fluctuation without information, thereby reducing the SNR. This is probably the major detrimental effect of scatter in digital mammography.

In screen-film mammography and in many digital mammography systems, scatter directed toward the image receptor is partially removed by an antiscatter grid. The grid is not efficient, in that it removes part (25–30%) of the useful directly transmitted "primary" X-ray beam, while rejecting most, but not all (80–90%), of the scattered radiation (Wagner 1991; De Almeida et al. 1999). The loss of both primary and scattered radiation reduces the number of X-rays recorded by the receptor. In film, both these must be replaced by increased exposure from the X-ray tube to ensure that the film is exposed to the proper level on its characteristic curve. The resultant increase in the tube output (and the radiation dose to the breast) is called the *Bucky factor* and this can be on the order of 2–2.5. In digital mammography, it is not necessary to compensate for the loss of scatter removed by the grid, as image brightness and contrast can be adjusted during image viewing. On the other hand, if there is to be no loss of SNR, then the exposure should be increased enough to make up for the loss of primary in the grid and maintain a desired value of n_d. In digital systems, it is also feasible to employ an alternative scanning approach to discriminate against scatter by acquiring the imaging in a geometry where the X-ray beam and detector are scanned across the breast and the beam is restricted to a slit or slot format. This is discussed further in Chap. 2.

1.7 Spatial Resolution

Before discussing the factors that affect spatial resolution in digital mammography, a brief introduction to the modulation transfer function, a valuable tool for quantifying resolution, is provided.

1.7.1 Modulation Transfer Function

In film-based imaging, spatial resolution is often assessed by determining the limiting resolution in terms of line-pairs/mm from a bar pattern. This is a subjective test, however, and is not very useful in the analysis of complex imaging systems.

Spatial resolution can be characterized quantitatively and more usefully through the *modulation transfer function* (MTF). The MTF describes how well the imaging system or one of its components such as the detector transfers the contrast of sinusoidal patterns from the incident X-ray pattern to the output. A sinusoid is a repetitive function, characterized as having a frequency (in this case a *spatial frequency* specified in cycles/mm) and an amplitude. The concept of spatial frequency can be visualized by considering ripples in a pond. Low spatial frequencies (long distance between wave peaks) represent coarse structures and high spatial frequencies (short wavelengths) describe fine detail.

Any pattern can be represented as a combination of sinusoidal shapes, each spatial frequency having a specific amplitude. The MTF simply describes how well each spatial frequency is transferred through a system, i.e., is simply the ratio of the amplitude of the sinusoid at the output of the system or a component to that at the input. The MTF of an imaging system is often 1.0 at very low spatial frequencies and falls with increasing spatial frequency. In a system containing several elements that affect the spatial resolution, the overall MTF is determined as the product of the MTFs of the individual components. For example, the MTF of a radiographic system is the product of that due to the focal spot, the detector, and any motion of the patient during the exposure. This is helpful in determining what part of the system is responsible for limiting its performance.

The MTF of a typical screen-film detector is shown in Fig. 1.7. As seen from the figure, it extends well beyond 20 cycles/mm. It is mainly determined by the screen, as the film has a very high MTF.

Several factors affect the spatial resolution in digital mammography. Some of them are identical to those that apply to film mammography, i.e., the focal spot must be sufficiently small to prevent excessive image unsharpness and relative motion of the patient, the X-ray source and the detector must be minimized during the exposure. The third factor affecting resolution is related to the lateral spread of signal (light photons or electronic charges) in the detector from

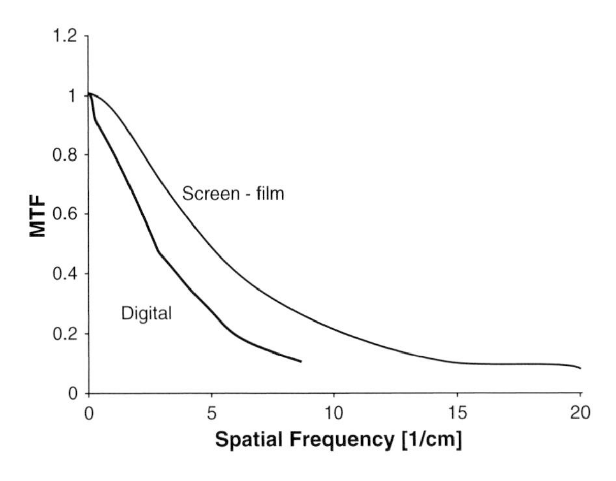

Fig. 1.7. MTF of a modern screen-film system (after BUNCH 1997) and of a typical detector used in digital mammography

Fig. 1.8. Illustrating the concept of aliasing (from PISANO et al. 2004. With permission)

the point where an X-ray is absorbed to the point where the signal is recorded. Generally, as the detector is made thicker to increase η, more blurring of this sort will occur and therefore there is a compromise between spatial resolution and η. As will be discussed in Chap. 2 there are opportunities in the design of detectors for digital mammography to overcome, at least partially, the need to make this compromise.

In digital mammography systems, an additional factor, the spatial sampling, affects resolution. The X-ray signal from each del is averaged over the aperture, d. This causes the MTF of the detector to decrease. For practical purposes, detectors for digital mammography are not designed to provide as high a spatial resolution as obtainable from a film mammography system and as seen in Fig. 1.7 its MTF is typically considerably lower. Other factors such as contrast and noise characteristics provide offsetting improvements in imaging performance. Typically, the size of the del ranges from 50 to 100 μm (0.05–0.1 mm). If d is expressed in mm and the del is a square, then the MTF falls to 0 at a spatial frequency of $1/d$ cycles/mm in both the x and y directions. For example, a del with $d = 100\,\mu m$, has an MTF that falls to 0 at 10 cycles/mm.

The spacing between dels or pitch, p is also an important factor. The larger that p is, the more information that will be lost through the sampling process and this restricts the maximum value of the spatial frequency of information in the image that can be represented accurately. The Nyquist theorem stipulates that for a pitch, p, the highest spatial frequency that can be accurately represented is:

$$f_N = \frac{1}{2p} \tag{1.13}$$

Higher spatial frequencies in the X-ray pattern will be mis-represented, a phenomenon known as *aliasing*. This is illustrated schematically in Fig. 1.8, where imaging of a pattern containing sinusoids with low spatial frequencies, high frequencies, and a mixture of the two are imaged with detectors having dels with fine and coarser pitch. Coarse sampling causes the erroneous creation of sinusoids of lower spatial frequency. So, undersampling both prevents reliable depiction of high spatial frequency information in the image, also because of interference due to these aliased low frequencies impairs the representation of the true low spatial frequency information in the image. According to (1.13), a detector with $d = 100\,\mu m$ and $p = 100\,\mu m$ will be susceptible to aliasing if there are spatial frequencies in the image above 5 cycles/ mm.

1.8 Detective Quantum Efficiency

The SNR or the SDNR are effective quantitative descriptors of the *quality* of the information carried by the radiological image. The larger the signal or signal difference is compared with the random fluctuation, the better the image is. As discussed above, SNR increases with increasing exposure and with higher values of η. It decreases when there are sources of noise other than quantum noise contributing to the image. The highest SNR occurs in the pattern of X-rays transmitted by the breast. This can be considered as the input signal. If the number of these X-ray quanta in a specified area was n_0, its value would be:

$$\mathrm{SNR}_{in} = \langle n_0 \rangle / \sqrt{\langle n_0 \rangle} = \sqrt{\langle n_0 \rangle}.$$

For a system that was perfect (i.e., no additional noise sources) except for incomplete absorption of all incident X-rays by the detector, the signal would be $\eta\langle n_0\rangle$ and the noise $\sqrt{\eta\langle n_0\rangle}$, giving a reduced SNR of $\sqrt{\eta\langle n_0\rangle}$.

We can characterize the performance of the imaging system by determining how efficiently it transfers the input SNR to the system output (i.e., the observer). The *detective quantum efficiency* (DQE) computes the ratio:

$$\mathrm{DQE} = \frac{\mathrm{SNR}^2_{out}}{\mathrm{SNR}^2_{in}} \quad (1.14)$$

For a perfect system, DQE would equal 1.0. Considering just the efficiency of X-ray interaction described above, DQE would be: $\eta\langle n_0\rangle / n_0 = \eta$, the fraction of incident X-rays used by the detector.

If there are other sources of noise, SNR_{out} will decrease below the value predicted by the number of interacting quanta so that DQE will fall below η. From the measurement of SNR_{out} it will appear that fewer X-rays have been used to form the image than has actually been the case and DQE is a measure of that apparent lack of efficiency. In fact, the quantity, SNR^2_{out}, which, in the absence of additional noise sources is just the number of X-rays detected, is known as the number of *noise-equivalent quanta* or NEQ.

It is common to present DQE and NEQ values as a function of spatial frequency (Bunch et al. 1987). DQE(f) specifies, at each level of detail, how well the system transfers the SNR information present at its input.

1.9 Energy Spectra for Digital Mammography

The X-rays emitted by the type of tube used in mammography are emitted over a range of energies, thus forming a spectrum. The shape of the energy distribution is controlled by the material forming the target of the X-ray tube, the type of metal foil used to filter the beam before it is incident on the patient, and the kilovoltage at which the tube is activated to make the exposure. In screen-film mammography, where there is little flexibility in controlling the aspect of the contrast provided by the film, the X-ray spectrum is chosen to provide the greatest practical contrast, C_{rad}. This tends to drive the choice toward relatively low energies, where the difference in attenuation coefficients between tissues is largest as was seen in Fig. 1.2, although the energy must be high enough to ensure that the breast is reasonably well penetrated. For an "average" breast, the examination is typically carried out at, e.g., 26 kV with a molybdenum target X-ray tube and a molybdenum filter placed in the beam as in Fig. 1.9. The use of a low energy causes the overall value of μ for breast tissue to be high, necessitating a relatively high radiation dose (1.10) to achieve an acceptable value of n_d (i.e., adequate SNR).

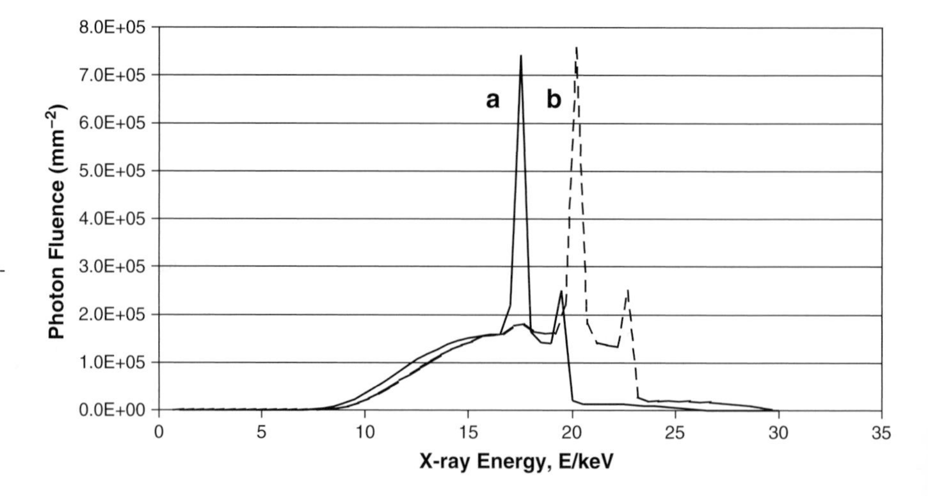

Fig. 1.9. X-ray spectra produced with (**a**) Mo target and Mo filter at 26 kV, (**b**) Rh target with Rh filter. With digital mammography the trend is toward use of more penetrating spectra

In digital mammography, obtaining an adequate value of SNR or SDNR is more important than achieving a specific value of C_{rad}, because the final contrast can be adjusted during image display. The SDNR does not change rapidly with X-ray energy and imaging with higher energies allows lower doses to be used and reduces the dynamic range requirements of the detector (Berns et al. 2003; Geertse et al. 2005; Lo et al. 2005; Bernhardt et al. 2006; Williams et al. 2008). Clinical practice in digital mammography is gradually shifting toward selection of higher kilovoltage and choice of rhodium rather than molybdenum filtration in Mo target systems or use of tubes equipped with rhodium or tungsten targets and appropriate filtration and kilovoltage to obtain a more penetrating beam than typically was used for screen-film mammography.

1.10 Clinical Dose Levels in Digital Mammography

The decoupling of image acquisition from display allows the brightness of the digital mammogram to be set at the desired level independently of the amount of X-ray exposure used to produce the image. Unlike the situation with film mammography, this provides great flexibility in choosing the dose for the examination. Nevertheless, there are important considerations in dose selection. Probably, the most important is to maintain an acceptable level of SNR or SDNR in the image. This implies that an adequate number of X-rays, n_d, are detected. In digital mammography, this can occur with a concommitent reduction in dose when compared with film if one or more of the following is true: (1) the DQE of the digital detector is higher than that of the screen-film image receptor, (2) the dose used for the film image is higher than that to achieve the required SNR, but is needed to produce a certain brightness and/or contrast, (3) a more penetrating X-ray beam can be used to produce the digital image so that less radiation is absorbed in the breast or (4) the scatter reduction mechanism for digital mammography is more efficient. In practice, for some digital mammography systems, dose reductions of between 25 and 30% compared with film mammography have been achieved. It is very important, however, that such reductions are obtained due to the above factors and not by degrading the SDNR at the risk of loss of diagnostic accuracy.

References

Barnes GT, Brezovich IA (1977) Contrast: effect of scattered radiation. In: Logan WW (ed) Breast carcinoma: the radiologist's expanded role. Wiley, New York, pp 73–81

Barnes GT, Brezovich IA (1978). The intensity of scattered radiation in mammography. Radiology 126:243–247

Barrett HH, Myers KJ (2001) Foundations of image science. Wiley, New York

Bernhardt P, Mertelmeier T, Hoheisel M (2006). X-ray spectrum optimization of full-field digital mammography: simulation and phantom study. Med Phys 33:4337–4349

Berns EA, Hendrick RE, Cutter GR (2003) Optimization of technique factors for a silicon diode array full-field digital mammography system and comparison to screen-film mammography with matched average glandular dose. Med Phys 30:334–340

Bunch PC (1997) The effects of reduced film granularity on mammographic image quality. In: Van Metter R, Beutel J (eds) Medical imaging 1997: physics of medical imaging. Proc. SPIE 3032, 302–317

Bunch PC, Huff KE, Van Metter R (1987) Analysis of the detective quantum efficiency of a radiographic screen-film combination. J Opt Soc Am A 4:902–909

De Almeida A, Rezentes PS, Barnes GT (1999). Mammography grid performance. Radiology 210:227–232

Geertse D, van Engen RE, Oostveen LJ, Thijssen MAO, Karssemeijer N (2005) Spectrum optimization for a selenium digital mammography system, Proceedings of the IWDM 2004, University of North Carolina, Chapel Hill, pp 116–122

Haus AG, Yaffe MJ (2000) Screen-film and digital mammography. Radiol Clin North Am 38(4):871–898

ICRU (2005) International Commission on Radiological Units. *Patient Dosimetry for X Rays Used in Medical Imaging*, ICRU Report 74, *JICRU* 5, Oxford University Press, Oxford, UK

Johns, PC, Yaffe, MJ (1987). X-ray characterization of normal and neoplastic breast tissues. Phys Med Biol 32:675–695

Lo JY, Samei E, Jesneck JL, Dobbins JT III, Baker JA, Singh S, Saunders RS, Floyd CE (2005) Radiographic technique optimization for an amorphous selenium FFDM system: Phantom measurements and initial patient results, Proceedings of the IWDM 2004, University of North Carolina at Chapel Hill,, pp 31–36

Wagner AJ (1991) Contrast and grid performance in mammography. In: Barnes GT, Frey GD (eds) Screen film mammography: imaging considerations and medical physics responsibilities Medical Physics Publishing, Madison, WI, pp 115–134

Williams MB, Raghunathan P, More MJ, Seibert JA, Kwan A, Lo J, Samei E, Ranger NT, Fajardo L, McGruder A, McGruder S, Maidment A, Yaffe MJ, Bloomquist A, Mawdsley G (2008) Optimization of exposure parameters in full field digital mammography. Med Phys 35(6):2414–2423

Pisano ED, Yaffe MJ, Kuzmiac CM (2004) Digital mammography. Lippincott, Williams & Wilkins, Philadelphia

Detectors for Digital Mammography

2

Martin J. Yaffe

CONTENTS

M. J. Yaffe, PhD
Imaging Research Program, Sunnybrook Health Sciences Centre, University of Toronto, 2075 Bayview Avenue, Toronto, ON, Canada M4N 3M5

KEY POINTS

The X-ray detector is the heart of a digital mammography system. Its improved characteristics of dynamic range and signal-to-noise ratio provide inherent advantages over screen-film technology. Detector technologies used for digital mammography can be distinguished by the acquisition geometry into scanning or full-field detectors, by energy conversion mechanism into phosphor-based and nonphosphor-based detectors and by how the detector signal is converted into an image value into signal-integrating and quantum-counting systems. Reading of the detector signal can be integrated into the detector assembly or the detector can be in the form of a sensitive plate in a portable cassette, which is moved to a separate device for readout. Detector performance characteristics vary among these different technological approaches. An understanding of the physics on which detector operation is based can help explain these differences. Various image processing operations can be carried out to correct for spatial nonuniformities in detector response and to improve the effective spatial resolution of the detector. In addition, use of a digital detector provides opportunities for more sophisticated automatic control of exposure factors for image acquisition.

2.1 Introduction

The detector is one of the defining features of a digital mammography system. The detector produces an electronic signal that represents the spatial pattern of

X-rays transmitted by the breast. The detector is designed to overcome several of the limitations inherent in the screen-film image receptor used in analog mammography, and in so doing, potentially provides improved diagnostic image quality and a reduction of dose to the breast.

The function of the detector can be described by a set of sequential operations that include:

(a) Interaction with the X-rays transmitted by the breast and absorption of the energy carried by the X-rays
(b) Conversion of this energy to a usable signal – generally light or electronic charge
(c) Collection of this signal
(d) Conversion of light to electronic charge (in the case of phosphor-based detectors)
(e) Readout of charge, amplification, and digitization

These operations must be optimized if the detector is to provide high-quality images at appropriate dose levels. Detectors are characterized by their quantum detection efficiency, sensitivity, spatial resolution properties, noise, dynamic range, and linearity of response.

As discussed in Chap. 1, digital images are acquired by sampling the pattern of X-rays transmitted by the breast. In practice, this is often accomplished using a detector that is constructed as an array of discrete detector elements or *dels*, each of which more or less independently measures the X-rays incident on it. The pitch or spacing between dels and the dimensions of the active portion of each del (aperture) in part determine the spatial resolution properties of the imaging system. The concept of spatial resolution and its quantification in terms of the modulation transfer function (MTF) were introduced in Chap. 1.

2.2 Geometric Considerations

A detector that is suitable for mammography must be able to capture the transmitted X-ray pattern from as much of the breast as possible. To satisfy this requirement, it must have spatial dimensions of approximately 24 × 30 cm. Positioning of a small breast can be facilitated if a smaller detector size (format of approximately 18 × 24 cm) is also available. Alternatively, the imaging system can be designed such that when a small breast is imaged on 24 × 30 cm detector, the collimated X-ray beam and the detector area over which the image is recorded can be shifted to cover the appropriate tissue in the axilla, depending on whether the left or right breast is being imaged.

It is important to be able to image the breast as close to the chest wall as possible. Therefore, there should be the minimum possible amount of insensitive material associated with the detector at this edge of the image. Finally, an excessively thick detector assembly can impede positioning of the breast for some views.

Two major types of digital mammography systems have been introduced. One uses a detector that is the full size of the field that is to be imaged. The other employs a detector array that is long in one dimension and narrow in the other. In the first type, "snapshot" imaging is performed by acquiring the X-ray transmission information from all parts of the breast simultaneously. As in film-based imaging, a radiographic grid is generally used in such systems to reduce the loss of quality caused by recording scattered X-rays.

In the second type of imager, the detector is scanned across the breast in synchrony with a long, narrow collimated X-ray beam to acquire the image progressively. Because the X-ray field is smaller, there is less scattered radiation recorded and these systems are operated without a radiographic grid.

2.3 Basic Physics of X-Ray Detectors

The initial stage in the detector, X-ray interaction is common to all detector technologies. This occurs at the level of individual atoms of the detector material. At the energies used for mammography, X-rays incident on the detector interact by one of the three mechanisms, elastic scattering, inelastic (Compton) scattering, or the photoelectric effect. Elastic scattering leaves no energy in the detector and produces no signal. Compton scattering results in part of the energy of the incoming X-ray being absorbed at the initial point of X-ray impact liberating an energetic recoil electron, but the remainder is carried away by the scattered quantum to be deposited elsewhere, resulting in the loss of spatial resolution. In a photoelectric interaction, the incoming X-ray knocks an electron out of an inner (K-shell or L-shell) orbital of the atom (Fig. 2.1) and much of the energy of the X-ray is transferred to this "photoelectron." When the vacancy is refilled by an electron from a more loosely bound shell, the remainder of the energy is

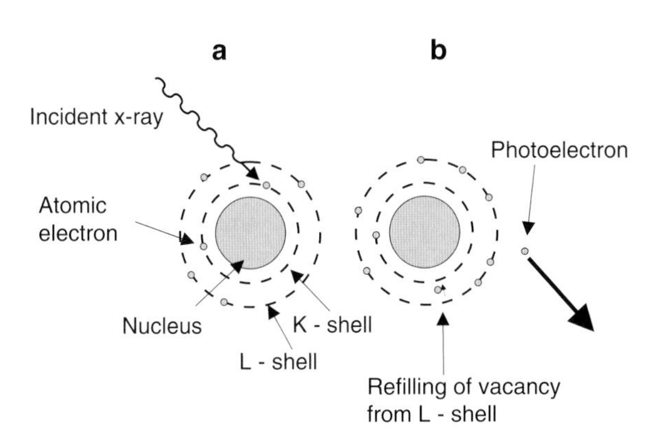

Fig. 2.1. X-ray interaction via photoelectric effect. (**a**) Incoming X-ray liberates inner shell electron. (**b**) Much of the energy of the X-ray is transferred to kinetic energy of this photoelectron. Vacancy is filled by an electron from a more loosely bound shell

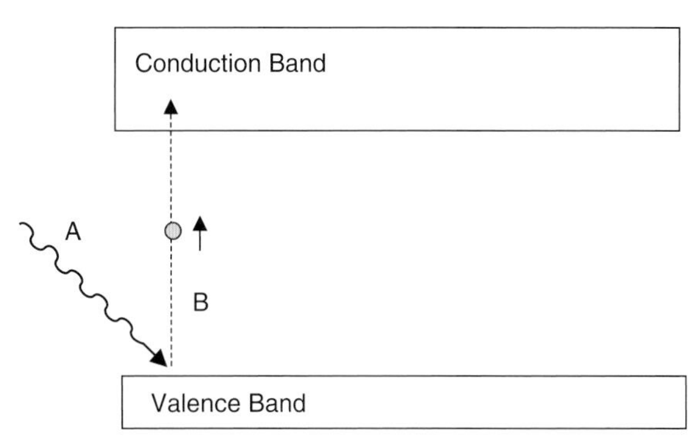

Fig. 2.2. Energy levels in a solid-state material. Interaction with incident X-ray (**A**) causes excitation of electron in molecule (**B**), leaving hole in valence band

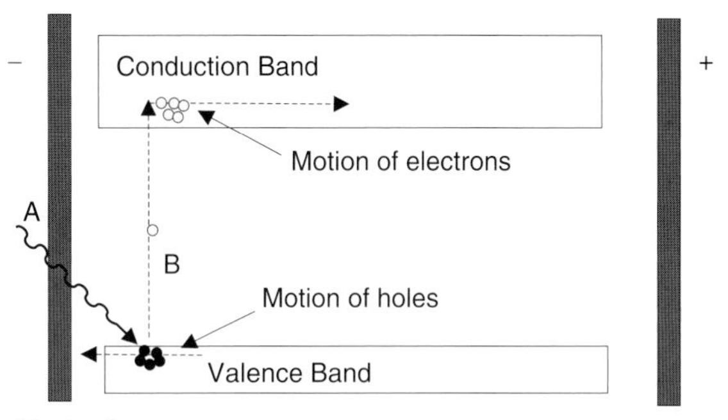

Fig. 2.3. Production of signal in a photoconductor detector. Electrons in conduction band and holes in valence band drift under influence of an externally applied electric field

transferred either to a second (Auger) electron or to a relatively low-energy fluorescent X-ray (which will be absorbed to liberate another electron). Therefore, when X-rays interact and lose energy through either the Compton or photoelectric mechanisms, this appears as kinetic energy of electrons. Because these are charged particles, they interact intensively with nearby atoms and their kinetic energy is lost within a short distance (<<1 mm) from the initial interaction site. This is desirable because it limits the blurring that would result from spreading of the energy and allows high spatial resolution to be achieved. Therefore, it is desirable that most interactions in the detector are of the photoelectric type and this can be achieved by employing a detector material having a relatively high atomic number. For example, for iodine and selenium, at 20 keV, 94% and 96%, respectively, of X-ray interactions will be by the photoelectric effect.

For all detector types, the transfer of energy to the detector material as the photoelectrons and recoil electrons slow down occurs through excitation or ionization of electrons in neighboring atoms within the detector structure. What occurs next depends on the specific molecular structure of the detector. We can distinguish between structures such as noble gases, photoconductors, fluorescent phosphors, and photostimulable phosphors.

In solids such as crystals and some amorphous materials, the orbitals associated with individual atoms become less apparent and the solid behaves as if the electrons (especially those in the outermost shells) exist in continuous bands, each occupying a range of energy levels, separated by unpopulated forbidden regions (bandgaps) as illustrated in Fig. 2.2. The electrons normally reside in the so-called "valence band." In nonmetallic solids, this band is separated by a bandgap of energy E_G from the "conduction band" in which the electrons are essentially free.

2.3.1 Photoconductors

In photoconductors, excitation (B in Fig. 2.3) of each electron in the molecular lattice to the conduction band results in a free electron and a "hole" left behind in the valence band. Because there are competing mechanisms of energy transfer, on average an energy $w \sim 3E_G$ is required to produce each electron–hole pair. Under the influence of an electric field, the electrons will drift in one direction in the conduction band and the holes in the other in the valence band, resulting in a movement of charge that is proportional to the energy deposited in the material by the X-rays (Fig. 2.3).

2.3.2 Phosphors

Phosphors are crystalline materials to which impurities have been added to create energy levels within the forbidden bandgap. Electrons excited into the conduction band (B in Fig. 2.4a) are captured by these impurity centers and rapidly de-excite (typically in nanoseconds), passing through these energy levels on their way down to their ground state (valence band). The energy levels of the impurities are chosen so that in transitioning between these levels, the electron causes energy to be emitted in the form of light (D). The energy of these light quanta is equal to the difference in energy levels, ΔE, for the transition and this determines the color (wavelength, λ) of the light by $\lambda = hc/\Delta E$, where h is Planck's constant and c is the speed of light in vacuum.

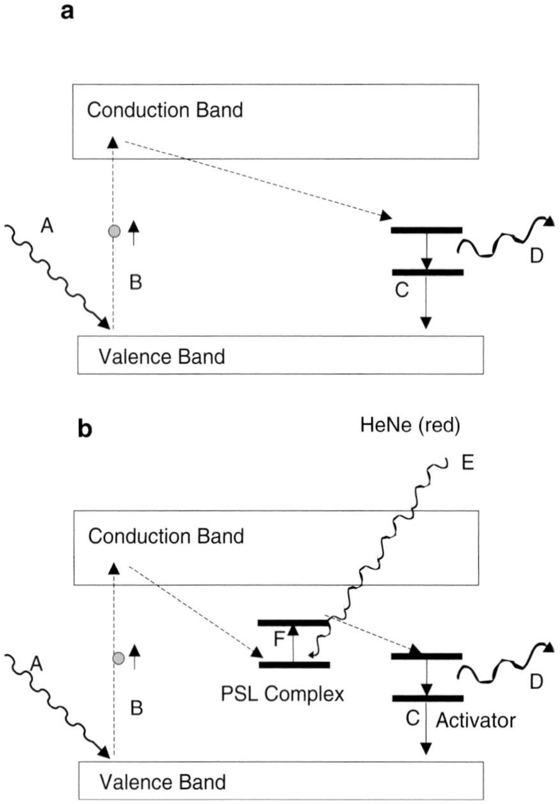

Fig. 2.4. (**a**) Operating principle of a conventional phosphor X-ray detector, (**b**) a photostimulable phosphor

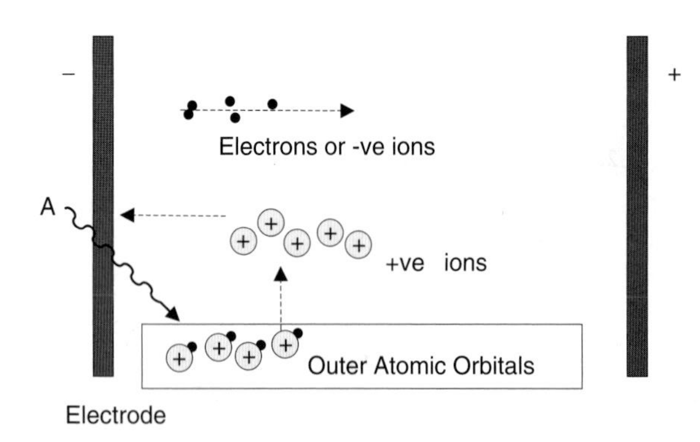

Fig. 2.5. Gas ionization detector

2.3.3 Photostimulable Phosphors

Photostimulable phosphors (Fig. 2.4b) are phosphors that contain a large number of electron trapping sites, also called f-centers or color-centers, located in the forbidden bandgap. Electrons excited to the conduction band have approximately equal probabilities of either producing light or being trapped at one of these sites, interrupting their de-excitation. Electrons can remain trapped for minutes to days until they are detrapped by exposure to a stimulating light (E). The stimulating energy is sufficient to cause them to be re-excited to the conduction band, where again they have a probability of de-excitation to the valence band, producing light (D) as in a conventional phosphor.

2.3.4 Noble Gases

In noble gases, energetic photoelectrons and recoil electrons cause electrons in outer atomic orbitals to be freed completely from the atom (ionized), resulting, for example in free electrons and positively charged gas ions (Fig. 2.5). These charged entities can be collected in an electric field to form a charge signal proportional to the amount of absorbed X-ray energy.

2.4 Aspects of Detector Performance

2.4.1 Quantum Detection Efficiency

As discussed in Chap. 1, quantum detection efficiency, $\eta(E)$, describes the fraction of the X-rays falling on the detector, that interact with it, producing at least some signal. Some calculated values of η for

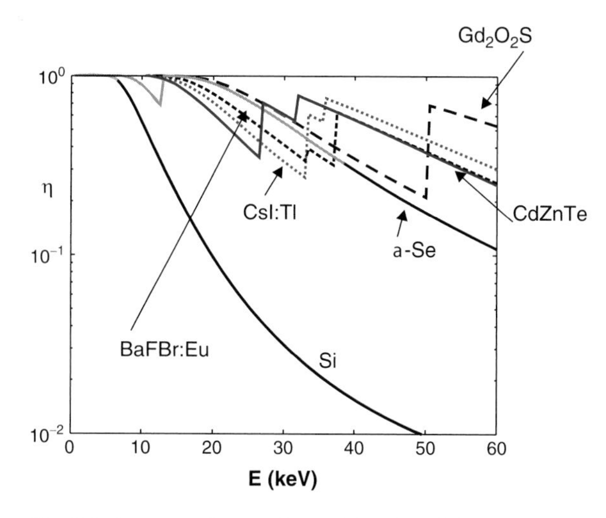

Fig. 2.6. Quantum interaction efficiencies of detector materials used in mammography for 0.1 mm thickness (redrawn from Yaffe and Rowlands 1997. With permission from IOP)

different detector materials are given in Fig. 2.6. In general, μ decreases as energy increases, causing η to do so as well. An exception occurs when the X-ray energy exceeds the absorption energy threshold for an atomic orbital (an "absorption edge") of the detector material. For example, as seen in Fig. 2.6 for the CsI phosphor, η increases dramatically at 33 keV because of the sudden increase in the attenuation coefficient of iodine at its K absorption edge. Some data on detector materials used for digital and film mammography are given in Table 2.1.

2.4.2 Sensitivity

Referring back to (1.2) in Chap. 1, for a certain mean number of X-rays, n_d, incident on the detector, the number, n_a, interacting with the detector will be:

$$n_a = n_d \eta(E) \tag{2.1}$$

The number of secondary light quanta or electrons produced will be:

$$n_{sq} = n_a g \tag{2.2}$$

Here, g is the gain. For the initial energy conversion in the detector (depending on detector design, more than one conversion stage may exist), $g = E_{abs}/w$, where E_{abs} is the energy absorbed in the detector per interacting X-ray and w is the amount of energy required to produce an element of signal (a light quantum or an electron, whichever is being measured). Here, the expression, E_{abs}, is used to reflect the fact that part of the energy, E, carried by the X-ray may not be absorbed due to escape of scatter and X-ray fluorescence.

The sensitivity of the detector, therefore, depends on: (a) $\eta(E)$, (b) g, (c) the efficiency of signal collection and measurement of the charge that is produced. Note that both η and g for the initial conversion stage, depend on E.

The X-ray interaction layer of the detector can be fabricated from materials with relatively high atomic numbers ($Z > 30$). At the low energies typically used in digital mammography (<50 keV), the majority of X-ray interactions in the detector will be through the photoelectric effect. The photoelectron will lose its energy within a very short distance of the point of initial interaction so that there will be little spread of charge. For example, for the cesium in CsI(Tl) phosphor, over 90% of interacting 30 keV X-rays are absorbed through the photoelectric effect. The remainder will interact through scattering, but there is a high probability that their energy will subsequently be absorbed locally. For incident X-rays above the K edge of the detector material, some of

Table 2.1. Characteristics of some common detector materials

Material	Z	E_K (keV)	W (eV)
CdTe	48/52	26.7/31.8	4.4
High purity Si	14	1.8	3.6
Amorphous selenium	34	12.7	50 (at 10 V/μm)
CsI(Tl)	55/53	36.0/33.2	19
Gd_2O_2S	64	50.2	13
BaFBr:Eu (as photostim. phosphor)	56/35	37.4/13.5	50–100
Xenon gas	54	34.6	

the energy will be re-emitted as X-ray fluorescence, although again, it may be reabsorbed, particularly if careful attention is paid to detector design, for example incorporating a choice of materials appropriate for the energy spectrum used for imaging and ensuring that one type of atom absorbs the fluorescence produced by another. In this way, it is possible to create a detector with $E_{abs} \sim E$.

For a particular X-ray energy, the amount of signal produced on interaction will be inversely related to the value of w of the detector material. Values of w (in electron volts) are given for various detector materials in Table 2.1. An X-ray quantum of energy 25 keV (20,000 eV) can potentially produce 25,000/50 = 500 electron–hole pairs in a selenium detector. From Fig. 2.6, a 0.1-mm thick Se detector will have $\eta \sim 90\%$ at 20 keV, so the sensitivity is 450 electron–hole pairs per incident X-ray.

2.4.3 Noise in Detectors

In Chap. 1, the basic concepts of noise in X-ray imaging were introduced and the dependence of one source, referred to as X-ray quantum noise, on the number of X-rays used to form the image, was discussed. In addition, several quantities used to describe imaging performance, namely the signal-to-noise ratio (SNR), the signal-difference-to-noise ratio (SDNR), the detective quantum efficiency (DQE), and the noise equivalent quanta (NEQ) were reviewed.

In an ideal imaging system, the only source of noise in the image should be quantum noise. Quantum noise is fundamental and unavoidable in an X-ray image, but for a given exposure, its effect should be minimized by ensuring that the quantum efficiency of the detector is as close to 100% as possible. In real imaging systems, there are also other noise sources, mainly associated with the detector, and efforts should be made in the system design to ensure that these are much smaller than the quantum noise.

One form of noise is the structural fluctuation in sensitivity over the area of the detector. In screen-film mammography, this effect cannot be removed, only minimized through design of the screens and film and tight quality control standards in manufacturing. In digital mammography, where the same detector is used repeatedly, if these del to del sensitivity differences remain constant, over time they can be considered as *fixed pattern noise*. Because the image is captured in digital form, the effects of fixed pattern noise can largely be removed in most digital detector systems. A *flat-fielding* or gain correction is applied to each acquired image.

All detectors convert the X-ray energy into a secondary signal such as light in a phosphor or electronic charge in a direct conversion type detector. These processes also introduce statistical fluctuation, i.e., noise over and above that caused by the primary quantum noise.

The fluctuation or noise in the number of secondary quanta from which the image signal is derived depends on both the Poisson fluctuation in n_a and in the fluctuation, σ_g in g. For a system with a single conversion stage such as that described by (2.2):

$$\sigma_{sq}^2 = \sigma_{n_a}^2 g^2 + n_a \sigma_g^2, \tag{2.3}$$

where the first term represents the quantum noise and the second the additional noise caused by fluctuation in the gain. The noise is then written as:

$$\sigma_{sq} = \sqrt{n_a (g^2 + \sigma_g^2)}. \tag{2.4}$$

There are various factors that can cause fluctuation in the gain. For example, because a polyenergetic spectrum of X-rays is used for imaging the breast, the absorbed energy resulting from an interacting X-ray will depend on the initial energy carried by that particular quantum. Therefore, as quanta of different energy in the spectrum are absorbed, the amount of light or charge produced will fluctuate from quantum to quantum, giving rise to an apparent fluctuation in g.

In addition, even for an absolutely constant amount of energy deposited by an X-ray quantum, the energy will be absorbed through a random chain of different types of subsequent interactions in the detector material and only some of these will give rise to emission of light. This variability will cause additional statistical fluctuation in the gain, a phenomenon described by Swank (1973). These additional sources of noise reduce the SNR for a given X-ray exposure to the detector, thereby reducing DQE.

For example, from (2.2) and (2.4), the SNR for a simple detector can be written as:

$$\mathrm{SNR} = \frac{n_{sq}}{\sigma_{n_{sq}}} = \frac{\sqrt{n_a}}{\sqrt{1+\left(\frac{\sigma_g}{g}\right)^{2}}}. \tag{2.5}$$

Therefore, SNR can be maintained at an appropriate level by ensuring that n_a is adequate and also that g is large when compared with σ_g. Frequently, this condition can be achieved by employing a fairly high level of gain.

Detectors often have multiple gain stages, and multiple stages of energy conversion, some where the gain is less than *1.0*, i.e., there is a loss of light quanta or electrons. An example of this is where light from a phosphor screen is collected with a lens to be recorded with an optical detector. This occurs in photostimulable phosphor detectors described below. If the number of secondary signal quanta collected and detected per interacting X-ray is not much larger than one, then we say that a "secondary quantum sink" exists. In this case, the statistical fluctuation in the detection of the secondary quanta becomes a significant noise source and will reduce the *SNR* and *DQE*. For this reason, it is important that the gain associated with the detector is large enough to offset any losses due to inefficiency in signal collection.

2.5 Detector Corrections

2.5.1 Uniformity Correction

The sensitivity of the imaging system should be spatially uniform so that any variations in the image signal can be attributed to structures in the breast.

In most types of digital mammography systems, an algorithm to correct for nonuniform sensitivity in the imaging system is applied to all images. The procedure is variously referred to as gain correction or "flat-fielding" and is illustrated in Fig. 2.7. If the detector has a linear response to X-rays, the response of each del can be described by a straight line having a slope, which represents its gain and an intercept representing the "dark signal" (detector output in the absence of radiation). Two such dels are illustrated in Fig. 2.7a. To perform the flat fielding correction, the slopes and

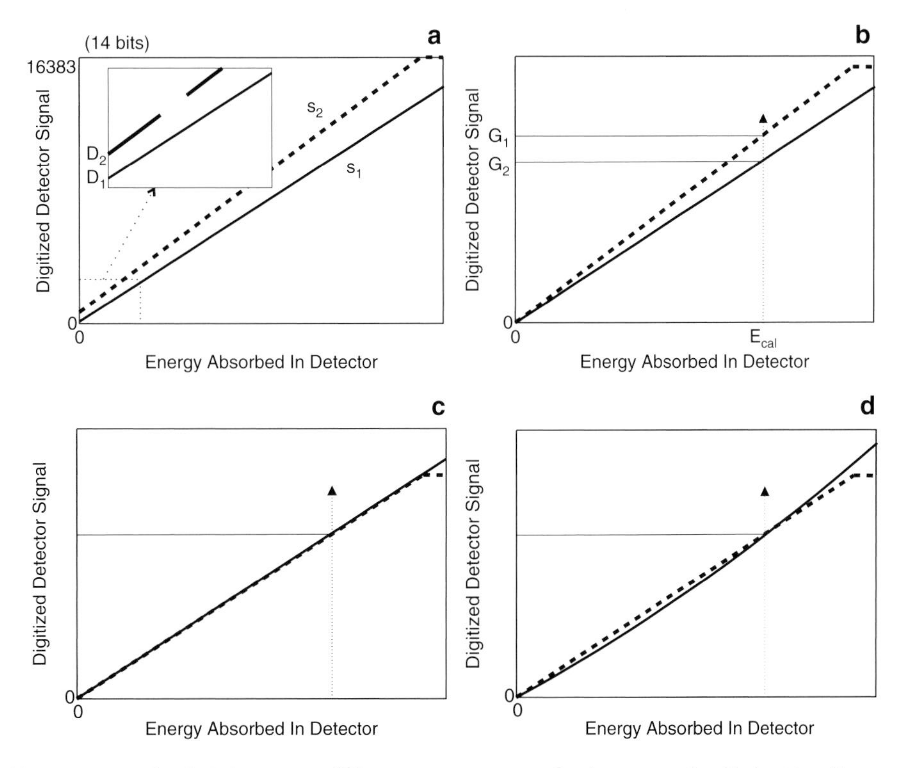

Fig. 2.7. Approach to flat field correction of a digital mammography detector. (**a**) A system with linear response illustrating two dels with different dark signals (intercepts) and gains (slopes). Enlarged view near the bottom end of the range illustrates dark signals, D_1 and D_2. These are measured by acquiring images without radiation. (**b**) Response has been corrected for different dark signals. Gain is still different. Exposure to a fixed amount of radiation E_{cal} allows determination of the slopes G_1/E_{cal} and G_2/E_{cal} for each del. (**c**) Response of the dels after correction. (**d**) Systems with nonlinear response cannot be completely corrected using this simple method (from Digital Mammography, eds. ED Pisano, MJ Yaffe, CM Kuzmiac. Lippincott, Williams and Wilkins, a Walters Kluwer Company, 2004. With permission)

intercepts describing every del in the detector must be measured. This is quite straightforward to do with a digital detector. The first step is that a "dark" image is obtained, by recording the detector response for the time equal to that of an X-ray exposure, but without X-rays. The pixels in this image consist of the intercepts D_1, D_2, etc. from all dels in the detector as shown in the inset to Fig. 2.7a) and these values are stored. In any subsequent image acquisition, these intercept values are subtracted from the measurement arising from each corresponding del, resulting in an image where it appears that the dark signals from all dels are zero. Such an offset correction can be made as frequently as necessary to compensate for temperature-related detector variations. At this point, it is possible to correct for differences from del to del in the slopes or sensitivities.

This is done by exposing the detector to an X-ray beam that has passed through a uniform attenuator. The constant exposure, E_{cal}, received by all dels will produce different signals according to the sensitivity of each del (Fig. 2.7b). This image, which is essentially a map of sensitivities, is stored and used to correct the response of each del in subsequent images, so that it appears that all dels provide uniform response (Fig. 2.7c).

An example of the effect of a flat-fielding correction is shown in Fig. 2.8. The interval between calibrations to measure flat-fielding correction constants depends on the temporal stability of the detector.

Because of the mask used for flat field correction is an X-ray image, it will contain quantum noise and this will be added to the digital mammogram when the correction is applied. If the digital mammogram and the mask image were produced with the same amounts of radiation, the standard deviation of the image pixels would be increased by ~ $\sqrt{2}$, i.e., about 40%. To avoid unnecessary increase in image noise, it is important that the flat-field mask be produced using a much larger amount of radiation than used for each individual mammogram. This can be most easily accomplished by averaging many acquired mask images together to form the working mask. For example, if the equivalent radiation for ten images were used to form the mask, the noise in the mask would be reduced by $\sqrt{10}$ or about threefold and the increase in noise due to flat-fielding would be $\sqrt{1.1}$ times that of the uncorrected image, i.e., virtually unchanged.

The flat fielding procedure described above essentially removes all spatial variation in what is assumed to be a uniform imaging field. Nonuniformities in signal that are not due to the detector, but are caused by such phenomena as heel effect of the X-ray tube, variation in X-ray path length through air (inverse square law), the beam filter, compression plate, and

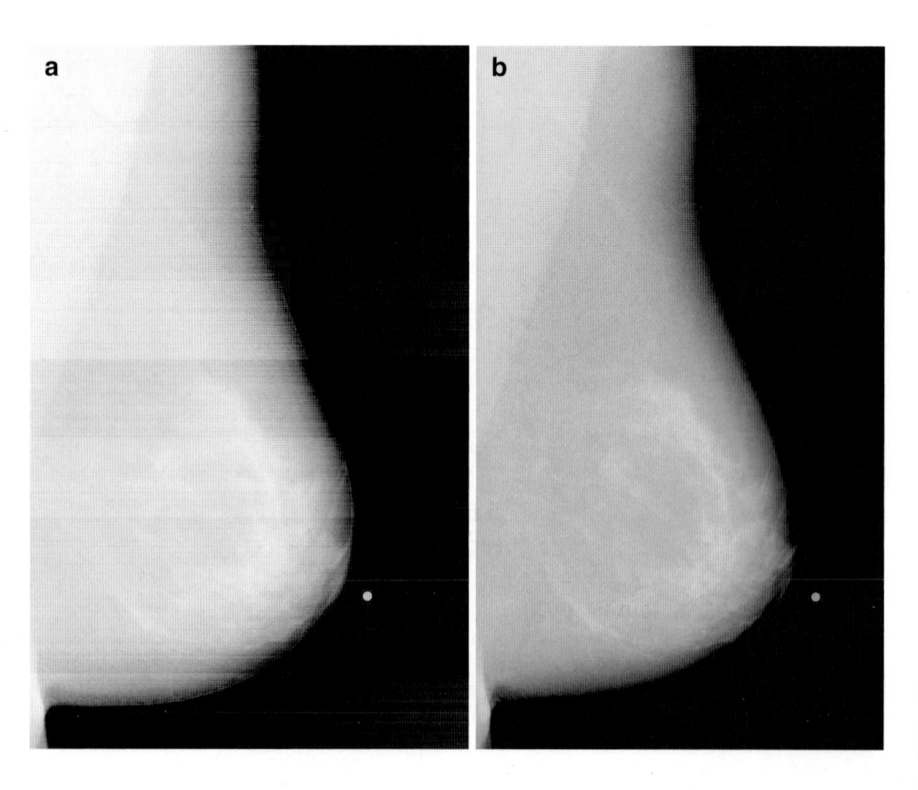

Fig. 2.8. Digital mammogram (**a**) before and (**b**) after flat-field correction (from Digital Mammography, eds. ED Pisano, MJ Yaffe, CM Kuzmiac. Lippincott, Williams & Wilkins, a Walters Kluwer Company, 2004. With permission)

antiscatter grid will also be completely or partially removed. If the flat fielding calibration is performed under one set of conditions, but imaging is done under another (e.g., a change of kV, target, or filter material, use of a compression plate of a different thickness or composition), the flat fielding procedure may generate artifacts.

The flat-fielding correction is generally based on an assumption that the detector responds linearly to radiation exposure. If some or all of the dels have nonlinear response, then a two-constant (slope and intercept) correction will only operate properly under exactly the exposure conditions used to obtain the calibration constants. Under other signal levels, different nonlinearities of the del responses will result in image nonuniformities (Fig. 2.7d).

2.5.2 Resolution Restoration

As discussed previously, there is a loss of spatial resolution in most detectors due to lateral spreading of signal. This causes the MTF to be reduced. But because the image is stored in digital format, it is possible to apply a correction to restore at least part of the drop in MTF and cause the image to be sharper. The method of implementing such corrections is proprietary to each manufacturer; however, a generic explanation can be offered. One approach to restoration is to perform a Fourier transformation of the image to represent it in the spatial frequency domain (what in MRI parlance is referred to as K-space). The MTF is a frequency-domain representation of how the imaging performance drops at each spatial frequency. By dividing the Fourier-domain image by the falling MTF, this drop is compensated. Then, by performing an inverse Fourier transform, the image is brought back to real-world coordinates, corrected for the resolution loss. This would be a perfect restoration except that images contain noise and when such a restoration is performed, the noise is amplified. For this reason, a modified restoration is generally performed. The noise level at each spatial frequency is automatically measured and a weighting function is established such that full restoration is applied at low spatial frequencies where the noise is low compared with the image signal and progressively less restoration is applied at higher spatial frequencies at which noise levels are relatively high.

An alternative, but entirely equivalent restoration can be applied without the need for the two Fourier operations by using a process called deconvolution.

2.6 Linear vs. Logarithmic Response

X-rays are attenuated in an exponential manner as they pass through matter. If we could image the breast with monoenergetic X-rays, the number of X-rays, n, that would arrive at the detector having followed a particular straight-line path through the breast would be:

$$n = n_0 e^{-\sum_{path} \mu(z)\Delta z},$$

where n_0 is the number of X-rays incident on the breast, $\mu(z)$ is the attenuation coefficient for a tissue element of size Δz at location z. A detector having linear response to X-rays produces a signal proportional to the number of X-rays transmitted and, therefore, exponentially related to the actual tissue properties. If instead, the detector signal was proportional to the logarithm of n instead of n itself the signal would be:

$$\log(n_0/n) = \sum_{path} \mu(z)\Delta z.$$

Such a signal would be more directly related to the tissue composition along the path. Logarithmic transformations can be applied to the detector data and this is done in photostimulable phosphor detector systems, with the effect of reducing the range of signal that must be digitized. Once the signal has been transformed in this way, the simple linear flat-field corrections described above can no longer be performed. Alternatively, logarithmic or similar types of transformations can be applied after flat-fielding, during image display, to compress the range of the data.

2.7 Detector Types

Several different types of detectors are used for digital mammography. These are briefly described here.

2.7.1 Phosphor-Flat Panel

Phosphor flat panel detector systems (Fig. 2.9) are based on a large-area glass plate. Using solid-state manufacturing techniques, a rectangular array of light-sensitive photodiodes is deposited onto the plate. These are interconnected with an array of control and data lines as well as a thin film transistor (TFT) switch adjacent to each photodiode. These electronic components are fabricated using amorphous silicon technology.

X-rays are absorbed by a layer of thallium-activated cesium iodide phosphor CsI(Tl) deposited onto the photodiodes. The physics of phosphors was described with reference to Fig. 2.4a. The photodiodes serve as the dels of the detector, detect the light emitted by the phosphor, and create an electrical charge signal that is stored on each del.

Because it can be manufactured to have a needle-like or columnar crystal structure, CsI can provide a better compromise between quantum efficiency and spatial resolution than is possible with the granular phosphors used in screen-film imaging. This is illustrated in Fig. 2.9b. In a conventional phosphor, the light quanta produced on X-ray absorption readily move laterally, leading to increased width of the line-spread function. The CsI crystals act as fiber optics or "light pipes" to reduce lateral spread. This allows the detector to be made thicker without as much resolution loss as would occur in conventional phosphors.

The arrangement of the individual dels with a photodiode and TFT switch is shown in the inset to Fig. 2.9a. Control lines for each row of the array are energized one at a time and activate all the switches in that row. A readout line for each column transfers the signal from the del at the activated row to an amplifier and digitizer. When a given row is activated, the signals from all of the dels on that row are collected along the readout line for all columns simultaneously.

In the system of this type, produced by General Electric Medical Systems (Milwaukee WI) (Fig. 2.10),

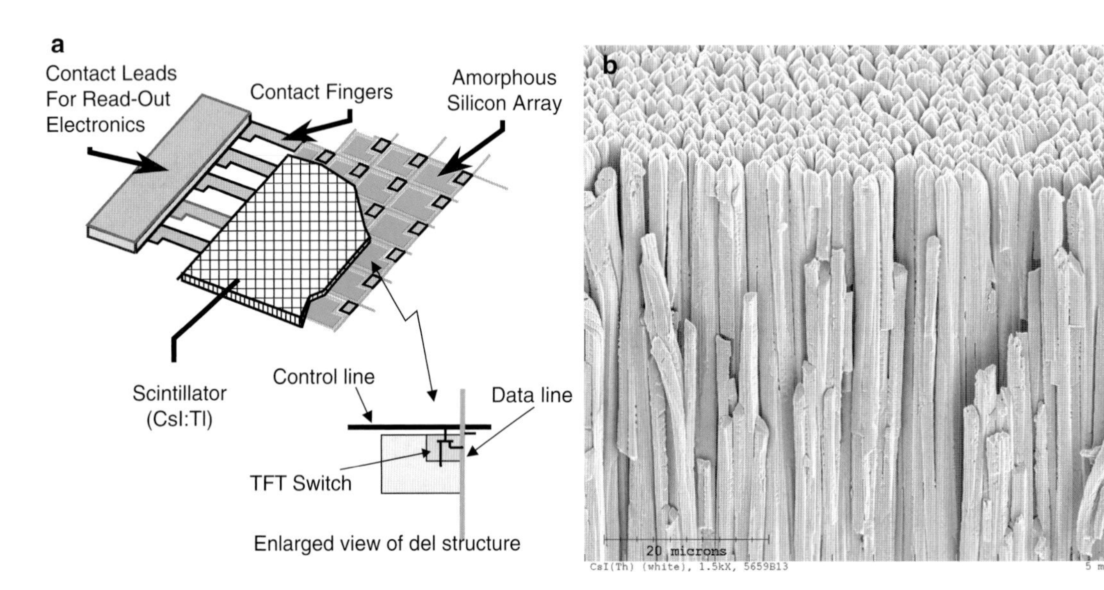

Fig. 2.9. Flat panel detector with CsI(Tl) absorber. (**a**) Detector with photodiode array. TFT readout element is shown in inset. (**b**) Structure of CsI:Tl needle phosphor (Reprinted from Enhanced a-Si/CsI-based flat-panel X-ray detector for mammography, by Jeffrey Shaw, Douglas Albagli, Ching-Yeu Wei, and Paul R. Granfors; Medical Imaging 2004: Physics of Medical Imaging. Proc. SPIE 5368, 370 (2004). With permission from SPIE)

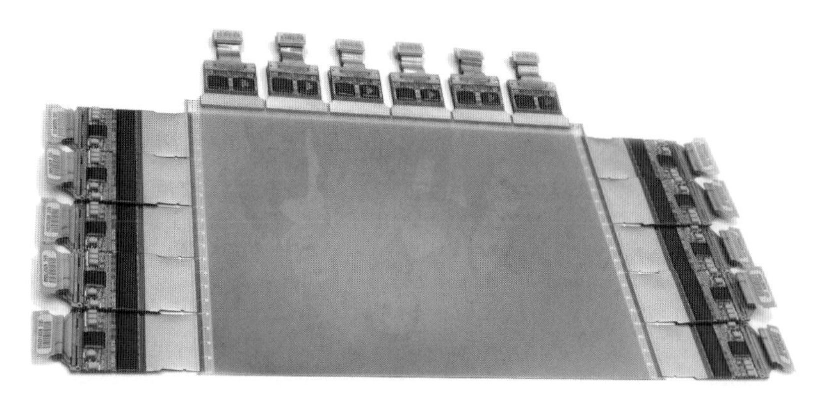

Fig. 2.10. Photo of flat-panel detector. (Courtesy, GE Global Research Center)

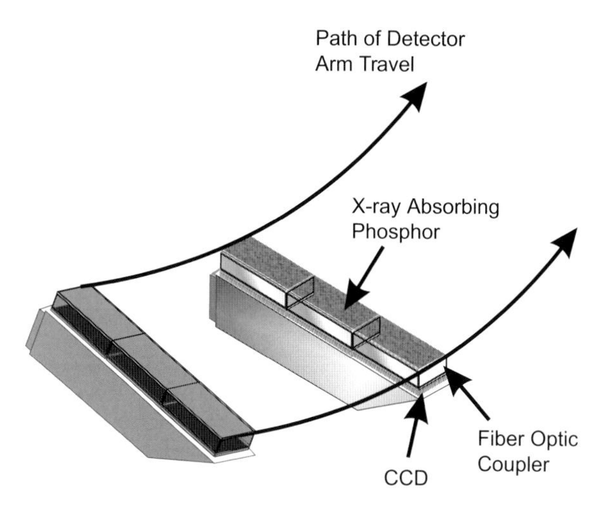

Fig. 2.11. Slot-format scanned CCD detector with CsI:Tl phosphor (Reprinted from Radiologic Clinics of N. America, 38:4 Haus AG and Yaffe MJ, Screen-Film and Digital Mammography, pp 871–898, (2000). With permission from Elsevier)

the del pitch is 100 μm, the field size is 24 × 30 cm and the digitization is carried out at 14 bits (Vedantham 2000). A comparison between the performance of the original detector system and one with an improved scintillating phosphor and reduced noise characteristics (Shaw 2004) was published by Ghetti (2008).

For flat-fielding correction, an offset value and a gain is measured for each del in the detector. Therefore, the number of such constants is equal to twice the number of dels in the detector, about 7.2 million values. It is typical to remeasure offset values between images; however, the gain matrix generally need only be measured occasionally.

2.7.2 Phosphor-CCD System

In this detector, an X-ray absorbing CsI(Tl) phosphor is deposited on a fiber-optic coupling plate, which conducts light from the phosphor to several rectangular *charge-coupled device* (CCD) arrays, arranged end to end. The fibers transmit the optical image from the phosphor to the CCD with minimal loss of spatial resolution. The CCD is an electronic chip containing rows and columns of light-sensitive elements. Light is converted in the CCD to electronic charge. The charge produced on each element in response to light exposure can be transferred down the columns of each CCD and read out by a single amplifier and analog-to-digital converter.

In the commercial implementation of this type of detector, the detector is rectangular with approximate dimensions of 1 × 24 cm. The X-ray beam is collimated into a narrow slot to match this format. To acquire the image, the X-ray beam and detector are scanned in synchrony across the breast (Fig. 2.11). Charge created in the CCD is transferred down the columns from row to row at the same rate, but in the opposite direction to the physical motion of the detector across the breast so that bundles of charge are integrated, collected, and read out corresponding to the X-ray transmission incident on the detector for each X-ray path through the breast. This is referred to as time-delay integration (TDI).

Scanning systems usually require longer total image acquisition time than full-field detectors. The slot collimators only allow use of a small portion of the total emission from the X-ray tube so that the overall heat burden for the tube for a scan is generally considerably higher than for full-field collimation. Because only part of the breast is irradiated at one time in scanning systems, the scatter-to-primary ratio is reduced. Collimation occurs before the breast so that transmitted X-rays are not lost. Normally, an antiscatter grid is not required with scanning systems while grids are used with full-field detectors. This provides a significant dose advantage for the former.

A slot-beam CCD-based scanning digital mammography system was originally marketed by Fischer imaging Inc (Denver CO). It employs dels of 54 μm. Over a limited portion of the detector, data can be read out at 27 μm intervals to provide a high-resolution mode. Digitization is performed at 12 bits.

2.7.3 Photostimulable Phosphor (PSP) System

PSP systems, which are often referred to by their trade name, "computed radiography" or "CR," have been widely used for many years in general radiographic applications. More recently, they were introduced for use in digital mammography. The operation of the detector in these systems is based on the principle of *photostimulable luminescence*, illustrated in Fig. 2.4b. Energy from X-rays is absorbed in a screen composed of a phosphor material containing a high prevalence of electron trapping sites. The absorbed energy causes electrons in the phosphor crystal to be temporarily freed from the crystal matrix and then captured in "traps" within the crystal lattice where they can be stored with reasonable stability for times ranging from seconds to hours. The number of filled traps in a particular location is proportional to the amount of X-ray energy absorbed in that location of the screen.

This analog image is then read by placing the screen in a reading device where it is scanned with a red laser beam. This causes the electrons to be freed from the traps and to return to their original state in the crystal lattice. In doing so, they may pass between energy levels in the crystal structure. These energy levels are defined by small amounts of specific elements deliberately incorporated into the crystal. The choice of these materials thereby determines the color of the light emitted (related to the difference in energy between the levels) as the electron makes its transition. A typical strategy is to design the crystal to emit blue light, so that this can be measured with an appropriate optical filter placed in a light-collecting system incorporating a sensitive photomultiplier tube (Fig. 2.12a), without interference from the red laser light. The amount of blue light measured is proportional to the energy of X-rays absorbed by the phosphor.

The phosphor plate is continuous and is not physically divided into dels. The laser beam is scanned across the plate along one dimension as the plate moves through the reader in the orthogonal direction and the location of the beam on the surface of the plate at each point in time is used to define the *x*–*y* coordinates of the image. The spatial sampling is determined by the size of the laser spot (aperture, d) and the distance between sample measurements (pitch, p).

A photostimulable phosphor system for digital mammography was originally introduced commercially by Fuji Film. Several other PSP systems are now available[1] and these are listed in Table 2.2. The dels are of a nominal size of 50 µm.

As discussed in connection with (2.2), (2.4) and (2.5), the gain, affects both the sensitivity and noise of the imaging system. It is important that the light produced from the phosphor is collected efficiently. If an inadequate amount of light is measured from each interacting X-ray, then the image will contain additional noise above and beyond the quantum noise, causing the SNR and DQE to be reduced.

[1]Note that due to national approval procedures, some digital mammography systems are currently commercially available in certain countries but not in others.

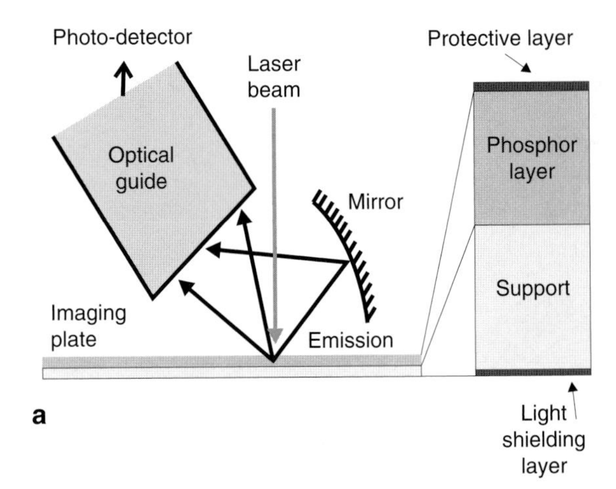

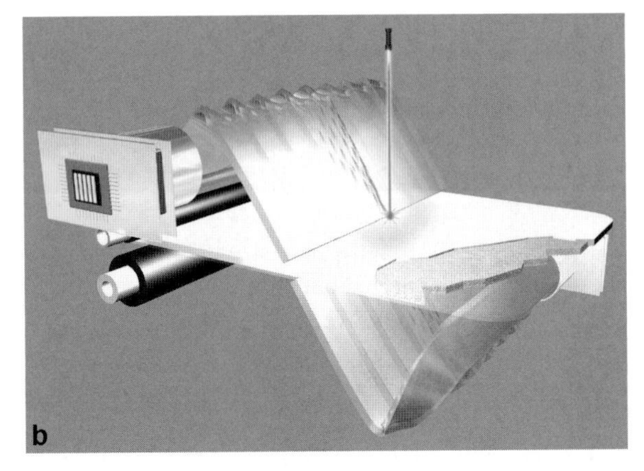

Fig. 2.12. Photostimulable phosphor system (**a**) mechanism of the PSP, (**b**) double-side d readout to increase efficiency (from Digital Mammography, eds. ED Pisano, MJ Yaffe, CM Kuzmiac. Lippincott, Williams and Wilkins, a Walters Kluwer Company, 2004. Used with permission)

To increase sensitivity and improve SNR, some photostimulable phosphor system manufacturers have refined their plate technology to reduce laser scattering and increased the efficiency of light collection by reading from both the top and bottom surfaces of the phosphor plate (Fig. 2.12b).

Unlike the other systems, this system employs removable cassettes, which can be used in the bucky tray of a standard mammography unit. While there are capital cost savings to this approach, it does require that phosphor plates be manually transported to the reader for processing. Because there are multiple detector plates, flat-field correction is normally not performed for the plates, but only for the plate reader. In principle, correction for nonuniformity of the individual plates could be done, but this would require precise registration within the reader and would be time-consuming.

Table 2.2. Current digital mammography systems

Manufacturer	Model	Del size (μm)	Detector dimensions (cm × cm)	Image matrix size	Bit depth	Technology	Grid
Flat panel detectors							
GE	Senographe 2000 D	100	19 × 23	1,914 × 2,294	14	CsI on a-Si	Y
GE	Senographe DS	100	19 × 23	1,914 × 2,294	14	CsI on a-Si	Y
GE	Senographe essential	100	24 × 31	2,394 × 3,062	14	CsI on a-Si	Y
Lorad/Hologic	Selenia	70	24 × 29	3,328 × 4,096	14	α-Se	Y
Siemens	Mammomat novation	70	24 × 29	3,328 × 4,084	14	α-Se	Y
Siemens	Inspiration	85	24 × 30	2,800 × 3,518	13	α-Se	Y
Planmed Oy	Nuance	85	17 × 24	2,016 × 2,816	13	α-Se	Y
			24 × 30	2,816 × 3,584			
IMS	Giotto	85	24 × 30	2,816 × 3,584	13	α-Se	Y
Fujifilm	AMULET	50	18 × 24	3,540 × 4,740	14	α-Se	Y
			24 × 30	4,728 × 5,928		with DOS technology	
Scanning systems							
Sectra	MDM L30	50	24 × 26	4,915 × 5,355	16	Si quantum counter	N
XCounter		50	24 × 30	4,800 × 6,000	16	Pressurized gas	N
Photostimulable phosphor (PSP) systems							
Fuji	Profect	50	18 × 24	3,540 × 4,740	12	BaF(BrI):Eu	Y
			24 × 30	4,728 × 5,928			
Carestream	DirectView CR950	50	18 × 23	3,584 × 4,784	12	BaFBr:Eu	Y
			23 × 29	4,800 × 6,000			
Agfa	CR 85 /35X	50	18 × 24	3,560 × 4,640	12	BaSrFBrI:Eu	Y
			24 × 30	4,760 × 5,840			
Konica	Pureview	43.8	35 × 43	~8,000 × 9,800	12	BaFI:Eu	Y
Konica[a]	Regius 190	43.8	18 × 24	~4,300 × 5,800	12	BaFI:Eu	Y
			24 × 30	~5,800 × 7,200			
Philips[b]	Cosima X Eleva	50	18 × 24	3,540 × 4,740	12	BaF(BrI):Eu	Y
			24 × 30	4,728 × 5,928			

[a]The Konica Regius 190 can use any of three possible plate types. Types RP-6M and RP-7M are based on BaFI:Eu whereas type CP-1M uses a CsBr needle phosphor

[b]The Philips CR unit uses the same plates as the Fuji CR unit

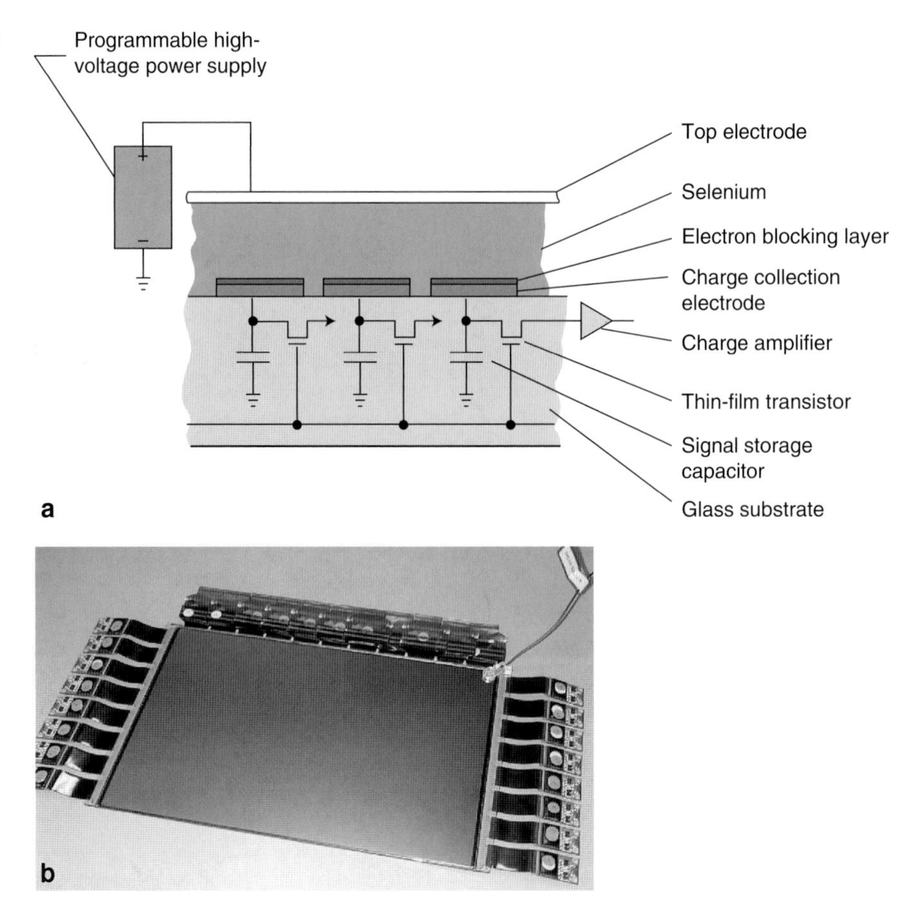

Fig. 2.13. Selenium system (**a**) schematic of detector, (**b**) photo of Anrad detector (from Digital Mammography, eds. ED Pisano, MJ Yaffe, CM Kuzmiac. Lippincott, Williams & Wilkins, a Walters Kluwer Company, 2004. With permission)

2.7.4 Selenium Flat Panel

In this type of detector, the X-ray absorber is a thin layer (100–200 μm) of amorphous selenium. When X-rays interact with the selenium and produce energetic photoelectrons, these lose their kinetic energy through multiple interactions with electrons in the outer orbitals of selenium atoms. The process causes some of these electrons to be liberated and the freed electron and the corresponding "hole" created by its departure, *i.e.*, the electron–hole pair, form the signal. An electric field applied between electrodes deposited on the upper and lower surfaces of the selenium as in Fig. 2.13a sweeps the charges toward the electrodes. One of the electrodes is continuous while the opposing one is formed as a large matrix of dels on a glass plate (Yorker 2002). The dels act as capacitors to store the charge. At the corner of each del is a TFT switch. Readout of charge from the dels is accomplished in the same manner as for the phosphor flat plate detector (Fig. 2.9), with control lines sequentially activating the TFTs for dels along individual rows. The signals from all activated dels are then simultaneously transmitted along readout lines adjacent to the columns of the matrix to be amplified and digitized. A detector of this type is produced by Hologic (Danbury CT). Dels are 70 μm, with 14-bit digitization. A selenium flat-panel detector system is also being produced by Anrad (St Laurent Quebec, Canada) with 85 μm dels (Fig. 2.13b) and this detector is currently used on the Giotto, Planmed, and Siemens systems. Some of its performance characteristics have been described by Bissonnette (2005).

Recently, an amorphous selenium detector that incorporates a different readout design (Fig. 2.14) has been introduced by Fuji. In this detector, there are two separate layers of selenium. The upper layer absorbs X-rays and produces electron–hole pairs similar to the operation of other selenium direct-conversion detectors. This charge is stored on the capacitance of each del. The lower selenium layer acts as an optically controlled switch that transfers the stored charge to a set of readout lines. This allows a del size of 50 μm to be achieved while avoiding the need for TFT switches, which would reduce the detector fill factor (Chap. 1)

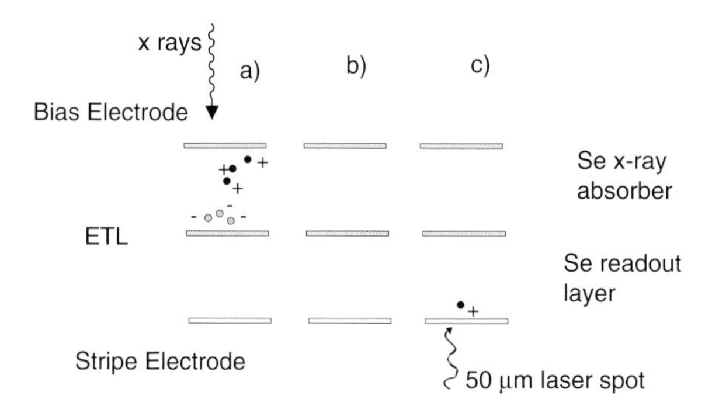

Fig. 2.14. Selenium flat panel detector with optically switched readout (Fuji Medical

in the conventional design and so reduce the geometric efficiency of the detector.

2.7.5 X-Ray Quantum Counting Systems

The detector systems described above operate by absorbing the energy from X-rays interacting with each del in the detector and accumulating the electronic signal produced by all the X-rays received during that measurement. This signal is then digitized to create the information corresponding to a pixel of the image. One aspect of these types of detectors is that higher energy X-ray quanta produce more signal in the detector than those of lower energy and this tends to weight the image signal to higher energy quanta. These carry comparatively lower image contrast than lower energy quanta.

Alternatively, the detector can be designed so that each del produces an electronic pulse every time an X-ray quantum interacts with it. The pulses are then counted to create the signal for that pixel. Pulse counting has several desirable features. Each interacting X-ray registers exactly one count regardless of its energy, so that the secondary noise sources discussed earlier associated with fluctuation in gain are eliminated. In addition, the equal weighting shifts the emphasis in the image signal away from the higher energies. Counting systems do not require the traditional analog-to-digital converter; however, it is important that the counting electronics is properly designed to handle the high rate of incident quanta, which can exceed 10^6 per second.

Currently, two quantum-counting systems have been introduced. Both use a set of multiple linear detectors, which are scanned across the image field beyond the breast during image acquisition in synchrony with an appropriate set of collimator blades located on the X-ray entrance side of the breast. A precise mechanical scanning system is require in order to avoid image artifacts.

The detector in the SECTRA system (Stockholm, Sweden) absorbs the X-rays in crystalline silicon (Fig. 2.15a). The electron–hole pairs produced from each interacting X-ray are collected in an electric field and shaped into a pulse, which is counted (Aslund 2007). The XCounter (Stockholm, Sweden) employs a pressurized gas as the X-ray absorber and pulses of ions created in the gas form the signal as illustrated in Fig. 2.15b (Thunberg 2002).

2.8 Spatial Resolution

Sample MTFs of several commercial digital mammography systems are compared with the MTF of a modern screen-film mammography system in Fig. 2.16. As will be discussed here, factors additional to the del aperture, d, also affect the MTF of detectors used for digital mammography. In some cases, these cause the ordering of the curves not to correlate with the del sizes.

The technical factors controlling spatial resolution differ among the types of detectors in use. In all phosphors, light is emitted from a small region near the point of X-ray interaction in the phosphor and tends to spread isotropically. In CsI(Tl) systems, where phosphor crystals are formed as columns (Fig. 2.9b), the crystals tend to guide the light down their length by total internal reflection and this considerably reduces spreading and allows the detector layer to be made thicker to obtain increased η. Even so, there is more spreading of light and a decrease in spatial resolution as the detector thickness is increased.

In photostimulable phosphor systems, the spread of light from its point of emission does not affect spatial resolution. The spatial localization is determined initially by the size of the laser spot that is used to read out the signal. As the laser light travels through the phosphor, it causes electrons trapped in the phosphor during the X-ray exposure to be freed from the traps. Some of these then produce light. The emitted light is collected by an optical system and a photodetector (Fig. 2.17a).

Sampling of the signal and the possibility of aliasing (discussed in Chap. 1) is controlled by the distance that the scanning laser spot moves relative to

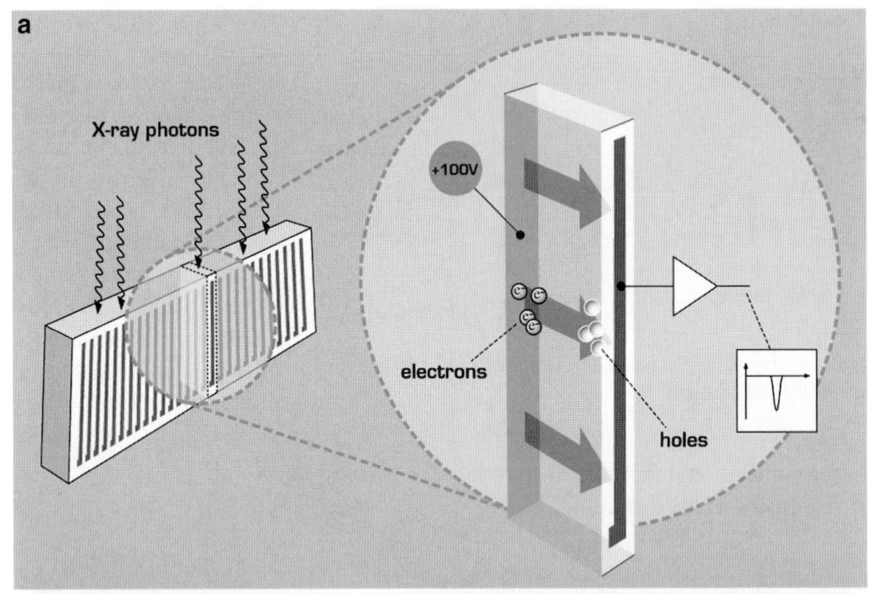

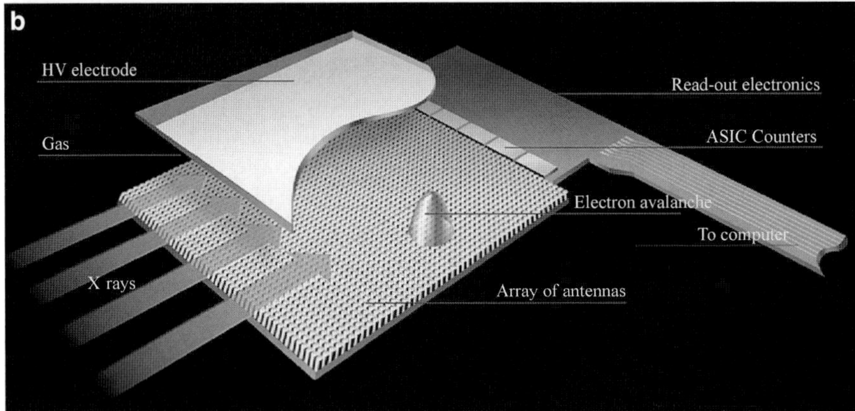

Fig. 2.15. (a) X-ray counting detector based on silicon (photo courtesy SECTRA), (b) high-pressure gas ionization detector (Xcounter. From Digital Mammography, eds. ED Pisano, MJ Yaffe, CM Kuzmiac. Lippincott, Williams and Wilkins, a Walters Kluwer Company, 2004. With permission)

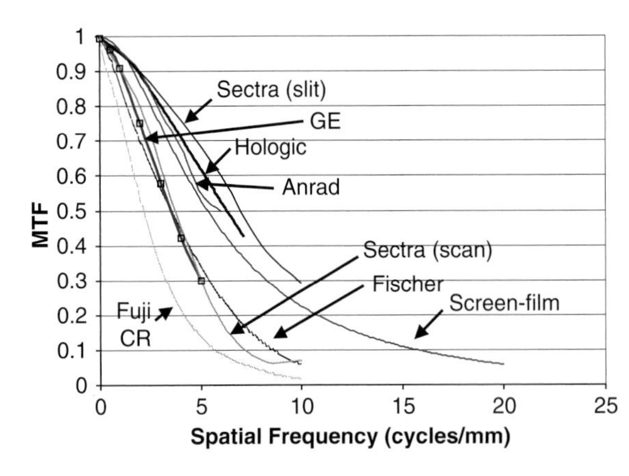

Fig. 2.16. MTFs of commercial digital mammography systems

the phosphor plate between light measurements. The laser light scatters in the phosphor so that traps are emptied, not only along the path on which the laser beam was directed, but also in adjacent regions of the phosphor corresponding to other areas of the image (Fig. 2.17b). This causes smearing of the signal and a loss of spatial resolution. The thicker the phosphor and the more scattering that takes place the greater the loss of resolution. Therefore, with these systems, there is a tradeoff between η and spatial resolution.

Lateral spreading of signal can be reduced in direct conversion detectors when compared with phosphors because the charge signal is quickly swept toward the collection electrode by the electric field before the charge has much opportunity to spread (Zhao 1997). This offers the possibility of excellent spatial resolution with a detector that is thick enough to obtain a high value of η.

A more comprehensive and probably more relevant measure of imaging performance is obtained from the graph of DQE (see Chap. 1) vs. spatial frequency. Recall that DQE describes the efficiency of

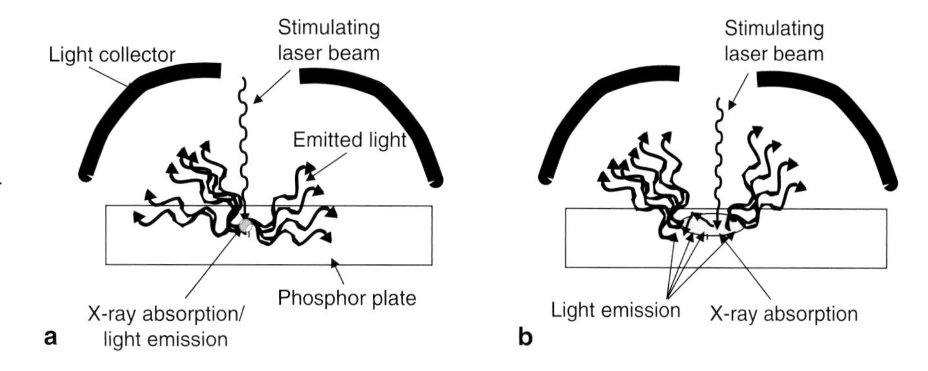

Fig. 2.17. Resolution loss in a photostimulable phosphor. (**a**) In the ideal system all emitted light arises from the point of X-ray interaction, (**b**) scattering of readout laser light within the phosphor material causes blur

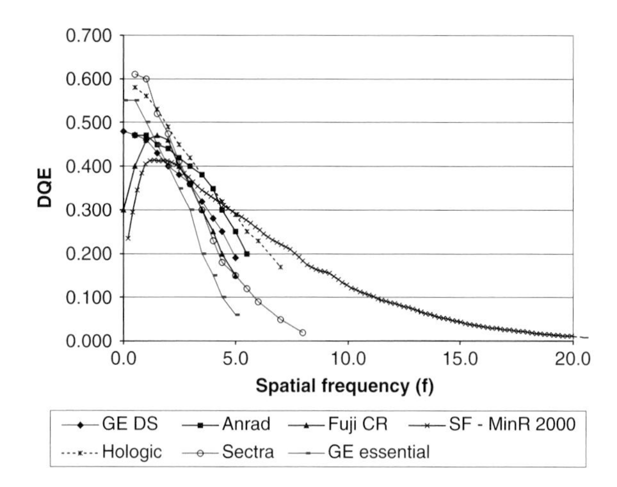

Fig. 2.18. Spatial-frequency dependent DQE of screen-film mammography and some digital mammography systems (Data on GE systems from Ghetti et al. (2008), Fuji and Anrad systems from Rivetti et al. (2006), Selenia from Lazzaria et al. (2007), Sectra from Monnin et al..(2007) and screen-film from Bunch (1999)

the system in transferring the signal-to-noise ratio in the X-ray beam transmitted by the breast to the recorded image. One of the performance advantages of digital over screen-film mammography is the improved DQE. The improvement can be due to better signal transfer or reduced noise and this varies according to system design.

In Fig. 2.18, the DQEs of several digital mammography systems are compared with that of a modern screen-film detector. As can be seen, the DQEs for the digital systems are higher than for the screen-film detector at low to mid spatial frequencies. Because of the intrinsically higher spatial resolution of the screen-film receptor its DQE persists to higher spatial frequencies; however, at very low values. In a complete analog, mammography system the overall DQE would be brought down by the effect of the X-ray focal spot size.

More importantly, as illustrated in Fig. 2.19, the DQE given for the film system occurs only at the optimum

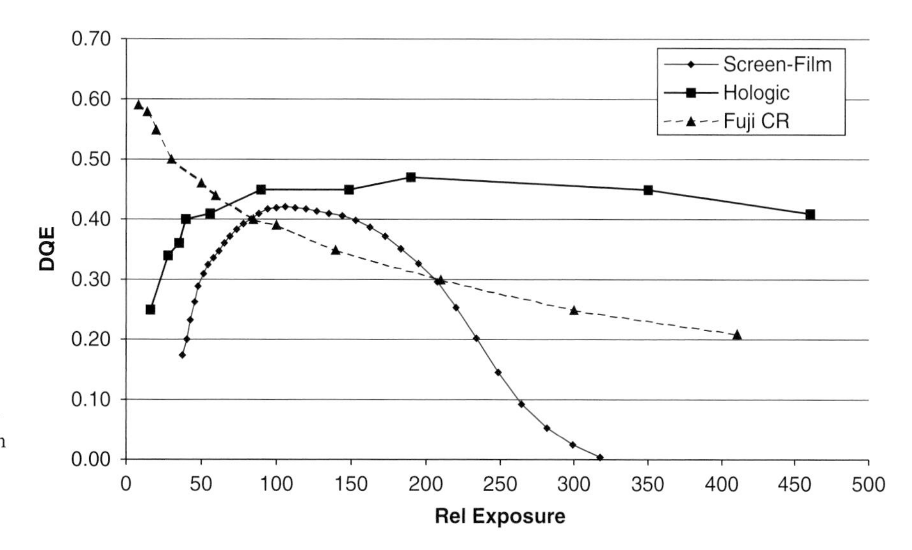

Fig. 2.19. Dependence of DQE on exposure to detector for a screen-film system (Bunch 1999), a flat panel detector and a CR system (Data from Monnin 2007)

exposure for the system. Ideally, the DQE of an imaging system should be independent of exposure to the detector, so that the value of SNR^2 should be directly proportional to exposure. For films systems, DQE is reduced considerably by the effects of film granularity (at low exposures) and nonlinearity of the response curve of the film (at both low and high exposures). For digital mammography detectors, there tends to be less dependence of DQE on exposure although there is generally some effect. For systems, where there is nonuniformity of sensitivity across the image field (e.g., CR systems where flat-field correction is not performed), DQE falls slightly toward the high end of the exposure range. For systems where there is a significant signal-independent level of electronic noise, DQE will fall at low exposures where the effect of this noise on the SNR is most important.

Several authors have performed comparison measurements of various performance indices such as MTF and DQE of commercial digital mammography systems, e.g., (Lazzaria 2007; Rivetti 2006; Monnin 2007).

2.9 Toward Smaller Dels

One feature of digital mammography that initially, at least, caused concern among radiologists, is that because of the del sizes used in the systems, the limiting spatial resolution would not be as high as when a high-quality screen-film receptor were used. For example, imaging a Pb line-pair test pattern with an excellent screen-film system might allow as many as 22 line-pairs/mm to be resolved. The basic design of digital systems suggests that maximum spatial resolution would be on the order of only 5 lp/mm for a 100 μm del and 10 lp/mm for a 50 μm del. Clearly, the expected advantages of digital mammography must come from something other than limiting spatial resolution, i.e., better contrast, image processing, quantitative information, improved archiving and retrieval, etc.

Nevertheless, one can ask, why not design a digital mammography system with smaller dels and obtain improved limiting spatial resolution? There are many different answers to this question. Perhaps, the first comes from asking another question: Is it necessary or desirable to have smaller dels? While the DMIST study illustrated a sensitivity advantage of digital mammography over film, primarily in women with dense breasts, many radiologists feel that in some cases, microcalcifications are not as well visualized with digital mammography. Magnification digital mammography, a technique accomplished by moving the breast closer to the X-ray source, causes the effective size of the del to be reduced with respect to the size of anatomical features and has been found to improve lesion conspicuity. This may also be the case with fine fibrils radiating from tumor masses. This suggests that with all other factors remaining equal, there may be an advantage to reducing the size of the del. Of course, in reality, the usual dilemma associated with any engineering problem then occurs, i.e., when one aspect of imaging is improved, another is generally degraded.

In flat panel systems, several characteristics of the performance of the detector change when the del size is reduced: (1) generally the fill factor of the del is reduced because an increased proportion of the del area must be used to accommodate switches and readout and control lines, (2) the charge storage capacity under the del is reduced, and related to this (3) the readout noise becomes larger when compared with the signal produced. In addition, there are nondetector factors that come into play and these may argue against reduction of the del size. For a given X-ray exposure to the breast, fewer quanta will fall on and be captured by a smaller del, causing the quantum SNR per del to be reduced. The amount of data produced per image increases inversely as the square of the linear dimension of the del (reduction of del dimension by a factor of 2 causes images to contain four times as many pixels). This increased amount of data must be read out at an acceptable rate and processed by the subsequent components of the imaging system. Finally, the production yield of detectors with an acceptable number of properly operating dels falls as the del size is reduced and this increases the cost of the detector. Nevertheless, many of these factors are technology-related rather than being fundamental and are therefore amenable to being overcome. The system described in Fig. 2.14 is an example of one approach to a solution.

2.10 Automatic Exposure Control

Digital image acquisition provides opportunities for major improvement in automatic optimization of image acquisition. For example, it is no longer necessary to have a separate AEC sensor as part of the system because the digital detector can serve as a multi-element sensor. This is currently not possible with the photostimulable phosphor systems because they are used with a conventional film mammgraphy machine and rely on the AEC that is part of the unit.

The optimal way to perform AEC in digital mammography is still a subject under study; however, several approaches have already been put forward. One method involves acquiring a test image of the breast at very low dose and using data gathered from this image to determine the exposure parameters that will be used for the main exposure. In the implementation used by Siemens, for example, the dose is kept very low for this test pulse by binning together data from many dels (e.g., 128 × 128) to create "super-pixels." After the exposure, either a manually preselected region of interest (ROI) from the detector data is used or else an algorithm is employed to segment the entire area of the breast from the image and this is used as the ROI. A goal is defined to specify the requirements for a high-quality image of the breast. For example, this could be that the mean pixel value over the ROI is at a specified level or that no pixel value in the ROI is less than some specified value, etc. From the statistics of the test image, the exposure requirements for the actual image can be inferred. These settings would be automatically selected immediately after the test exposure and the main exposure would then occur.

To use the detector in this way, it is necessary that the detector can be read out quickly to determine what the optimum exposure factors should be. In addition, the detector response must be stable at the low exposures used for the test shot and the detector response at different X-ray fluence levels must be sufficiently predictable to allow scaling of the settings from those used during the test shot to those required for the main exposure. This may impose difficulty for some of the current detectors and creative ways of accomplishing this will have to be found.

One approach to automatic optimization of exposure is to compute one or more appropriate statistics that reflect the attenuation of the breast from a brief, low-dose test exposure obtained immediately prior to the actual image acquisition. One possible statistic might be the minimum signal. This would arise from the most attenuating region of the breast. The algorithm could then set the parameters for the actual exposure so that the image signal in this region would exceed some preset value. Because, in a digital image signal is less important than SDNR, it might be more useful to have the algorithm compute the SDNR from the test exposure and then ensure that some minimum acceptable value of SDNR is exceeded in every part of the image. Additional data on the compression thickness and force can also be used to refine the algorithm. There is likely considerable opportunity to improve imaging performance in mammography through design of "smarter" AEC systems.

References

Åslund M, Cederström B, Lundqvist M, et al (2007) Physical characterization of a scanning photon counting digital mammography system based on Si-strip detectors. Med Phys 34(6):1918–1925

Bissonnette M, Hansroul M, Masson E, Savard S, et al (2005) Digital breast tomosynthesis using an amorphous selenium flat panel detector. In: Flynn MJ (ed) Medical imaging 2005: physics of medical imaging, Proceedings of SPIE Vol. 5745. SPIE, Bellingham, WA, pp 529–540

Bunch PC (1999) Advances in high-speed mammographic image quality. In: Proceedings on Medical Imaging 1999, Physics of Medical Imaging, 3659.SPIE, Bellingham, WA, pp 120–130

Fetterly KA, Schueler BA (2003) Performance evaluation of a dual-side read dedicated mammography computed radiography system. Med Phys 30:1843–1853

Ghetti C, Borrini A, Ortenzia O, et al (2008) Physical characteristics of GE Senographe Essential ` DS digital mammography detectors. Med Phys 35:456–463

Lazzaria B, Belli G, Gori C, et al (2007) Physical characteristics of five clinical systems for digital mammography. Med Phys 34:2730–2743

Monnin P, Gutierrez D, Bulling S, et al (2007) A comparison of the performance of digital mammography systems. Med Phys 34:906–914

Rivetti S, Lanconellia N, Campanini R, et al (2006) Comparison of different commercial FFDM units by means of physical characterization and contrast-detail analysis. Med Phys 33:4198–4209

Shaw J, Albagli D, Wei C-Y, et al (2004) Enhanced a-Si/CsI-based flat panel X-ray detector for mammography. In: Yaffe MJ, Flynn MJ (eds) Medical imaging 2004: physics of medical imaging, Proceedings of SPIE Vol. 5368 SPIE, Bellingham, WA, pp 370–378

Swank RK (1973) Absorption and noise in X-ray phosphors. J Appl Phys 44:4199–4203

Thunberg S, Francke T, Egerström J, et al (2002) Evaluation of a photon counting mammography system. In: Antonuk LE, Yaffe MJ (eds) Medical imaging 2002: physics of medical imaging, Proceedings of SPIE Vol. 4682

Vedantham S, Karellas A, Suryanarayanan S, et al (2000) Full breast digital mammography with an amorphous silicon-based flat panel detector: physical characteristics of a clinical prototype. Med Phys 27(3):558–567

Yaffe MJ, Rowlands JA (1997) X-ray detectors for digital radiography. Phys Med Biol 42:1–39

Zhao W, Rowlands JA (1997) Digital radiology using active matrix readout of amorphous selenium: theoretical analysis of detective quantum efficiency. Med Phys 24(12):1819–1833

Yorker JG, Jeromin LS, Lee DL, et al (2002) Characterization of a full-field digital mammography detector based on direct X-ray conversion in selenium, Proceedings of SPIE 4682, pp 21

Quality Control in Digital Mammography

3

Kenneth C. Young, Ruben Van Engen, Hilde Bosmans, Jurgen Jacobs, and Federica Zanca

CONTENTS

Kenneth C. Young, PhD
National Coordinating Centre for the Physics of Mammography, Royal Surrey County Hospital, Guildford GU2 7XX, UK
Ruben Van Engen
LRCB, Radboud University Nijmegen Medical Centre, Weg door Jonkerbos 90, 6532 SZ Nijmegen, The Netherlands
Hilde Bosmans, PhD
Federica Zanca
Jurgen Jacobs, MSC
Department of Radiology, University Hospitals Leuven, campus Gasthuisberg, Herestraat 49, 3000 Leuven, Belgium

KEY POINTS

An effective quality control system for digital mammography needs to evaluate the status of each stage of image formation – acquisition, processing and display. Such quality control benefits greatly from the ability to make more precise and reproducible measurements than was possible with film-screen systems. On the other hand, the greater variety of system designs and general lack of experience with different digital systems has complicated the introduction of quality control (QC) procedures. Those with extensive experience of QC in digital mammography have stressed the importance of checking regularly for artifacts in images of uniform test blocks for the early detection of any problems arising in the image acquisition stage, e.g. the detector. Although the tests for the subsequent stages of image processing and display are less well developed, they are of considerable importance and will be the focus of further work. Digital technology makes possible the automation of routine QC procedures and a method of doing this is described.

3.1 Introduction

It is widely recognised that mammography must be of high quality in order to allow the earliest detection of breast cancers. To ensure that quality, a number of documents providing quality control (QC) guidance have been developed.

Procedures for quality control of screen-film mammography systems have been in use for many years but vary from one jurisdiction to another. In the United

States, the guidance from the professional organisations American Association of Physicists in Medicine (AAPM 1990) and American College of Radiology (ACR 1999) have been used to develop the MQSA act that gives QC procedures a legal framework. The introduction of a national breast cancer screening programme in the UK in 1988 was the stimulus that brought about guidance from the medical physics profession in the UK. The current guidance for the QC procedures followed by physicists in the UK (Moore et al. 2005) was developed jointly by the Institute of Physics and Engineering in Medicine (IPEM) and the NHS Breast Screening Programme. In several European countries, the development of breast cancer screening also saw the development of national QC protocols. In an effort to standardise Quality Assurance and QC procedures for breast cancer screening, the European Commission has published comprehensive guidance for all the professional groups (EC 2006). Until recently, most QC guidance related to screen-film systems. However, with the advent of digital mammography, there is now guidance for these systems in the latest version (4th edition) of the European document. In the UK, the NHSBSP have provided additional guidance for physicist's procedures for conducting QC of digital systems (Workman et al. 2006). Separate guidance is provided for the routine and more frequent tests that are conducted by the users of the equipment (Burch et al. 2007).

It is important to distinguish between the following different types of QC procedures

(a) "Type testing" of new designs of mammography systems: These are tests on a sample model of a new design and are similar to acceptance testing but may be more rigorous. Examples of this are the technical equipment evaluations published by the NHSBSP on their website (www.cancerscreening.nhs.uk). There is also a protocol for type testing against European Guidelines on the EUREF website (www.euref.org).
(b) Acceptance testing and follow-up tests at 6 or 12 monthly intervals. In highly developed countries, these tests may be conducted by medical physics services.
(c) "Routine" frequent (e.g., daily or weekly) testing by users – typically radiographic staff (Burch et al. 2007).

It is a complication that there is no single international source of guidance on QC procedures that covers the whole world and the International Atomic Energy Authority (IAEA) is attempting to address this by producing documents that provide harmonised guidance.

Digital mammography comprises three distinct stages – acquisition, processing and display (Fig 3.1). For an effective QC system, the user must be able to evaluate the status of the system at each stage. It is usual to conduct measurements on each of these stages separately. Thus, to assess the acquisition stage quantitative measurements are made on the unprocessed images. Measurements on this stage are fairly

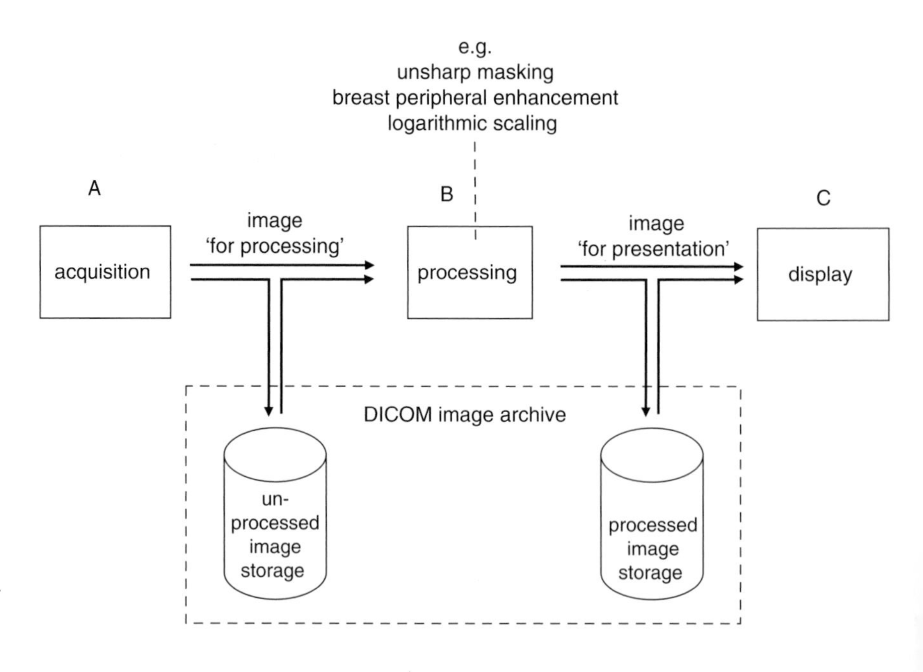

Fig. 3.1. Image formation stages in digital mammography

well developed and examples are provided of the appropriate measurements in the UK and European protocols. Measurements on the unprocessed images are sensitive to variations in the detector design, tube, beam quality, scatter removal by the grid, and the radiation dose. The QC of the image processing stage is much less developed and is discussed later in this chapter. Tests for the display stage are independent of the first two stages and generally rely on electronic test patterns as discussed later.

QC in digital mammography benefits greatly from the ability to make more precise and reproducible measurements on digital images than was possible with film-screen systems. On the other hand, the greater variety of system designs and general lack of experience with different digital systems has complicated the introduction of QC procedures with digital systems. Thus, for example, the fact that some systems have a flat breast support table and some curved is a variation not seen with traditional film screen systems. In general, it is thought to be desirable to have standard QC protocols that are applicable to all types of digital mammography system. One accepted variation in protocols is that special tests are introduced for computerised radiography (CR) systems. Many manufacturers have developed QC procedures designed specifically for their systems and indeed this is a requirement in the United States. However, we believe that it is preferable to work towards standard QC protocols applicable to all manufacturers' systems.

3.2 Image Quality

In mammography, it has been traditional to use test phantoms to make an assessment of image quality. A wide variety of such phantoms have been used and most rely on subjective judgements of the visibility of details simulating masses, calcifications or spicules. However, such a simple and non-quantitative approach is unlikely to be adequate to detect subtle changes in a system's quality or differences between systems. A more quantitative approach is desirable and possible. The European guidelines for the QC of mammography include minimum standards for the image quality of digital mammography systems based on contrast-detail measurements. The method involves the determination of threshold object thickness visibility using circular gold discs with diameters from 2.0 mm down to 0.1 mm using the CDMAM

Fig. 3.2. Photograph of the contrast detail phantom CDMAM

phantom (version 3.4, Artinis, St. Walburg 4, 6671 AS Zetten, The Netherlands) (Fig 3.2). The minimum standards were set to ensure that digital systems are as good as or better than film screen systems (Young et al. 2005). In practice, contrast detail measurements require a large number of observer readings. This procedure suffers from two main disadvantages. One is the presence of significant inter-observer error, which undermines the reliability and confidence in the measurements. A possible solution using automated software to assess the phantom images has been described by Young et al. (2008).

Although measurements of threshold object thickness are difficult to conduct reliably, one may take advantage of the fact that on a given system, they are simply related to the contrast-to-noise ratio (CNR) of a large area object (sometimes called signal-difference-to-noise-ratio SDNR) as shown in (3.1).

$$\text{Threshold contrast} = \frac{\lambda}{\text{CNR}} \tag{3.1}$$

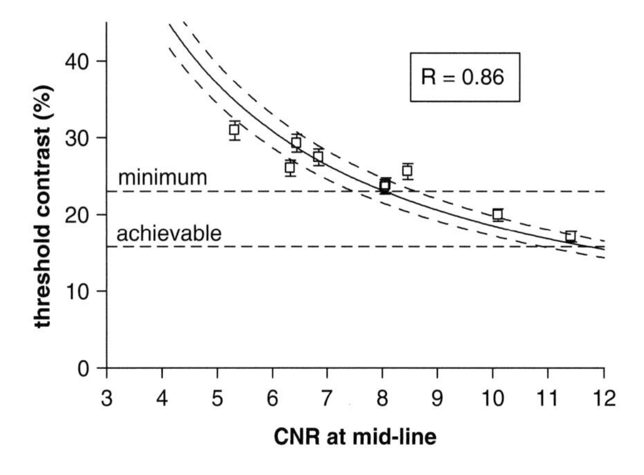

Fig. 3.3. CNR and threshold contrast for 0.1 mm details using three human readers. The fitted curve is in the form of (3.1) along with 95% confidence limits. Also shown are the minimum and achievable standards for threshold contrast in the European protocol (Young et al. 2006)

This is likely to be true as long as the system MTF does not change and has been demonstrated experimentally by Young et al. (2006) (Fig 3.3). The advantage of the CNR measurement is that it is easy to conduct accurately and reproducibly while being highly relevant to image quality in phantom or clinical images. In particular, CNR is sensitive to changes in noise, e.g. due to a change in dose, or changes in object contrast, e.g. due to changes in beam quality. For this reason, the CNR measurement is very useful for assessing the performance of automatic exposure control systems and can be related to the effect on threshold object thickness on a given system. In the European protocol, the object used is a 10 × 10 mm square of Aluminium 0.2 mm thick used with different thicknesses of PMMA and combined with a dose measurement. In other protocols, other objects may be used but should ideally be made of a material with attenuation characteristics that adequately mimic breast tissue. One potential source of confusion is that while a high CNR leads to better image quality on any particular system, it is not a valid measure for comparing systems that have different MTFs. This is because the systems with the best MTFs need a lower CNR for a given level of image quality. The CNR values required to achieve the minimum and achievable levels of image quality as defined in the European protocol are reported in the technical evaluations published by the NHS Breast Screening Programme and summarised in Table 3.1.

3.3 Image Noise

Noise in digital mammography images represents a fundamental limitation in image quality and therefore

Table 3.1. CNR at minimum and achievable image quality levels for different systems tested by the NHS Breast Screening programme

Manufacturer	System	CNR at minimum image quality	CNR at achievable image quality	NHSBSP equipment report
Hologic	Selenia Mo	4.3	6.2	0701
Hologic	Selenia W	4.1	6	0801
Siemens	Novation	4.6	6.7	0710
Siemens	Novation	4.2	6.1	0710
Sectra	MDM-L30	3.1	4.6	0805
GE	Essential	11	16	0803
IMS	Giotto	6.8	9.8	0804
Agfa CR	CR: MM3.0 plates	12	17.5	0707
Kodak	CR: EHR-M2	8.3	12.1	0706
Konica	CR: RP-6M plates	9.9	14.4	0806
Konica	CR: RP-7M plates	7.6	11.1	0806
Konica	CR: CP-1M plates	5.7	8.3	0806

an important part of QC procedures. Image noise can be assumed to consist of three components: electronic, quantum and structured noise. Electronic noise is assumed to be independent of the exposure level and may arise from a number of sources: dark noise, readout noise, amplifier noise. Quantum noise arises due to the variations in X-ray flux and (if present) secondary photon flux and is proportional to the square root of the exposure. Structured noise is caused by spatially fixed variations of the gain of an imaging system and is proportional to the exposure. In practice, it is usual to observe all three types of noise in varying proportions according to the dose absorbed by the detector. This difference in behaviour with dose makes it possible to separate these noise components.

If high electronic or structured noise is present on images of a particular system, image quality may be lower than expected. Therefore, the magnitude of additional noise sources is an important parameter to check in QC testing. However, at present a thorough analysis of noise may only be part of type testing or acceptance procedures rather than routine QC. A methodology for such an analysis is provided in the NHSBSP technical evaluations and the EUREF type testing protocol and is summarised as follows.

Standard deviation (SD) in a standard region-of-interest (ROI) is taken as a measure of noise. This standard ROI is chosen to be 0.6 × 0.6 cm and the ROI is located 6 cm from chest wall side and laterally centred. The total SD in an ROI can be written as squared sum of noise components:

$$SD^2 = SD_e^2 + SD_q^2 + SD_s^2 \tag{3.2}$$

SD is the standard deviation in the standard ROI
SD_e is the standard deviation in the standard ROI due to electronic noise
SD_q is the standard deviation in the standard ROI due to quantum noise
SD_s is the standard deviation in the standard ROI due to structured noise

To facilitate comparison, this equation can be rewritten in terms of relative noise (SD/P):

$$\frac{SD^2}{P^2} = \left(\frac{k_e}{E}\right)^2 + \frac{k_q^2}{E} + k_s^2 \tag{3.3}$$

SD is the standard deviation in a standard ROI
k_e is the electronic noise coefficient
k_q is the quantum noise coefficient
k_s is the structured noise coefficient
P is the average pixel value in a standard ROI
E is the exposure level at the detector surface in μGy

The set-up of the measurement is identical to that of the NPS measurement in the IEC 62220–1–2 standard (IEC 2007). A 2-mm thick aluminium attenuator is positioned as closely to the X-ray tube as possible to minimise the influence of scatter and all removable parts are taken out of the X-ray beam, including the anti-scatter grid and compression paddle. By making images at different detector exposure levels, and fitting (3.3), the presence of additional noise sources can be evaluated as shown in Fig. 3.4.

To check whether the system is operating at a dose level at which quantum noise is the largest component, the response function can be used to translate typical pixel values (in the unprocessed image) to a detector dose level. It is also possible to evaluate the noise components in the frequency domain using the Noise Power Spectra. However, these results are more difficult to interpret and for QC purposes the analysis in the spatial domain is sufficient (Ravaglia et al. 2009).

3.4 Homogeneity and Artifacts

Homogeneity is an important measurement to check regularly in digital mammography, because the majority of problems encountered with digital systems have been caused by either incorrect flat field calibration or artifacts caused by image receptor defects. Both can be evaluated in a homogeneity test (Van Engen et al. 2006b). It is recommended that this test is performed weekly and before and after detector calibration.

In the homogeneity test, a homogeneous block of PMMA, which covers the whole detector, is imaged in full automatic mode. In the European Guidelines, a PMMA thickness of 4.5 cm has been chosen because this approximates the attenuation of an average breast. The resulting unprocessed image is divided into ROIs with a size of 0.5 × 0.5 cm. For each ROI, the average pixel value, standard deviation and variance are determined. Signal-to-noise ratio (SNR) is calculated as pixel value over standard deviation. The pixel value, SNR and variance in each ROI should be plotted in surface plots as a function of position on the detector, which can be used to determine homogeneity in pixel value and SNR (Fig. 3.5). Because of the flat field calibration, which is performed on DR systems, the pixel value will normally be approximately equal over the imaging field and SNR will drop towards nipple side due to the heel effect and

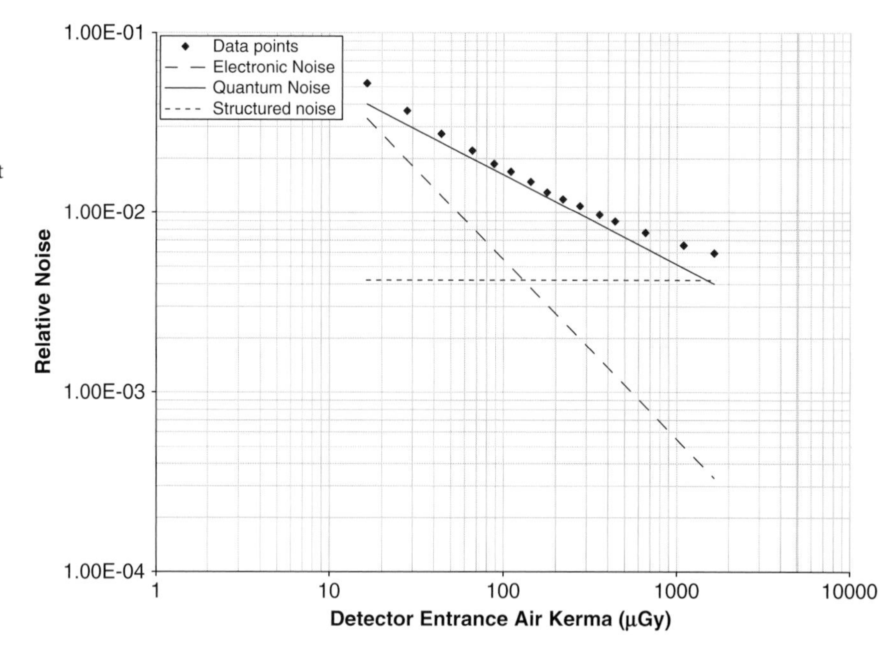

Fig. 3.4. Example of results from the noise analysis in spatial domain. In the dose range between 10 and 1,500 μGy, detector dose quantum noise is the largest noise component

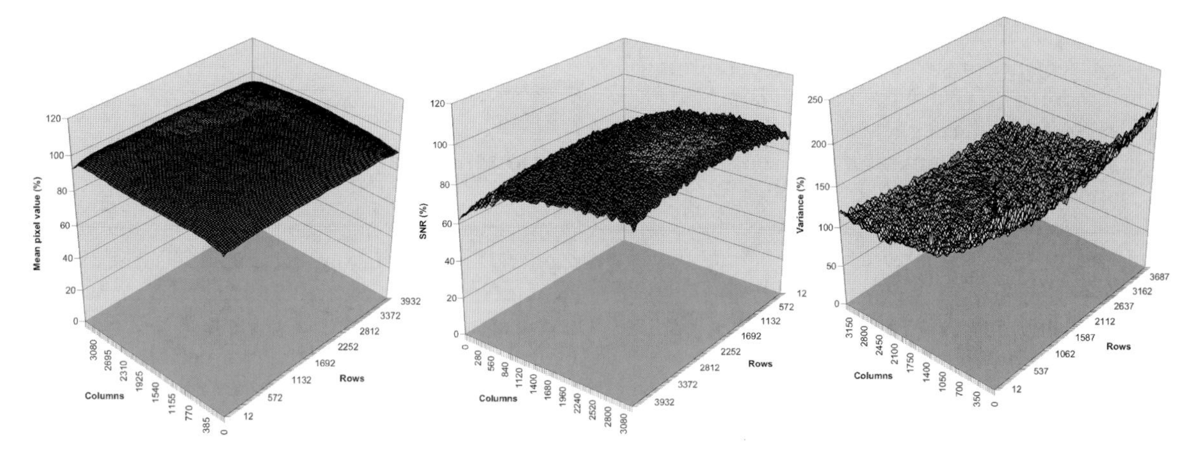

Fig. 3.5. Pixel value, SNR and variance for each position on the detector. The signal has been normalised to the standard ROI

geometric factors. Spikes in the SNR and variance graphs indicate the presence of artifacts on the image.

This approach is an excellent means of identifying artifacts. These can be divided into categories: image receptor problems, detector calibration problems, and other problems as discussed below.

3.4.1 Artifacts Due to Problems with the Image Receptor

Image receptor problems can have multiple causes. For example, in DR systems the long-term stability of the detector might be insufficient or artifacts might be caused by the specific architecture of the detector. Examples of this last group of artifacts are geometric distortion due to incorrect stitching of sub-images and inhomogeneities towards the lateral sides of the image. For CR systems, the screens can develop physical defects that mimic micro-calcifications. Examples are shown in Figs. 3.6–3.9.

3.4.2 Artifacts Related to Detector Calibration

In all DR systems, imperfections and differences in gain and offset of individual parts of the detector are reduced by the detector calibrations. For some systems,

Fig. 3.6. Example of an artifact due to insufficient long-term stability

Fig. 3.7. Example of an artifacts: inhomogeneities due to detector problems

Fig. 3.8. Examples of an artifacts: defects in a CR screen mimicking microcalcifications

Fig. 3.9. Examples of an artifacts: inhomogeneities caused by the long-term use of a CR cassette (ghost)

this detector calibration is performed by service engineers. For other systems, the user has to perform the flat field calibration. Faults may occur in performing this calibration. Besides this, artifacts may occur due to a shift in position of an object in the X-ray beam after the flat field calibration has been performed. For example, inhomogeneities of the X-ray filter may become visible. Examples are shown in Fig. 3.10.

3.4.3 Artifacts Due to Other Problems

The other causes of artifacts are due to a diverse range of problems and include issues related to non-moving anti-scatter grids, collimation of the X-ray beam and alignment of the compression paddle. Examples are shown in Figs. 3.11–3.13.

3.5 Dosimetry

In many ways, the QC procedures for measuring dose are the same for digital mammography systems as for analogue film screen systems. However, there are some subtle differences. Most dosimetry relies on using a simple phantom to represent the breast. In the original European protocol for analogue systems, an exposure is made of a slab of PMMA with a thickness of 45 mm under automatic exposure control. The incident air kerma is measured and used to estimate the mean glandular dose for a standard breast model with a 53-mm overall thickness with a central

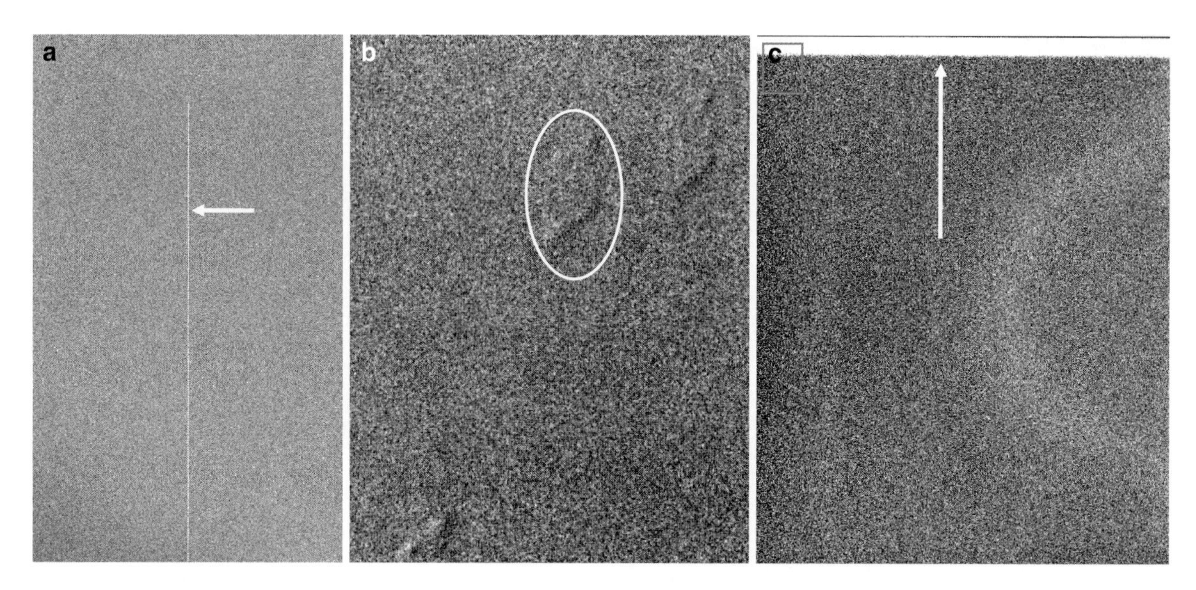

Fig. 3.10. Examples of an artifacts related to the flat field procedure, (**a**) not corrected defective column on a homogeneity test image, (**b**) artifact caused by the movement of an object in the X-ray beam after the flat field calibration on a homogeneity test image, (**c**) artifact caused by an incorrect flat field procedure: the PMMA did not cover the whole detector during part of the flat field calibration procedure

Fig. 3.11. Example of an artifact: the X-ray beam is collimated to 18 × 24 cm while the image has the size of the whole image receptor

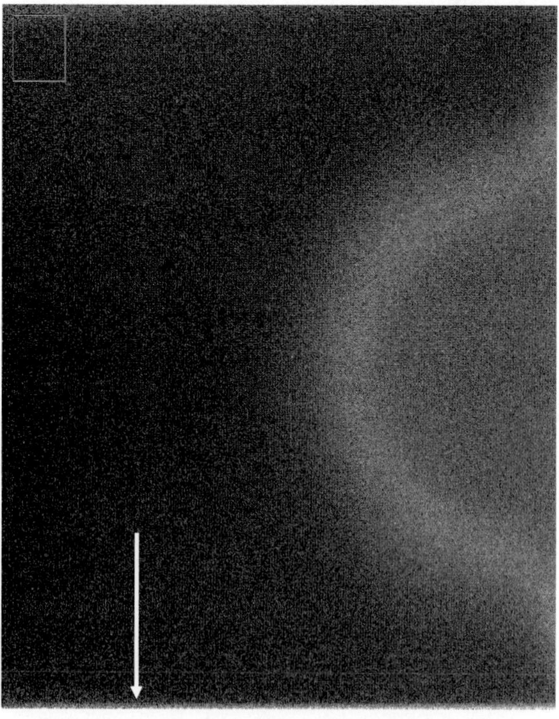

Fig. 3.12. Example of an artifact: at the lateral side of the image, an underexposed edge is visible caused by the compression paddle. On this image a ghost image is also visible

Fig. 3.13. Example of an artifacts: the anti scatter grid is visible on the image

Table 3.2. Dose limits in European guidelines

Thickness of PMMA (cm)	Thickness of equivalent breast (cm)	Acceptable level (mGy)	Achievable level (mGy)
2.0	2.1	<1.0	<0.6
3.0	3.2	<1.5	<1.0
4.0	4.5	<2.0	<1.6
4.5	5.3	<2.5	<2.0
5.0	6.0	<3.0	<2.4
6.0	7.5	<4.5	<3.6
7.0	9.0	<6.5	<5.1

layer with 29% glandularity. This is typical of the breasts of women attending for screening in the age range 50–64 (Dance et al. 2000). However, when digital systems were introduced, it was decided that a single dose measurement at one thickness was insufficient and dose limits in the European and UK protocols were provided across a range of simulated breast thicknesses (Van Engen et al. 2006; Workman et al. 2006). This was because the AECs on digital systems can be easily adjusted to give different doses at different thicknesses. The dose limits in European guidelines increase with breast thickness as shown in Table 3.2.

Another issue is that some modern digital mammography systems have introduced target filter combinations (e.g. W/Ag and W/Al) that were not considered in the original dosimetry protocol (Dance et al. 2000). However, further guidance has been published to provide factors to enable the protocols to be extended to cover these target filter combinations (Dance et al. 2009).

Another issue to consider is that the AECs on digital mammography systems are becoming much more sophisticated than those used with analogue systems. Thus, on some manufacturer's systems, the AEC will adjust the exposure to ensure adequate detector dose below the densest part of the breast being imaged. This means that doses estimated using a uniform block of material are no longer simply related to the dose expected for an inhomogeneous real breast. In general, we can expect the dose to a real breast to be higher. This makes it particularly important to also estimate the dose from samples of exposures recorded for real breasts.

3.6 Quality Control of Image Processing

Image processing aims to visualise radiographic (raw) images in "the best possible way". The necessity of this part of the imaging chain is immediately obvious from just one practical mammogram. Figure 3.14a shows a so-called raw patient image. Subtle X-ray contrasts that are possibly present in this image can neither be seen nor radiologically interpreted. Figure 3.14b shows the same image after applying image processing.

Image processing is subject to task-driven requirements. In breast cancer screening, the ultimate goal for image processing algorithms is to visualise small masses and subtle microcalcifications in such a way that they can be easily perceived by the radiologists. In addition, lesions should be visualised without deformation. Lesions should be faithfully represented, without any artificial enhancements, and the breast glandularity should be visible from the images. Other requirements for image processing are more general:

(1) Robustness with regard to type of breast

The effect of the image processing on the global appearance should not be overly sensitive to breast. Otherwise, the global appearance between different views of the same breast or between left and right breasts may be too different for comparison during reading.

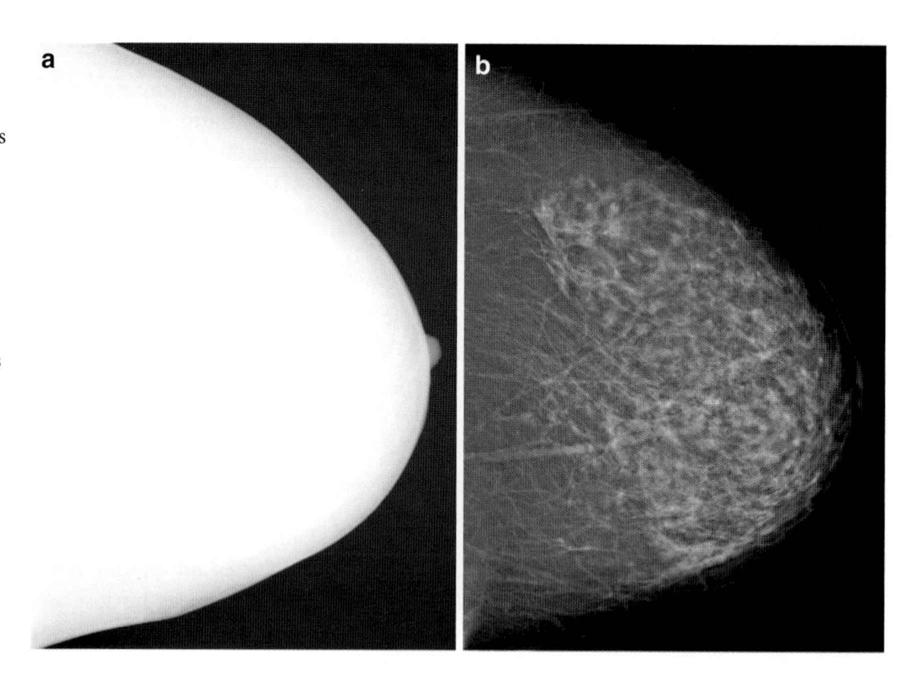

Fig. 3.14. (a) Unprocessed, "raw" mammogram ofa normal patient. Visualisation of contrasts is suboptimal. (b) Processed mammogram of the same patient using routinely applied image processing. Image contrasts can be appreciated. This image can be presented to radiologists for reading purposes

(2) Robustness with regard to dose level or dynamic range

The local dose level at the detector should not have too much influence on the processing. Otherwise, the software may react differently in different parts of the breast or produce different appearances if in successive acquisitions slightly different doses were used.

(3) Reproducible

A series of raw images of the same views of the breast should provide similar processed images.

(4) No or minimal need for extra window-levelling (or any other manipulation)

For practical reasons, image processing should prepare the visualisation of the images such that no extra window-levelling is needed. This is very beneficial for the workflow.

(5) Radiologists should have confidence in the images

The reality today is that the appearance of digital mammograms is determined by the type of detector and the vendor-specific processing that has been applied. This is a particular issue in centres conducting second reading, where images acquired with different mammographic units are read. In these centres, radiologists may have to train themselves with images having very different appearances due to the different technology used rather than the breast characteristics.

Image processing consists essentially of mathematical operations performed on radiographic images. It is important to notice that in current image processing algorithms, the results of the processing are determined by the content of the raw image. As an example, the histogram of a small dense breast can be very different from a larger breast. Characteristics of this histogram can drive subsequent parameters. As a result, simple applications of the image processing software on high spatial resolution objects or specific contrasts all inserted in a phantom with a homogeneous background may not be representative of the quality of real processed mammograms. Are all processing algorithms equally faithful? Would they all pass a minimal standard?

From the above scene setting, it can be understood that evaluation of image processing algorithms should include both a clinical part (evaluating the apparent global image quality) and a technical part (evaluation of contrasts and sizes of just visible lesions).

Currently, quality assurance protocols do not provide tools to assess image processing algorithms. Algorithms could be applied repeatedly on the same set of raw clinical input images and this action could then be followed by an evaluation process. Although this would not require any additional exposure (or ethical consideration), such systematic studies have not been reported.

3.6.1 Radiological Evaluation

Some studies have scored the quality of mammograms (with specific processing algorithms) in terms of the radiologist's performance in clinically relevant tasks (Pisano et al. 2000a, b; Cole et al. 2005; Sivaramkrishna et al. 2000). This way of assessing image quality is very direct but is often time-consuming and expensive. In these studies, there was no preference seen for particular image processing algorithms.

A possible method of evaluating the global quality of an image, including image processing, consists of the use of "image quality criteria". This approach to assessing the image quality of radiological images has a long tradition in Europe (EC 2006). The basic concept behind these criteria is that images showing all normal structures with high detail and contrast will perform similarly for (subtle) lesions. Van Ongeval et al. (2008) proposed such a set of criteria for the evaluation of the global quality of mammograms (Tables 3.3 and 3.4). These authors distinguish between strict anatomical criteria (Table 3.3) and more physico-technical characteristics (Table 3.4). The criteria are applied to a representative sample of images. The absolute scores can be evaluated with an observer who judges a mammogram to an internal reference. On the other hand, comparative tests can be performed during which raw images processed with different processing algorithms or different parameter settings are shown side by side and in which the quality of each mammogram is ranked. Such a comparative evaluation may be easier than giving absolute scores but other statistics have to be applied. This approach to testing image processing remains subjective. It is not proven that radiologists' confidence with the apparent image quality correlates with the quality of the processing for the visualisation of lesions.

Table 3.3. Radiological image quality criteria, score 1 = fulfilled, score 0 = not fulfilled

Visualisation of the skin line	1 or 0
Visualisation of the vascular structures through the dense parenchyma	1 or 0
Sharp visualisation of vascular and fibrous structures and pectoral muscle	1 or 0
Sharp reproduction of skin structures along the pectoral muscle	1 or 0
Visualisation of the Coopers ligaments and vascular structures in the white areas	1 or 0
Visualisation of the Coopers ligaments and vascular structures in the dark areas	1 or 0
In the white areas, no disturbing noise (pseudomicrocalcifiations due to noise)	1 or 0
In the dark areas, no disturbing noise (pseudomicrocalcifiations due to noise)	1 or 0
Enough contrast in the white areas	1 or 0
Enough contrast in the dark areas	1 or 0
The glandular tissue is white	1 or 0
The background is dark	1 or 0

Copied from Van Ongeval et al. (2008)

Table 3.4. Physical characteristics of an image

Contrast	−2, −1, 0, +1, +2
Sharpness	−2, −1, 0, +1, +2
Saturation in white regions	0, +1
Saturation in dark regions	0, +1
Artifacts	If present, please list
How confident are you with the representation of the microcalcifications?	−2, −1, 0, +1, + 2
How confident are you with the representation of opacities?	−2, −1, 0, +1, +2
How confident are you with the representation of the image?	−2, −1, 0, +1, +2

Copied from Van Ongeval et al. (2008)

3.6.2 Quality of Image Processing Algorithms in Terms of Detectability of Lesions

An ideal test methodology for image processing algorithms would deliver a clinically relevant result, like the detectability of particular lesions, the risk for false-positive lesions, the time to read the images, etc. A large database of raw clinical images with well-circumscribed lesions could be the basis for such a methodology. Raw images could be processed with an algorithm to be tested and the number of detected lesions could be obtained from different readers scrutinising these images. If the same database of images were always used, reference curves could even be established. In practice, the time and costs to prepare such a database are huge as all lesions are to be described and/or proven. To mimic screening conditions, large numbers of (proven) normal images

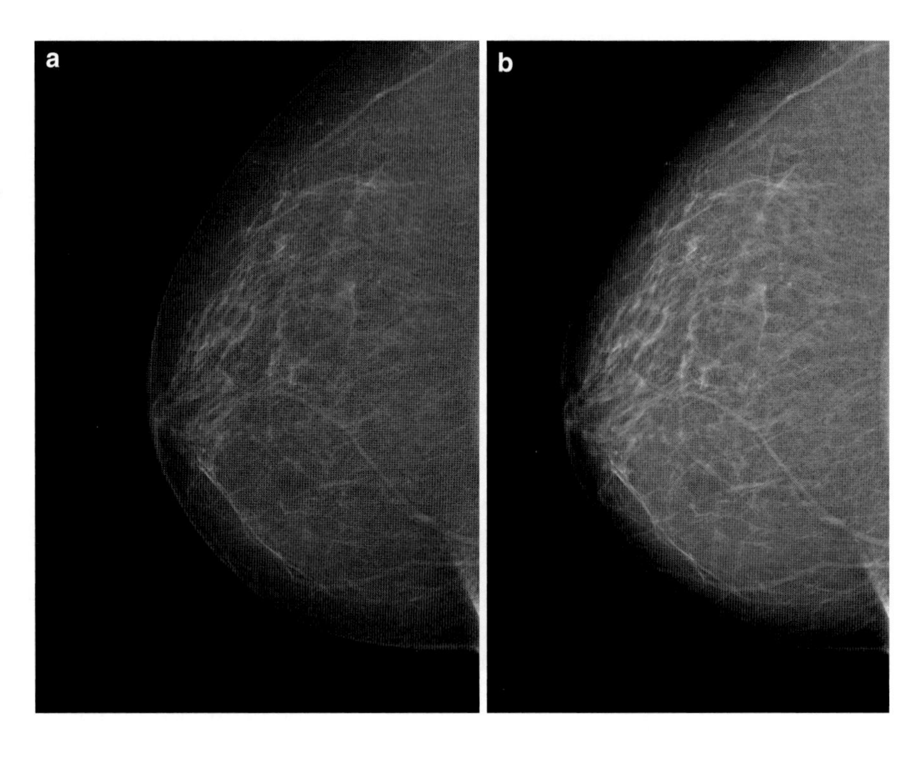

Fig. 3.15. (**a**) Raw image that has been reprocessed using Siemens Opview 1. (**b**) Same raw image as in Fig 3.15a that has been reprocessed using Siemens Opview 2. Copied from Zanca et al. (2009)

would be required. In addition, there is a fundamental problem with databases of real images – namely it is difficult to make databases that can serve to test image processing algorithms that are potentially much better than the ones that you started with. For example, what if suddenly subtle lesions started to appear? Would they be real lesions or artificially enhanced noisy structures?

An alternative and more practical approach exploits the possibility of simulating lesions. Templates of lesions can be multiplied with the pixel values in raw images to simulate specific lesions at specific places. The basic physics behind this principle is very simple: if a template represents the X-ray attenuation of a particular lesion, multiplying this template with raw images will result in an image in which an artifact shows up that has modulated the X-ray beam in much the same way as the lesion. If this process is done carefully, the observer may not notice the difference between the "artifact" and the real lesion (Carton et al. 2003; Samei et al. 2007; Ruschin et al. 2007; Zanca et al. 2008). This approach is appropriate for small lesions and has been worked out for comparative studies of image processing by Zanca et al. (2009). This experiment used a set of 200 unprocessed normal images. In 100 of these images, clusters of microcalcifications were simulated. The clusters were characterised in terms of contrast value, size, LeGal type of the individual microcalcifications in the cluster, their position in the images as well as the glandularity of the local background. Subtle and more obvious clusters were simulated via the introduction of contrast values that could be freely tuned. All images were processed with five processing algorithms used in routine radiological practice and provided by the manufacturers. An example of an image reprocessed with the two of five algorithms is shown in Fig. 3.15. Four observers took part in an observer performance experiment to detect the clusters. This study was practically feasible; thanks to a dedicated software tool (Jacobs et al. 2008a). Statistical analysis was performed with FROC (Chakraborty and Berbaum 2004). A result from this study showing a statistically significant difference in the detectability of clusters for two algorithms is shown in Fig. 3.16. All readers had been confident with the different image processing presentations, yet their scores were different for the different algorithms. This result suggests that a careful check of image processing algorithms is necessary and the paper describes a methodology for doing this, which should be considered in future QA protocols.

3.7 Quality Control of Monitors

Digital images can be viewed using a monitor and workstation (soft copy viewing) or using a printed film and light box (hard copy). Advantages and

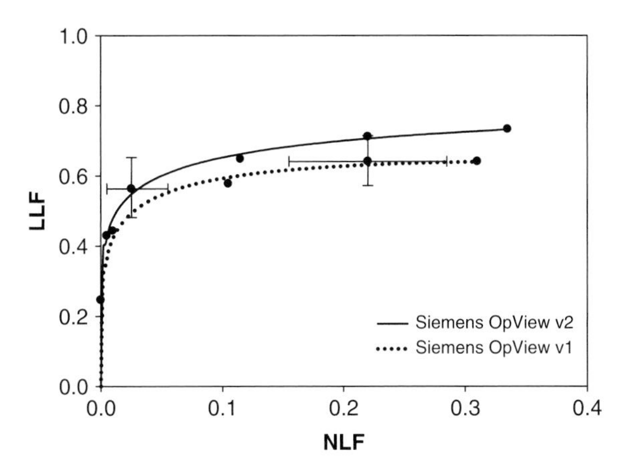

Fig. 3.16. FROC curve for one reader (Reader 1) of an observer performance experiment with simulated clusters of microcalcifications. Error bars are calculated using bootstrap method. Images were reprocessed with 2 versions of the Siemens image processing algorithms Opviewl and Opview2. Superiority of Opview 2 to Opview 1 (in terms of microcalcification visibility) is demonstrated. Figure copied from: Zanca et al. (2009)

disadvantages of both viewing modes as well as requirements for optimal workflow using workstations are discussed in Chap. 7. In this section, the Quality Assurance issues relating to viewing digital images are discussed.

Image viewing is an important and integral part of the normal workflow in the radiology department. Viewing modalities are, however, not always included in QA procedures, as they are not directly linked with X-ray exposure and therefore not always included in safety procedures. Viewing of images is however indirectly linked with exposure, or at least with image quality. QA guidelines therefore include them in test protocols. The American DMIST manual and the European Guidelines suggest daily tests of display monitors.
Display systems can be tested in different ways:

- Using luminance meters on the one hand and test patterns on the other hand
- By human observation or using automated computerised methods
- By trained physicists (on half-yearly basis) or by the local personnel (more frequently)
- Using static test patterns or variable patterns

Testing of viewing monitors and viewing conditions includes a careful evaluation of the ambient light in the viewing room and ideally also of the configurations of the monitors relative to each other and to any viewing boxes used for film.

3.7.1 Physics Tests

The AAPM task group 18 (TG18) has published a very comprehensive document on the assessment of monitors as well as an executive summary of this document (Samei et al. 2005). The characteristics, tested in conjunction with specially designed test patterns (i.e. TG18 patterns), include reflection, geometric distortion, luminance, the spatial and angular dependencies of luminance, resolution, noise, glare, chromaticity and display artifacts. The most important QA test on a monitor is probably the verification of the luminance response. The luminance response refers to the relationship between displayed luminance and the input values. The response can be measured using a calibrated luminance meter and the TG18-LN test patterns. The luminance L in the test region is measured in each pattern and the measured values related to the targeted luminance response for the image display device. The slope of the luminance response should remain within set intervals. In mammography protocols, the Gray Scale Display Function (GSDF) curve is usually imposed with limiting values. The shape of the GSDF was derived from measurements on the response of the human visual system (Barten 1992). The GSDF is defined as a table of luminance values such that the luminance change between any two sequential values corresponds to the peak-to-peak relative luminance difference, dL/L, predicted by the Barten model. It is believed that this curve makes optimal use of the luminances over the complete dynamic range for an average person. In Fig. 3.17, the luminance responses of two monitors are shown. The upper curve corresponds to one that passed the limiting values dictated by the GSDF curve and the lower one that failed.

In a given ambient lighting, the luminance varies between the minimal luminance, L'min, and the maximal luminance, L'max. The ratio L'max/L'min is the luminance ratio. In order to have similar image appearance with respect to contrast, all display devices should have the same luminance ratio and the same display function. If more than one monitor is used in a workstation, as is usually the case, the luminance responses should be compared and found to be similar within specified limits.

Ambient light levels are also subject to limiting values. In the European QA protocol, it is specified that the ambient light should be below 5 lux. However, Pollard et al. (2009) have discussed whether these strict requirements should be applied to the viewing by LCD monitors.

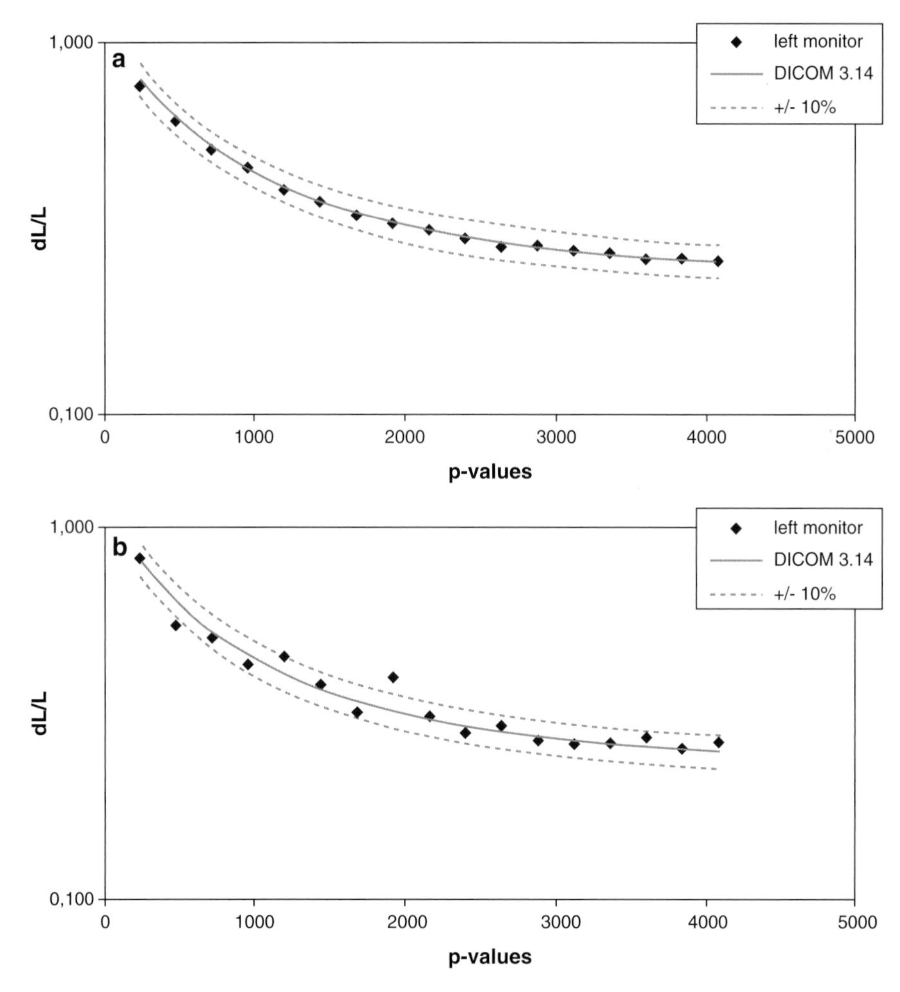

Fig. 3.17. Examples of a good (**a**) and a bad (**b**) contrast response (5MP LCD monitor used in mammography screening)

3.7.2 Human Reading of Test Patterns

Frequent, periodic tests are usually performed using specific test patterns. The AAPM TG18 proposes the use of a pattern called "TG18-QC" for constancy checks (Fig. 3.18). This test pattern is readily available and allows for a complete test of a monitor. However, some disadvantages of this design are that the test pattern is fixed with the consequence that users may become so familiar with the pattern that they do not examine the image carefully; the test patterns are rather complicated for personnel not well trained for monitor QA; the relative importance of monitor problems is not documented and therefore it is difficult for the user to judge the relevance of an observed abnormality.

An alternative would consist of variable test patterns with a series of simple questions for the observers. Such a test pattern "MoniQA" was proposed by Jacobs et al. (2007) and a typical example of a test pattern is shown in Fig. 3.19. In the University Hospital Leuven, daily QC of monitors is performed in all mammographic units involved in screening using this variable (MoniQA) pattern and the results of the reading are transferred to a QC supervision centre. Figure 3.20 shows a typical follow-up of a monitor over time. We presume that an even more simplified pattern that consists of low contrast circles on a homogeneous background could detect the following major problems: apparent changes in luminance because the ambient light level is at the moment of the test too high; finger prints; calibration problems. A further analysis of the results may determine the appropriate frequency for these tests.

3.7.3 Fully Automated Procedures

The digital nature of the equipment has also been exploited to facilitate QC. Most vendors of monitors

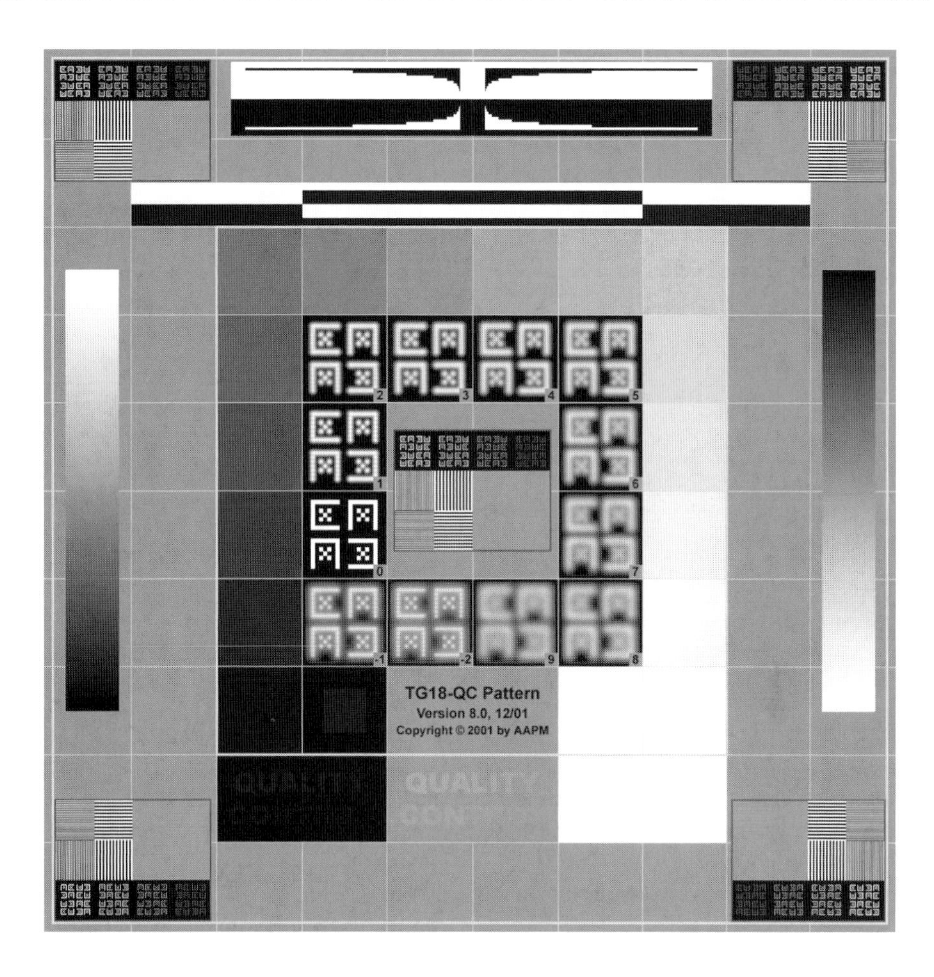

Fig. 3.18. The AAPM-TG 18 QC comprehensive test pattern

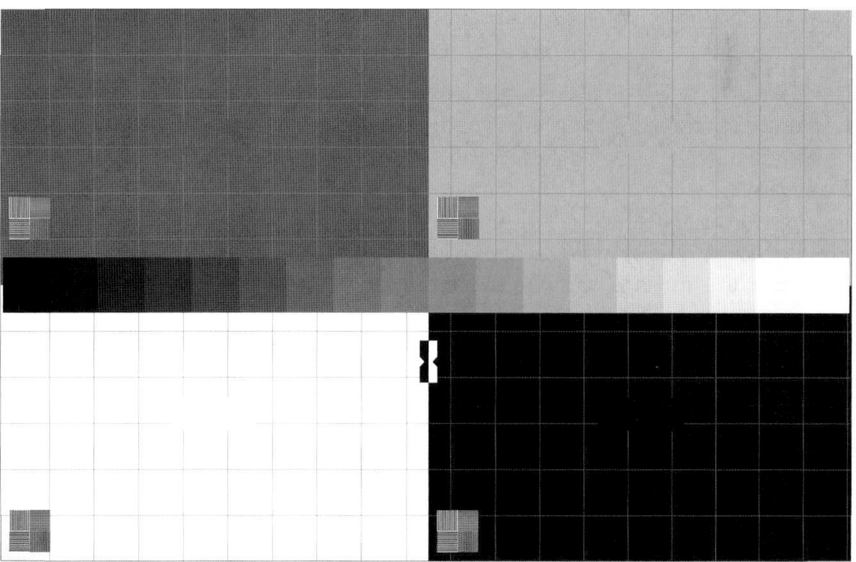

Fig. 3.19. An example instance of the variable test pattern MoniQA

offer their solution as an option to their product. Overnight, the GSDF curve of the monitor is typically verified at one reference point by means of a built-in luminance meter. Deviations from set curves can be calibrated automatically and immediately. Next, the scores can be supervised centrally, e.g. in the QC department of the vendor. The amount of information to be gained on stability is enormous. There is

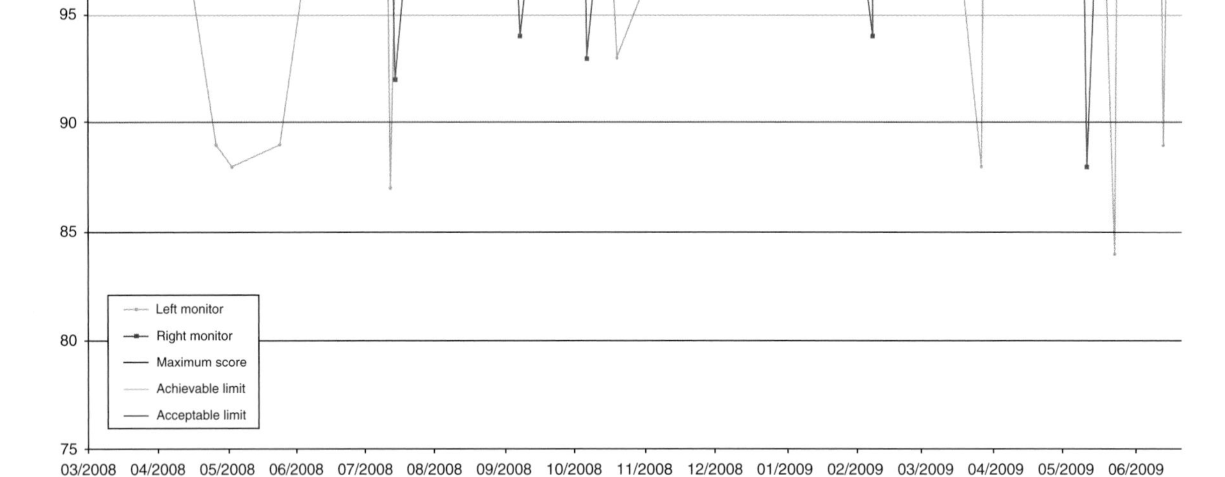

Fig. 3.20. Typical follow-up of the quality of the monitors of a dual-monitor workstation using the MoniQA score

only one major shortcoming to this system: one of the most important parameters, namely ambient light during the reading of images, is not taken into account with this nightly procedure. The effect of a deviation of the perception by the reader is also not known.

3.7.4 Conclusion

It is generally agreed that viewing conditions should be included in a QC protocol. However, there are not many practical tools available that test all aspects of viewing or that easily link with perception in radiological practice. Further research is needed with regard to the most appropriate parameters to be tested.

3.8 Routine Quality Control Tests and Their Automation

Quality can be measured at several stages of the imaging process, by means of a large variety of methods and with different frequencies. This section considers the frequent tests that can be executed by local staff and of which the parameter analysis can reveal the state of the equipment, image quality and dose levels. The role of periodic QC is to ensure constant imaging systems such that high-quality mammograms can be acquired in all women (Van Engen et al. 2006a).

Clinical image quality is multifactorial with some factors potentially very variable and patient-dependent. A rigorous QC management system will ensure good control of patient-related aspects (e.g. positioning, artifacts in clinical images) and technical issues. Positioning and other patient- or lesion-related issues can be controlled as in the film-screen environment. It can be hoped that in the near future, software tools become available that will allow automation of these types of analysis (Jacobs et al. 2008c). Artifacts in clinical images will be discussed in Chap. 4. It is very important that tools are foreseen to explain and overcome the artifacts and to quantify the amount of problems. The remainder of this paragraph focuses at the strictly technical aspects of image quality.

Simple test objects can be used for frequent periodic testing of systems. Some protocols list items to be tested (Van Engen et al. 2006a). Some vendors also prescribe how their system should be tested and also provide limiting values for these measurements. Straightforward parameters for regular QC of digital mammography are: the exposure settings (dose level) from an AEC-controlled acquisition of a standardised test block, the SNR in a region of interest of a raw image, the CNR generated by a thin Al object and the homogeneity of an image of a uniform test block. The DICOM header of (DR) images contains extra information that can be valuable such as detector temperature, image processing software versions, etc.

There are a series of unsolved questions regarding routine QC in practice. First, there is not yet a consensus regarding algorithms to calculate the parameters. Second, there is an absence of clinically validated limiting values. The following fundamental questions have as yet no answer:

- What amount of fluctuation in SNR is acceptable?
- Which local inhomogeneities pose clinical problems?
- Which clustering of defective pixels would obscure a subtle lesion?
- Do we detect all clinically relevant technical problems with our current approaches?

Although effective routine QC requires easy access to unprocessed images for analysis and timely reports, this facility is frequently not available.

In the remainder of this chapter, we illustrate a practical approach to daily QC as worked out in the screening unit of Leuven, Belgium. Currently, QC data for 50 digital mammography systems are collected daily and centralised data supervision is performed (Jacobs et al. 2008b). Based on this experience, we believe that periodic QC is an important and necessary part of the management of image quality. Numerous problems have been detected that would have remained largely undetected if QC was restricted to investigations on a half-yearly or yearly basis.

3.8.1 A Practical Example of Periodic Technical Quality Control

A very simple test object is chosen, namely a homogeneous block of PMMA that is large enough to cover the complete detector (with the largest compression paddle). Two images are acquired every day, using the same fully automatic acquisition mode that is routinely used on patients. The first image is acquired, then the slab is rotated by 180° and a second acquisition is performed with an identical AEC setting.

Access to raw data is enabled in two steps: the vendor enables a node on the X-ray system to which raw data can be sent, and at this node, a DICOM receiver (Jacobs et al. 2006) receives the QC image and extracts a series of data out of all relevant incoming raw data. By convention, the patient name tag is used to distinguish between an image for QC purposes and any other image. In practice, patient dose data analysis can also be performed on routine patient images using the same DICOM receiver software tool.

Image analysis takes place as follows:

- The two images are subdivided into overlapping ROIs that are equally spaced over the image and with a left–right and under–upper overlap of 50%.
- In all these ROIs, mean pixel value, maximum and minimum pixel value, standard deviation, variance and deviating pixels are calculated.
- Thumbnail images are created out of all these data in which every pixel is the parameter from a ROI.
- These thumbnail images can be colour coded for easy visual inspection.

Instead of sending the two original images for analysis to a remote centre for quality supervision, a smaller package of data is sent, which includes the SNR in a specific ROI, the thumbnail images, a selection of DICOM header tags (for DR systems) or manual input such as anode/filter combination, peak tube voltage and tube current for CR with incomplete DICOM headers, a few line profiles and noise power spectra of six specific ROIs.

In the centre for daily QC supervision, these data are analysed. Special points of interest are drifting of parameters over time and focalised artifacts.

Examples of the type of problems detected by this method are shown in Figs. 3.21–3.25.

3.8.2 Conclusion

Periodic QC can be performed by means of very simple test objects. Centralised supervision is possible and offers the advantage that a lot of experience can be gathered, of different brands of systems, of different centres, using the same IT technology. However, more research is needed to determine the clinically relevant limiting values.

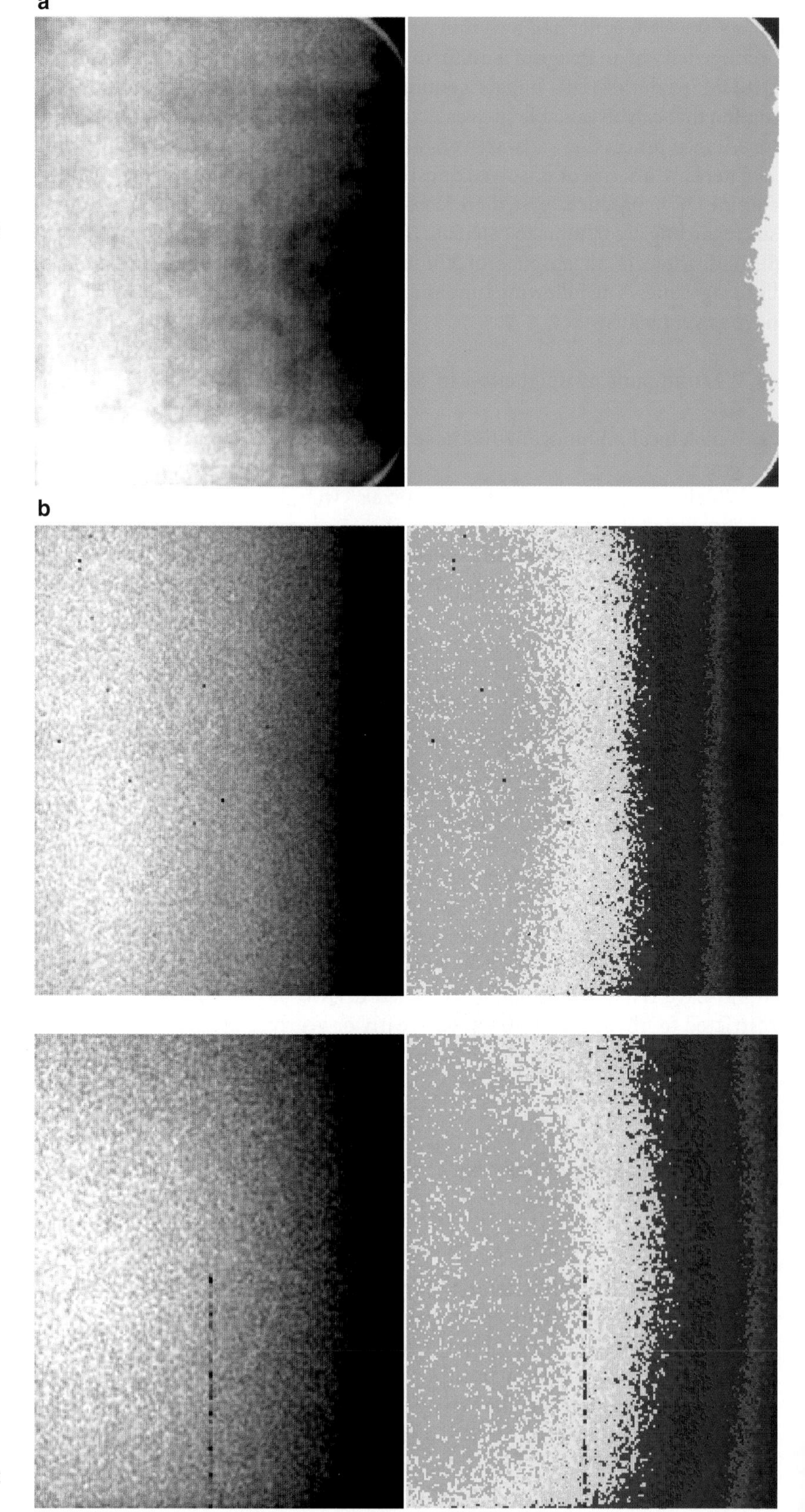

Fig. 3.21. Defective pixels in a DR detector: (**a**) mean thumbnail image (monochrome and colour deviation map) indicating no defective pixels and an almost perfect homogeneity. (**b**) SNR thumbnail image (monochrome and colour deviation map) indicating defective pixels. For both cases chest wall side is at the left

Fig. 3.22. Defective row in a DR detector: variance thumbnail images (monochrome and colour deviation maps) obtained from a system in which a column of pixels is not shown with the proper data

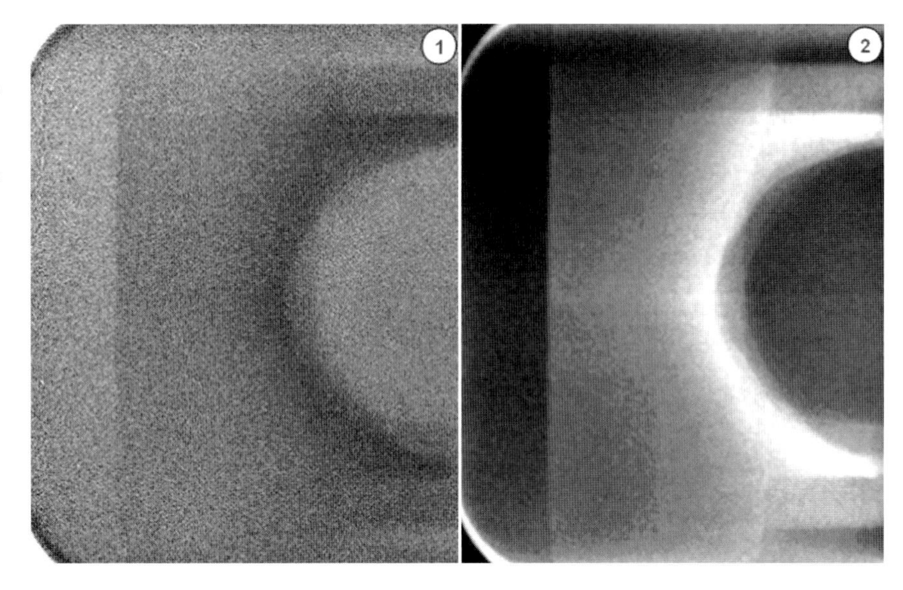

Fig. 3.23. Lag ghosting in a DR detector: phantom image (*1*) thumbnail image of the mean values in different ROIs (*2*) obtained from a DR system in which a typical lag ghost can be seen

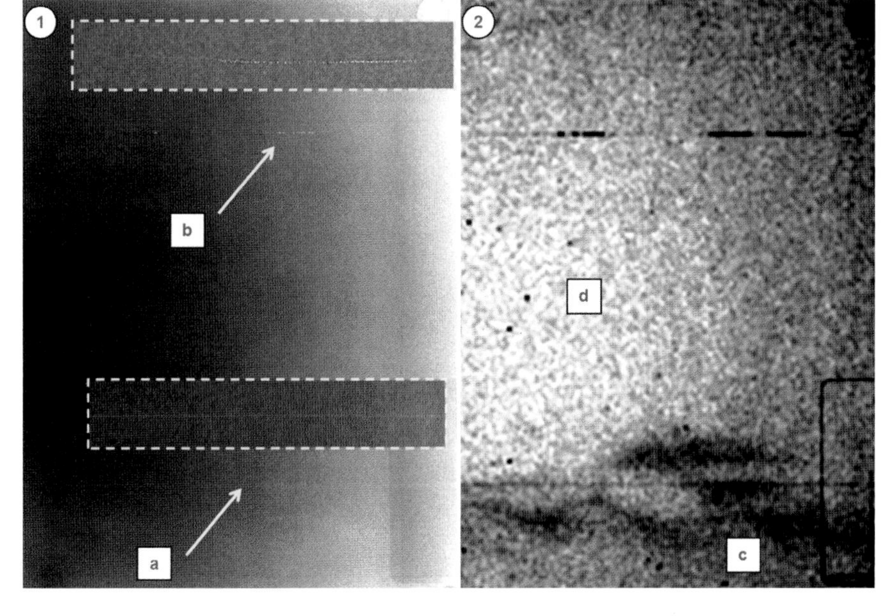

Fig. 3.24. Scan line artifacts in a CR system: phantom image (*1*) thumbnail image of the SNR in different ROIs (*2*) obtained from a CR system in which a row of pixels is not shown with the proper data due to a scan line artifact in the CR reader (**a**). Additional artifacts that can be seen are (**b**) scratches on the Imaging Plate (IP), (**c**) a bad phosphor uniformity and (**d**) local point artifacts due to phosphor damage

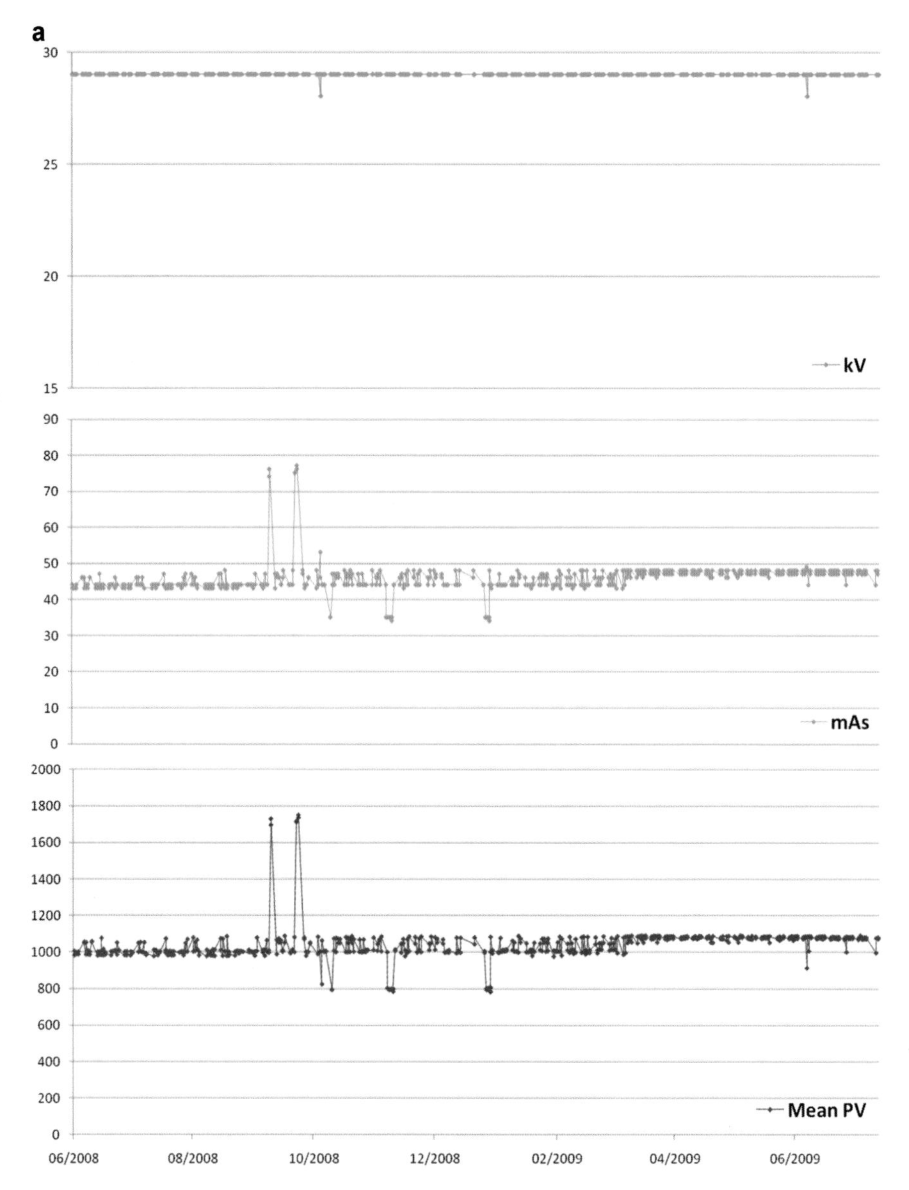

Fig. 3.25. Drift of exposure parameters over time. (**a**) kV, mAs and signal-to-noise ratio of a system that was stable in terms of dose (kV, mAs) over a period of 1 year. (**b**) shows a system where the mAs gradually increased but the kV and mean PV remained constant. (**c**) Evaluation of the mAs during daily quality control. Every change in mAs indicated by an arrow is due to interventions by service engineers

Fig. 3.25. (Continued)

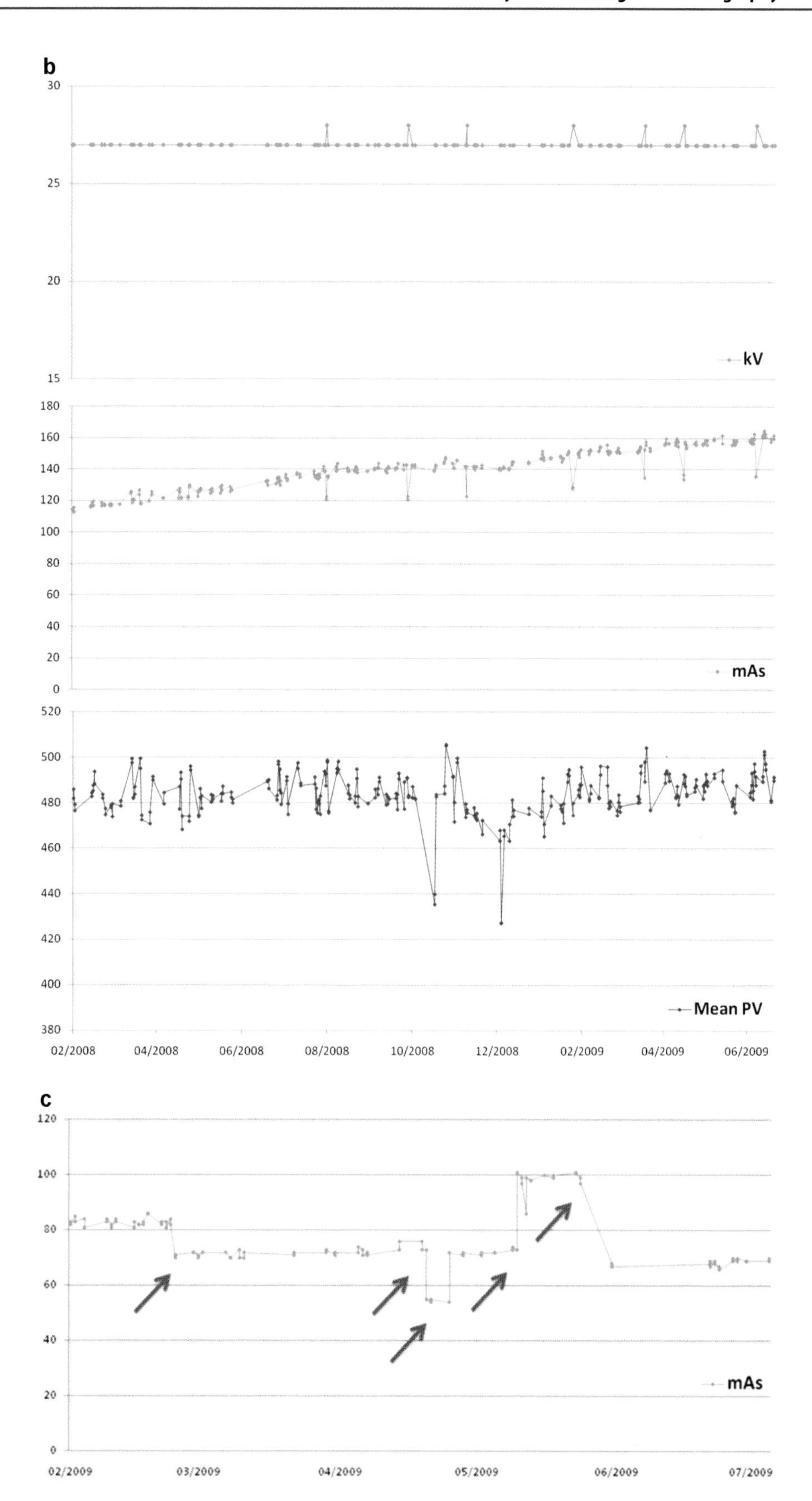

References

American Association of Physicists in Medicine (1990) Equipment Requirements and quality control for mammography, AAPM Rep. 29, New York

American College of Radiology (1999) Mammography quality control manual. Reston, VA

Barten PGJ (1992) Physical model for contrast sensitivity of the human eye. Proc SPIE 1666:57–72

Burch A, Whelehan P, Young KCS (2007) Cush, Routine quality control tests for full field digital mammography systems. NHSBSP Equipment Report 0702

Carton AK, Bosmans H, Van Ongeval C, et al (2003) Development and validation of a simulation procedure to study the visibility of micro calcifications in digital mammograms. Med Phys 30:2234–2240

Chakraborty DP, Berbaum KS (2004) Observer studies involving detection and localization: Modeling, analysis and validation. Med Phys 31:2313–2330

Cole EB, Pisano ED, Zeng D, et al (2005) The effects of gray scale image processing on digital mammography interpretation performance. Acad Radiol 12:585–595

Dance DR, Skinner CL, Young KC, et al (2000) Additional factors for the estimation of mean glandular breast dose using the UK mammography dosimetry protocol. Phys Med Biol 45:3225–3240

Dance DR, Young KC, Van Engen RE (2009) Further factors for the estimation of mean glandular dose using the United Kingdom, European and IAEA breast dosimetry protocols. Phys Med Biol 54:4361–4372

European Commission (1996) European guidelines on quality criteria for diagnostic radiographic images. EUR 16260: (Brussels EC)

International Electrotechnical Commission (2007) Determination of the detective quantum efficiency – detectors used in mammography. IEC 62220-1-2

Jacobs J, Deprez T, Marchal G, et al (2006) The automatic analysis of digital images for quality control purposes made easy with a generic extendable and scriptable DICOM router. Presented at the Annual meeting of the RSNA, Chicago

Jacobs J, Deprez T, Van Steen A, et al (2008c) Simplifying image oriented quality control in digital mammography. Int J CARS 3(suppl 1):S3–S10

Jacobs J, Lemmens K, Nens J, et al (2008b) One year of experience with remote quality assurance of digital mammography system in the Flemish Breast Cancer Screening Program. In: Krupinski EA (ed) Proceedings of the 9th International Workshop on Digital Mammography. Springer, Berlin, Germany, LNCS 5116, pp 703–710

Jacobs J, Rogge F, Kotre J, et al (2007) Preliminary validation of a new variable pattern for daily quality assurance of medical image display devices. Med Phys 34:2744–2758

Jacobs J, Zanca F, Marchal G, et al (2008a) Implementation of a novel software framework for increased efficiency in observer performance studies in digital radiology. Presented at the Annual meeting of the RSNA, Chicago, p 888

Moore AC, Dance DR, Evans DS, et al (2005) The commissioning and routine testing of mammographic X-ray systems. Institute of Physics and Engineering in Medicine, IPEM Report No 89

Pisano ED, Cole EB, Hemminger BM, et al (2000a) Image processing algorithms for digital mammography: a pictorial essay. Radiographics 20:1479–1491

Pisano ED, Cole EB, Major S, et al (2000b) Radiologists' preferences for digital mammographic display. The international digital mammography development group. Radiology 216 (3):820–830

Pollard BJ, Samei E, Chawla AS, et al (2009) The influence of increased ambient lighting on mass detection in mammograms. Acad Radiol 16:299–304

Ravaglia V, Bouwman R, Young KC, et al (2009) Noise analysis of full field digital mammography systems. Proc SPIE Med Imaging 72581B:1–11

Ruschin M, Timberg P, Båth M, et al (2007) Dose dependence of mass and microcalcification detection in digital mammography: free response human observer studies. Med Phys 34:400–407

Samei E, Badano A, Chakraborty D, et al (2005) Assessment of display performance for medical imaging systems: executive summary of AAPM TG18 report. Med Phys 32:1205–1225 (AAPM TG18 document can be downloaded from http://deckard.mc.duke.edu/~samei/tg18)

Samei E, Saunders RS Jr, Baker JA, et al (2007) Digital mammography: effects of reduced radiation dose on diagnostic performance. Radiology 243:396–404

Sivaramakrishna R, Obuchowski NA, Chilcote WA, et al (2000) Comparing the performance of mammographic enhancement algorithms: a preference study. Am J Roentgenol 175:45–51

Van Engen RE, Swinkels MMJ, Oostveen LJ, et al (2006b) Using a homogeneity test as weekly quality control on digital mammography units. In: Proceedings of the 8th International Workshop on Digital Mammography, Manchester, UK. pp 259–265

Van Engen R, Young KC, Bosmans H, et al (2006a) The European protocol for the quality control of the physical and technical aspects of mammography screening. In: European Guidelines for Quality Assurance in Breast Cancer Screening and Diagnosis, 4th Edition. European Commission, Luxembourg

Van Ongeval C, Van Steen A, Geniets C, et al (2008) Clinical image quality criteria for full field digital mammography: a first practical application. Radiat Prot Dosimetry 129: 265–270

Workman A, Castallano I, Kulama, et al (2006) Commissioning and routine testing of full field digital mammography systems. NHSBSP Equipment Report 0604

Young KC, Alsager A, Oduko JM, et al (2008) Evaluation of software for reading images of the CDMAM test object to assess digital mammography systems. Proc SPIE Med Imaging 69131C:1–11

Young KC, Cook JJH, Oduko JM (2006) Use of the european protocol to optimise a digital mammography system. In: Astley SM, Bradey M, Rose C, Zwiggelaar R, editors. Proceedings of the 8th International Workshop on Digital Mammography. Springer, Berlin, Germany, LNCS 4046, pp 362–369

Young KC, Johnson B, Bosmans H, et al (2005) Development of minimum standards for image quality and dose in digital mammography. Proceedings of 7th International Workshop on Digital Mammography, Berlin, Germany. pp 149–154

Zanca F, Chakraborty DP, Van Ongeval C, et al (2008) An improved method for simulating microcalcifications in digital mammograms. Med Phys 35:4012–4018

Zanca F, Jacobs J, Van Ongeval C, et al (2009) Evaluation of clinical image processing algorithms used in digital mammography. Med Phys 36:765–775

Classification of Artifacts in Clinical Digital Mammography

4

Chantal Van Ongeval, Jurgen Jacobs, and Hilde Bosmans

CONTENTS

Chantal Van Ongeval, MD
Hilde Bosmans, PhD
Jurgen Jacobs, MsC
Department of Radiology, University Hospitals Leuven, Herestraat 49, B-3000 Leuven, Belgium

KEY POINTS

Digital mammography is now fully introduced in the radiological examination of the breast.

This was only possible after the setup of a dedicated quality control program of the digital mammography systems. Quality assurance not only includes the technical quality procedures, but also all the activities of the radiologists and technologists to ensure high-quality of the mammogram. The detection and registration of artifacts are part of this quality assurance. The artifacts are divided into patient-related artifacts, technologist-related artifacts, mammography unit related artifacts, software-related artifacts, and viewing conditions related artifacts. As some artifacts can cause interpretation problems, early recognition is important, and this is only possible with a close collaboration between all persons involved in the quality assurance of digital mammography.

4.1 Introduction

Quality assurance (QA) refers to all systematic activities undertaken by the breast imaging staff to ensure high-quality mammography (Hogge 1999).

Physical–technical quality control (QC) procedures are often a very visible part of QA (Bloomquist et al. 2006a; Yaffe et al. 2006; Euref 2006), but next to the strictly technical issues, clinical images deserve investigations too. Radiologists and technologists should actively participate in QA. Only then, a strictly implemented QA routine can guarantee optimal radiological images. Part of the QA routine is the evaluation of the clinical images for artifacts.

We refer to BASSETT (1995) for defining an artifact: any variation in mammographic density not caused by true attenuation differences in the breast. It is important to learn about artifacts to recognize them on mammograms and to avoid diagnostic mistakes. In film-screen mammography (FSM), most artifacts were linked to film or developer. Digital mammography has different physical–technical characteristics, and surely other artifacts are associated with the new technology. Some artifacts are not well understood yet.

Artifacts can originate in the different steps of the imaging chain but also in the handling of the system. Physical–technical periodic QC based on simple test object measurements can certainly detect some of them, but other artifacts may go undetected. Until now, we know one report that discusses the different kinds of artifacts in digital mammography (AYYALA et al. 2008).

In this chapter, we will describe different artifacts that appeared in our images over the last years, when digital mammography was introduced for both diagnostic and screening purposes.

4.2 Classification

Following the classification proposed by HOGGE et al. (1999) for FSM, we divided all artifacts into five different categories: patient-related artifacts, technologist-related artifacts, mammography unit related artifacts, software-related artifacts, and artifacts related to the viewing conditions. We give a short description of each category:

1. *Patient-related artifacts* are independent of the technology that is used and can be found in standard text books. Therefore, we will not discuss them any further. Typically, they include hair or traces of deodorant on the image and patient movement.
2. *Technologist-related artifacts* are caused by an incorrect handling of the system by the technologist. Examples are an improper cassette handling, an inadequate screen-cleaning procedure, a wrong setting of the imaging parameters (e.g., the choice of a wrong automatic exposure program), or an incorrect use of the mammography equipment (like deactivation of the grid).
3. *Mammography unit related artifacts* are divided into artifacts concerning direct radiography systems (DR) and computed radiography systems (CR). For the DR systems, the artifacts are often related to the detector or the readout electronics whereas for CR systems the imaging plate (IP) and the plate reader are the main cause of problems. Artifacts introduced by the X-ray tube are also included in this category.
4. *Software-related artifacts* include processing-related artifacts that are introduced by malfunctioning or badly configured image-processing algorithms as well as artifacts introduced by software during the acquisition.
5. *Viewing condition related artifacts* include problems with the calibration of the monitor toward the Grayscale standard display function (GSDF) (NEMA 2000), artifacts residing on the viewing station (like pixel defects or dirt), bugs in the viewing software, and problems with the environment wherein the images are read. We refer to Chap. 7 for more detailed discussions.

It is not always possible to categorize an artifact in one specific group, as several items may be combined. Examples are artifacts that occur with specific breast sizes or glandularities only.

The origin of some of the artifacts is still not known and nobody has encountered today already all possible artifacts. It is very important, however, that artifacts are explained to all end-users.

Because we focus on the clinical presentation of the artifacts, we will not work out the physical–technical cause of the artifact, nor discuss the best solution(s). The aim of this chapter is to illustrate the effect or the appearance of artifacts on the clinical image.

4.2.1 Technologist-Related Artifacts

The most frequent artifacts seen on CR images are *white dots* due to dust and localized problems in the coating or phosphor layer of the IP. In the images, the dots are often whiter than microcalcifications, but in some cases they can hardly be differentiated (Figs. 4.1 and 4.2). In our experience, a large amount of these problems can be avoided by having an accurate cleaning protocol and visual check of the cassette. In case this artifact appears every day, even after careful cleaning, the company needs to be contacted to change the cassette or redesign the problematic layer of the IP. In this case, the artifact is related to the mammographic unit. When parts of the IP are cracked (of wrecked), *scratches* will become visible on the image (Fig. 4.3).

Both on CR and DR systems, *noise* is often seen when the imaging parameters are not properly set or when the duration of the exposure was not long enough (Fig. 4.4a, b). This can be the result of an improper use or incorrect preprogrammed settings. A system setup at very low dose level can occasionally provide images with an excessive noise level, while the imaging parameter setting is normal. It is important to determine whether the setting is too low (to be solved by the company) (SAUNDERS et al. 2007) or whether wrong preprogrammed settings were used (which can be overcome with better local instructions). Possibly, the settings are too low for particular categories of breasts only (like very small breasts). Because there is no definition of the parameter "disturbing noise" in a clinical mammogram, the

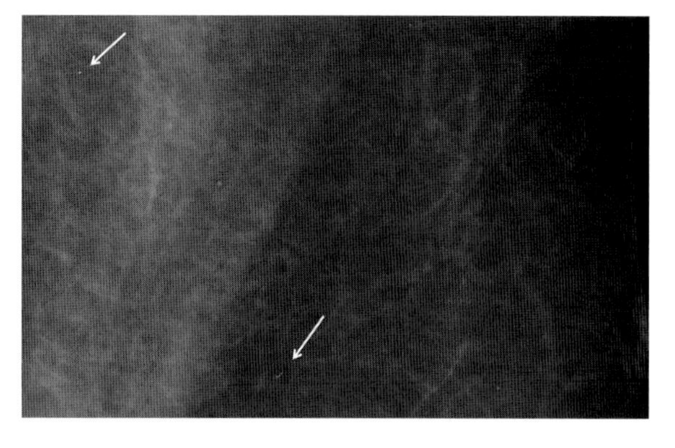

Fig. 4.2. *White dots* and *white fiber structures (white arrows)* in a CR image (Agfa CR85)

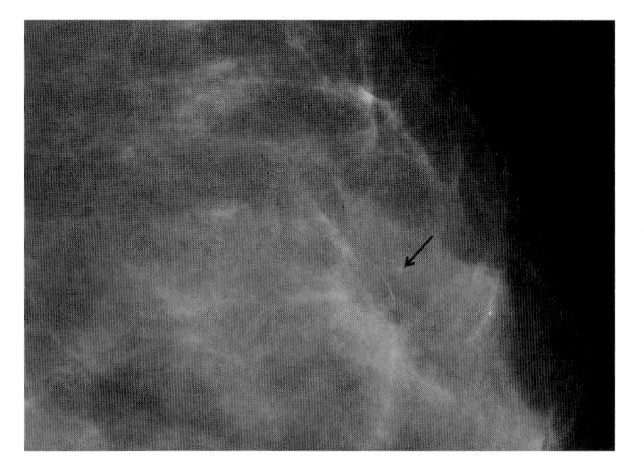

Fig. 4.1. *White curved line (arrow)* in a CR image (Agfa CR85) due to a small fiber of the protective layer of the imaging plate (IP)

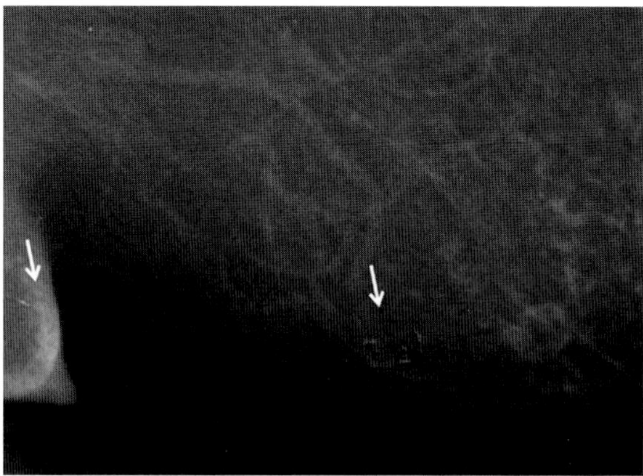

Fig. 4.3. Oblique view of the left breast: *white structures* (arrows) in the inferior part of the breast due to scratches in the phosphor of the IP (Fuji Profect CR)

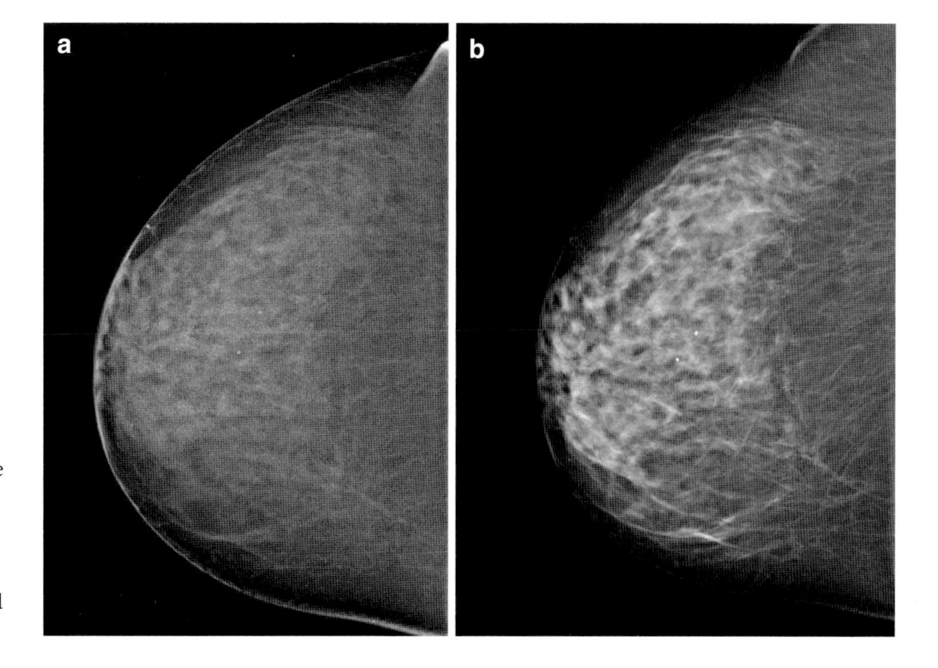

Fig. 4.4. (**a**) CC view of the right breast with noise due to wrong acquisition parameters, (**b**) the image was retaken with new acquisition parameters and the noise disappeared (Siemens Inspiration DR)

radiologist has to judge the noise levels in a subjective way. The number of failures or retakes due to noise should be registered and if necessary the automatic exposure control has to be readjusted.

The presence of *gridlines* in the image is also part of the artifact spectrum (Fig. 4.5). This artifact is also well known in FSM and is often due to a wrong exposure setting (example: relatively high tube potential and low mAs with total exposure times that are very short when compared with one period of the grid motion).

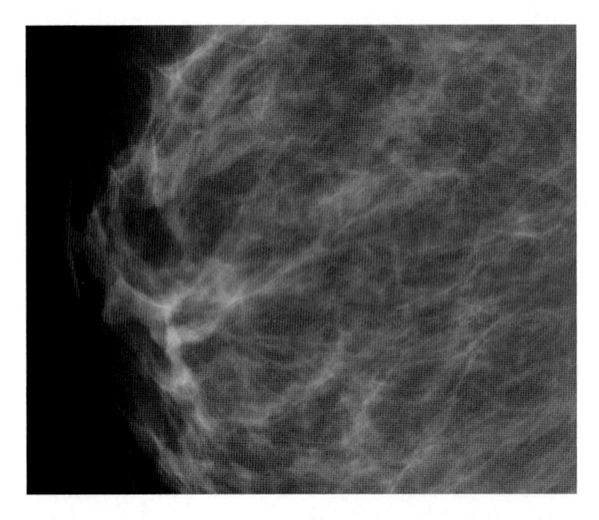

Fig. 4.5. Segment of the oblique view of the right breast. Presence of a grid structure on the image (Siemens Novation DR)

Another artifact in this group is an incorrect *positioning of the cassette* in the bucky of the mammography system. In one CR system, this results in the serial number that will superimpose on the mammogram (Fig. 4.6).

4.2.2 Mammography Unit Related Artifacts

For CR systems, the artifacts reside often in the IP reader. A typical example here is dust on the readout laser of the IP reader. This results in *white lines and/or black lines*, which follow the readout direction and are perpendicular to the mammogram (Fig. 4.7). Because of the linear structure, this artifact can be recognized easily. The composition of the lines and the different contrasts of the lines relate to the type of dust and the effect of processing on the final image after it passed the readout. Although the internals of IP readers can often be cleaned easily, care should be taken towards the placing of the reader. If this is foreseen in a busy environment with a lot of people passage, the equipment will suffer more from intruding dust.

Similar to dust, an *incorrect alignment* of the IP reader (Fig. 4.8) can also have a negative influence on

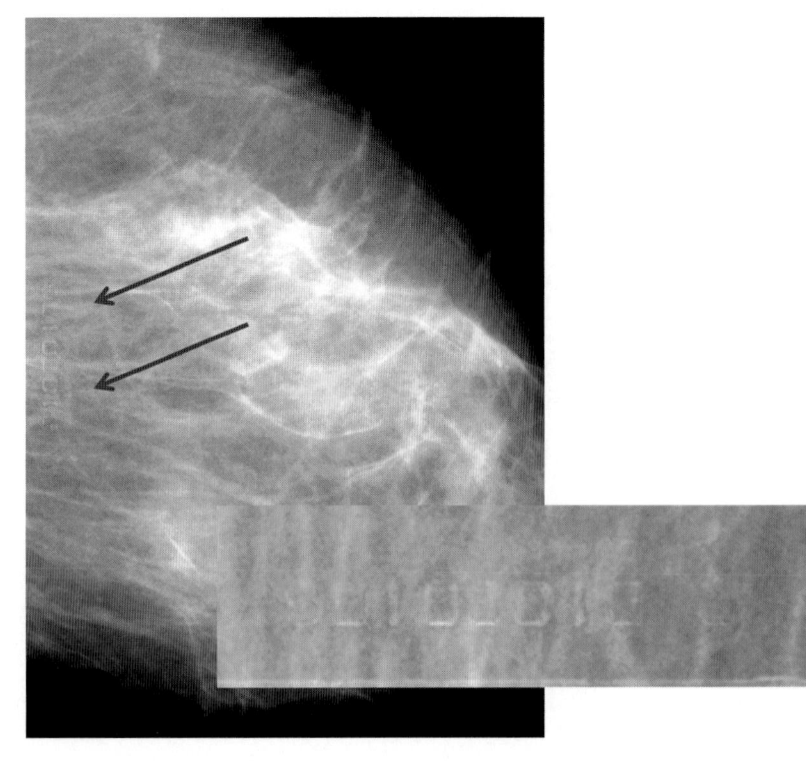

Fig. 4.6. Oblique view of the left breast. The identifier of the IP becomes visible in the pectoral part of the breast due to an incorrect manipulation of the IP by the radiographer (arrows) (Fuji Profect CR)

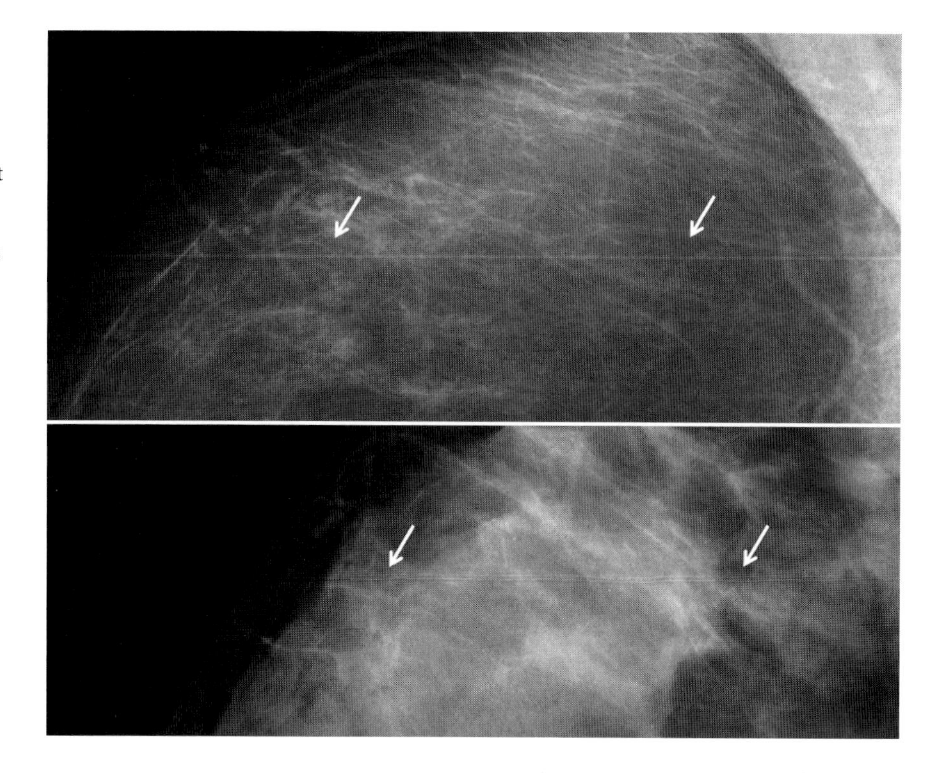

Fig. 4.7. Two examples of a scanline artifact. The *white lines* (arrows) are caused by dust in the CR reader. The different appearance of the artifact in the two images can be explained by the impact of processing on the lines (Agfa CR85)

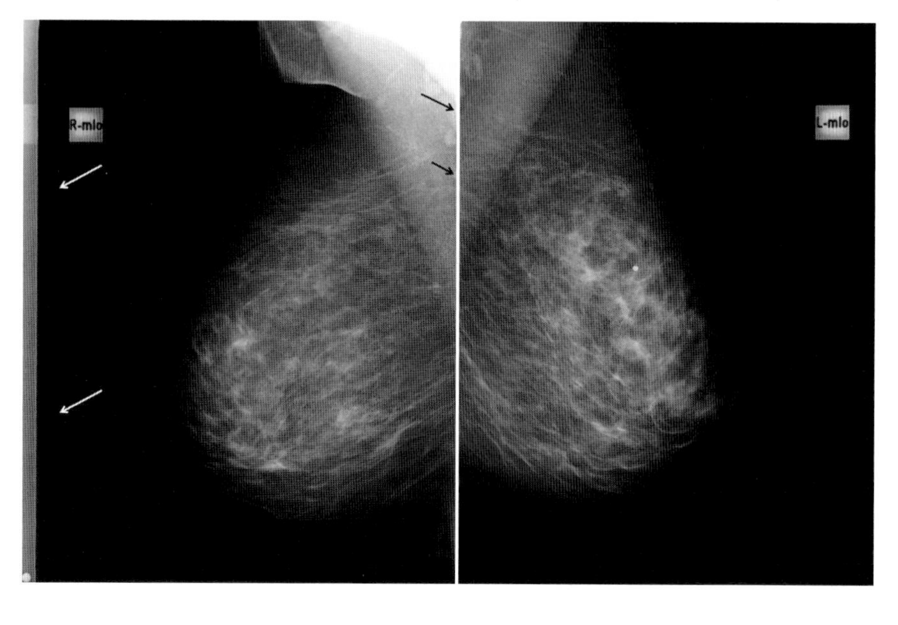

Fig. 4.8. Oblique view of the right and left breast. There is a *white area* at the pectoral side of the right breast: a small pectoral part of the right breast (*black arrows*) is missing and a *gray vertical line* is present (*white arrows*) due to alignment problems of the CR reader due to bad positioning of the IP in the CR reader (Fuji Profect CR)

the quality of the readout and can possibly result in missing breast tissue. The recognition of this artifact is of high importance for the clinical interpretation of the mammograms. A specific test for this is foreseen in the physics protocol.

Deterioration of the detector can appear under different circumstances.

When parts of the images of a-Se detector detectors start to be blurred, this may be caused by crystallization of the selenium (Marshall 2006). As this blurring starts at the axillary side or the bottom of the image, it can take some time before it can be recognized and before it starts affecting image quality. This crystallization is progressive and, in our experience, after a period of one year, the blurring becomes visible in the breast image (Fig. 4.9).

Also typical for DR systems are *defective pixel artifacts*. These can occur in different forms. The

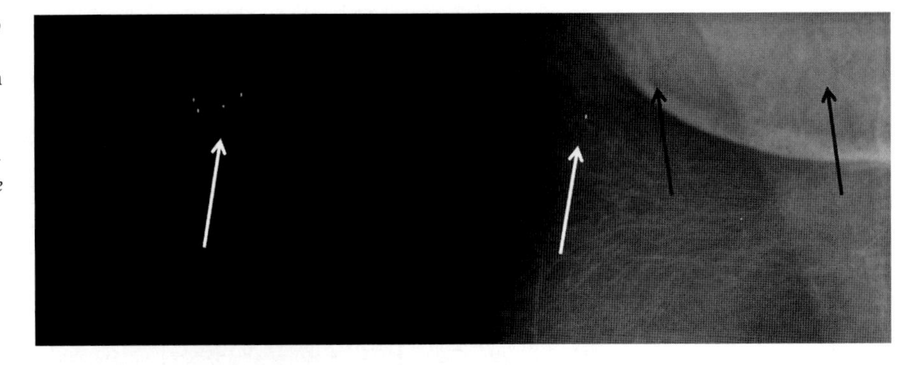

Fig. 4.9. Blurring at the top and bottom edges of the image due to crystallization (*black arrows*) of the a-Se detector (Agfa DM1000 DR). Also, clusters of defective pixels are visible (*white arrows*)

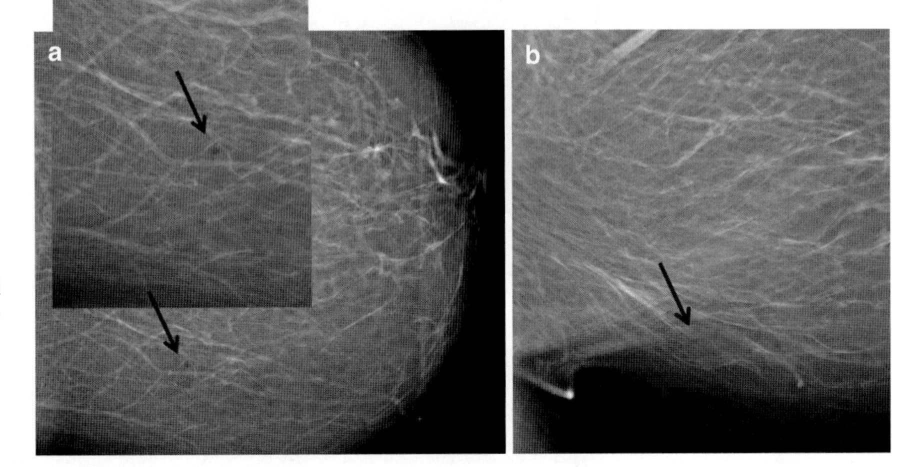

Fig. 4.10. Segments of the cranio-caudal view (**a**) and oblique view (**b**) of the left breast. Presence of a *black circular artifact (black arrows)* at the same position in both images (Siemens Novation DR)

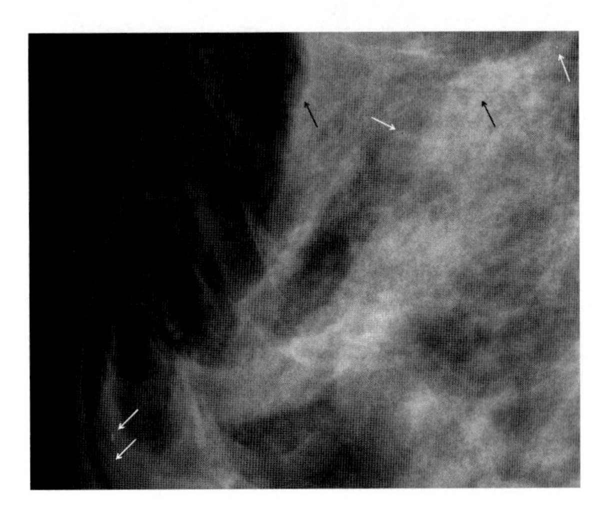

Fig. 4.11. Segments of the oblique view of the right breast. Presence of *white* and *black dot* artifacts (*white arrow*) and horizontal line artifacts (*black arrows*) due to bad flat fielding (Philips Mammo Diagnost DR)

most well-known form is that of a single defective pixel (Figs. 4.9–4.11). The clinical impact of the artifact is often neglectable, but care should be taken if the number of defective pixels becomes too large or if adjacent defective pixels start to form clusters. If these clusters become too large, they could have the shape of pathology-like microcalcifications. Vendors have usually specifications regarding the acceptable number and clustering of these dead pixels.

Another manifestation of defective pixel is blooming (Fig. 4.12). These are the result of an early discharge of a single detector element that subsequently appears as a white spot in the image (as all the signal has disappeared at the data collection moment). The charge of this detector element increases the charge of neighboring elements resulting in a black halo around the white spot.

Electronic artifacts can present as a waveform pattern (example: when there is a problem with the speed of the cooling fans) or as white lines, perpendicular to the pectoral side (when there is a problem with the readout of one column in the detector) (Fig. 4.13). In clinical images, these are often not visible as straight lines but as dashed lines. This is due to the interpolation of the image-processing algorithms. These artifacts are known as "vibration artifacts" (Ayyala et al. 2008).

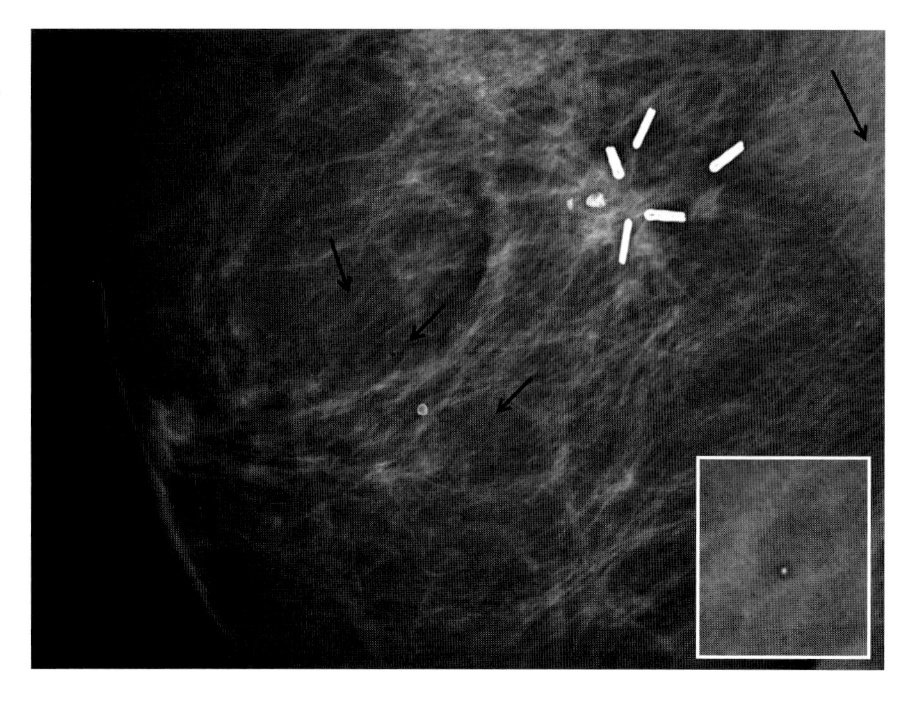

Fig. 4.12. Oblique view of the right breast with multiple *black dot (arrows)* artifacts ("*blooming artifact*") at variable places in the image (Siemens Novation DR)

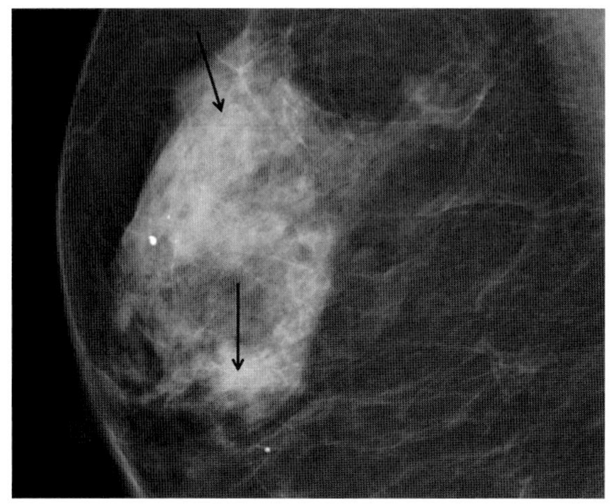

Fig. 4.13. Oblique view of the right breast with vibration artifacts visible at two different places (*black arrows*) (Siemens Inspiration DR)

Ghost or lag ghost artifacts introduced by an incorrect electronic clearing of the detector is a typical problem for systems using an amorphous selenium detector (Fig. 4.14a, b) (Bloomquist et al. 2006b). Although regularly seen during the physical–technical constancy checks of these systems, the influence of this artifact on the interpretation of clinical data is not very clear. It is recommended, however, to avoid this type of artifact.

Besides artifacts related to the specific technology used in DR systems, several artifacts can also be introduced in the system, mainly due to *bad calibration procedures*. The proper handling of several DR systems requires a calibration following a strict time schedule. In Fig. 4.11, we show images of a system where this calibration had not been performed: in the homogeneous image, several pixels with reduced sensitivity can be observed. The same pixels show also reduced pixel values in the clinical images. The calibration procedure aims to acquire the template that multiplication with an image of a flat field object provides a homogeneous image. It compensates for both sensitivity problems of the detector and for the heel effect. Some systems perform thickness and beam quality specific flat fielding. Multiplication with these templates is performed with all clinical images. If this calibration is performed incorrectly, artifacts are introduced, instead of being solved. On systems that suffer from ghost problems, this calibration should not be performed immediately after a clinical acquisition as the calibration algorithm will tend to compensate for that temporary ghost on all successive images (until a new calibration is performed several days later). Also, artifacts visible on the calibration phantoms (Fig. 4.15) or dust in front of the filter of the tube during calibration (Fig. 4.16) can result in errors in the calibration result. Owing to the influence of the calibration results in the next acquisitions, these artifacts appear in all clinical data. Therefore, radiologists should be able to recognize them and give instructions to recalibrate the system.

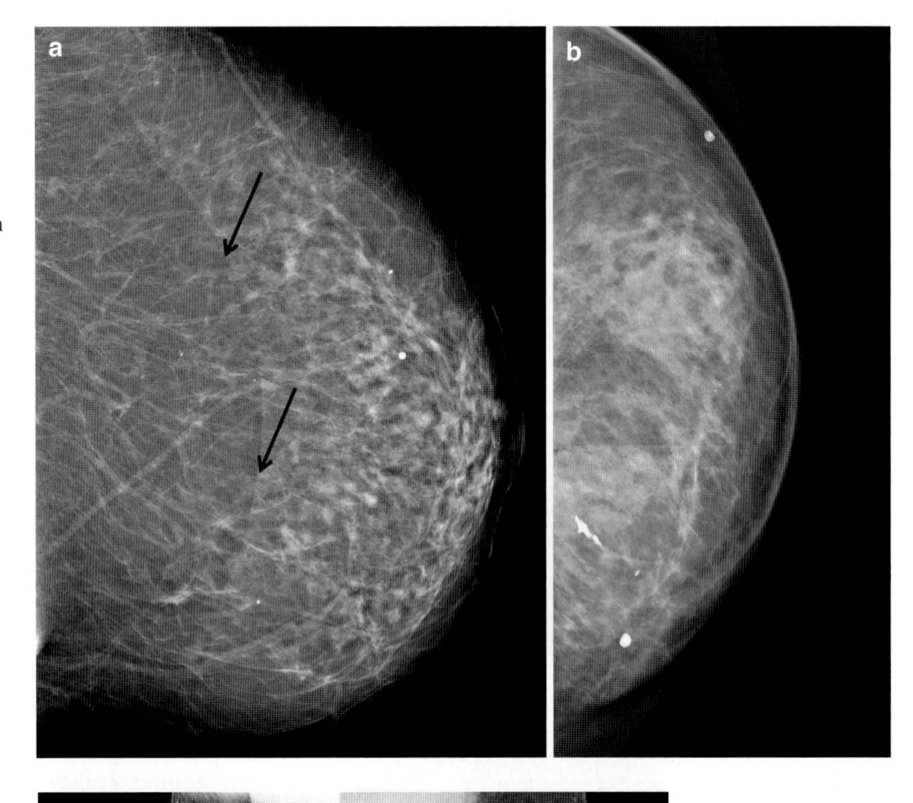

Fig. 4.14. Cranial caudal view of the left breast (**a**) contains a lag ghost (*arrows*), which is the electronic rest of the cranial caudal view of the right breast, (**b**) (horizontally flipped for comparison reasons)

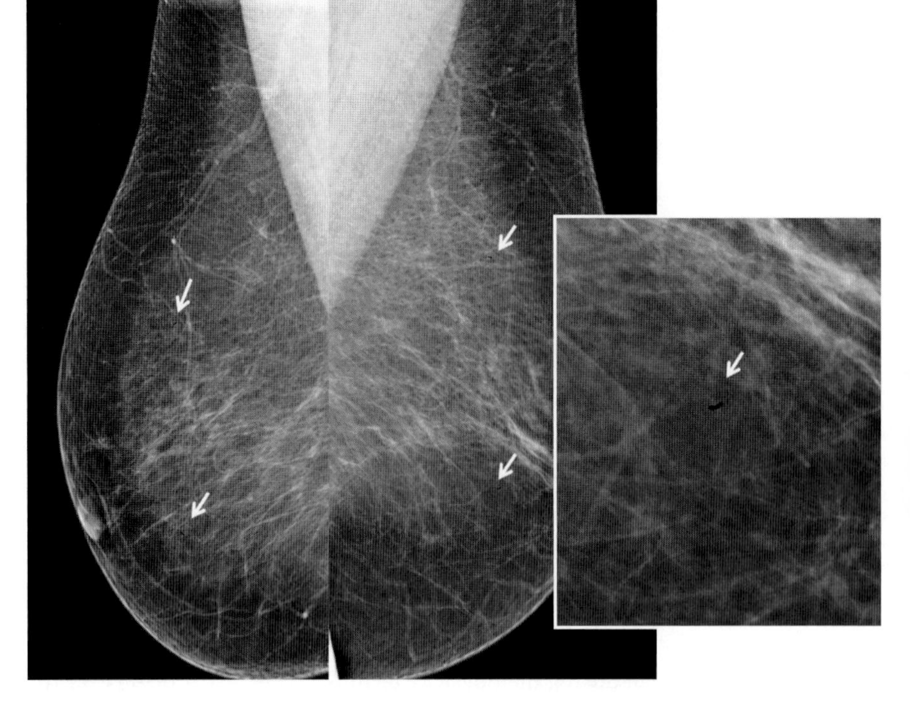

Fig. 4.15. Oblique view of the right and left breast. *Black dots (white arrows)* are visible on the same place in the images because of one defect in the calibration phantom, which causes multiple calibration errors due to the handling of the phantom during the calibration procedure (Hologic)

4.2.3 Software-Related Artifacts

Software-related artifacts can occur at several points during the acquisition process.

We have observed the following artifacts: (1) the system reads out the wrong detector size: the complete detector instead of only the smaller part is read out and then scaled to the smaller size, resulting in random border artifacts, loss of image quality, and

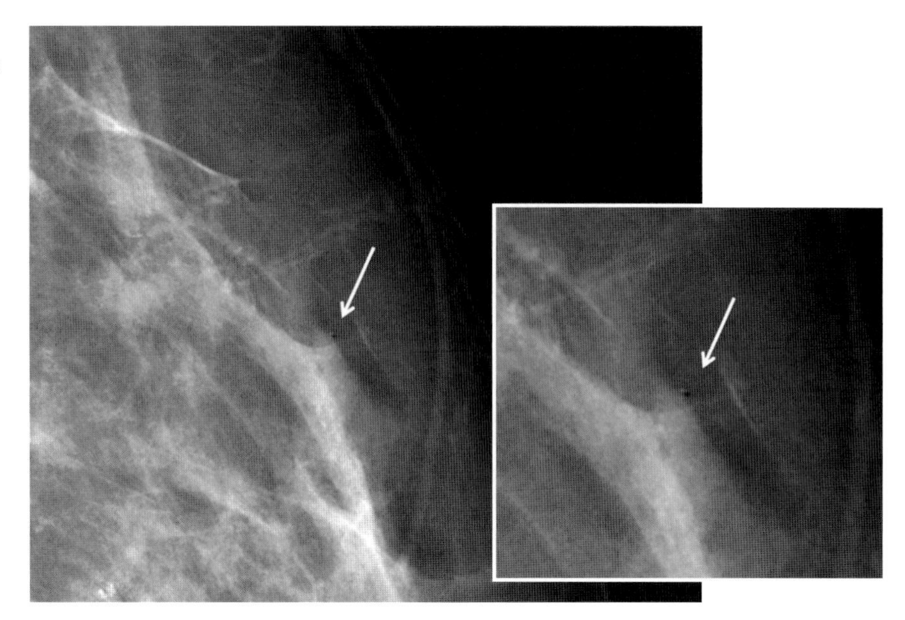

Fig. 4.16. *White* and *black* artifact (*arrow*) can be seen due to a combination of dust particles on the filter during calibration and the impact of the image processing on this artifact (Siemens Novation DR)

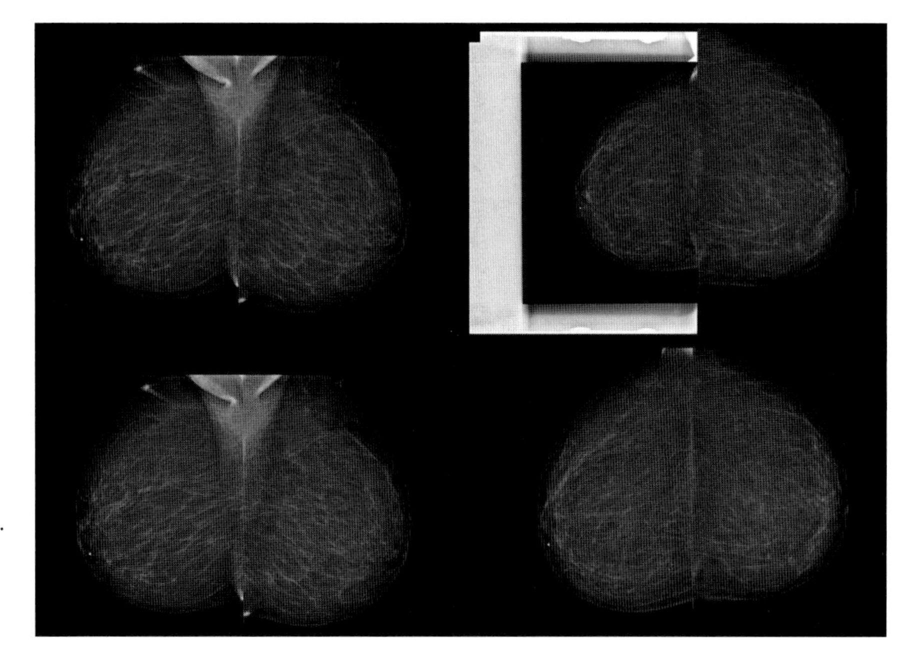

Fig. 4.17. Because of a software error, the system reads out the wrong detector size (large detector size) and then scales back to the correct detector size. This artifact is visible in the cranial caudal view of the right breast (Siemens Inspiration DR)

wrong distance measurements (Fig. 4.17), (2) horizontally flipped images, or (3) shifted parts of clinical images (Fig. 4.18). These artifacts happen just as a very infrequent coincidence and are so obvious that a repeat image is taken and bug report forms are completed.

Image processing related software artifacts can have different causes. Algorithms used during the processing of the clinical data may be very sensitive to high contrasts introduced by foreign materials or by very high contrast structures like large calcifications in the breast and therefore could result in the use of incorrect image processing parameters or could introduce artifacts that overlay potentially important information. Typical examples are halo artifacts around biopsy needles or implants (Fig. 4.19a, b).

Another cause could be the use of incorrect image-processing parameters. The reason could be a poor processing algorithm, badly deployed software installation, or upgrade or incorrect configuration settings (Figs. 4.20a, b and 4.21).

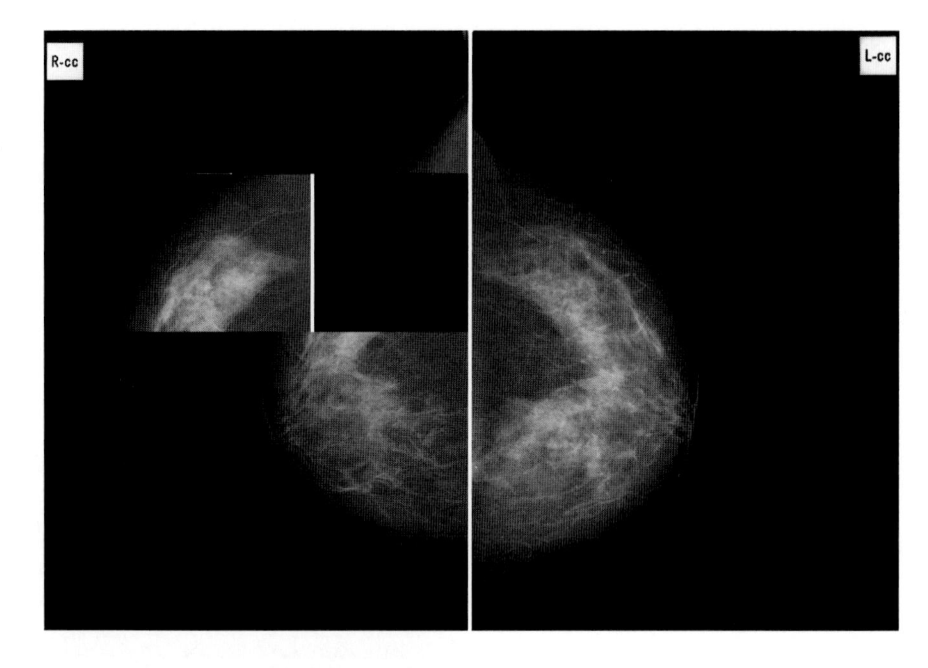

Fig. 4.18. Cranial caudal view of the right breast shows a deformation of a part of the image after transportation to the second reader center (Fuji Profect CR, Agfa IMPAX)

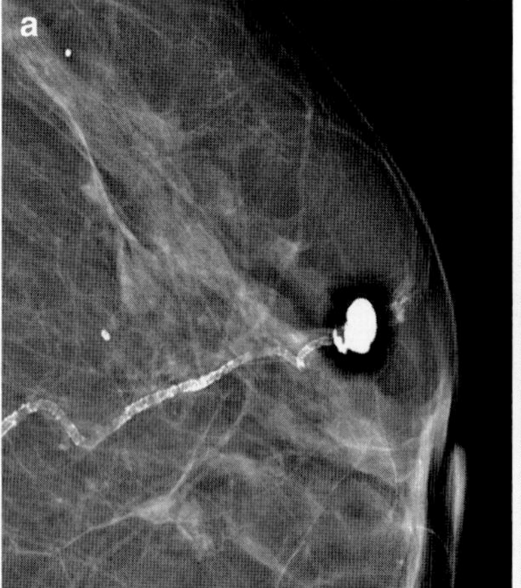

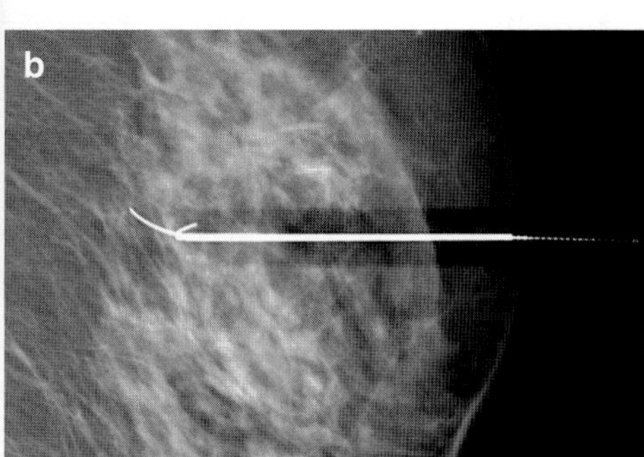

Fig. 4.19. (**a**) A large calcification with a black halo in the left breast, (**b**) a black halo surrounding the dense needle. Both artifacts are due to bad image processing (Siemens Novation DR)

Fig. 4.20. Same image of the right breast (**a**) in 2006, (**b**) in 2008. (**a**) Shows *black areas* in the subcutaneous fat on an image of 2006 while (**b**) shows a gray aspect of this subcutaneous fat due to different processing algorithms (Siemens Novation DR)

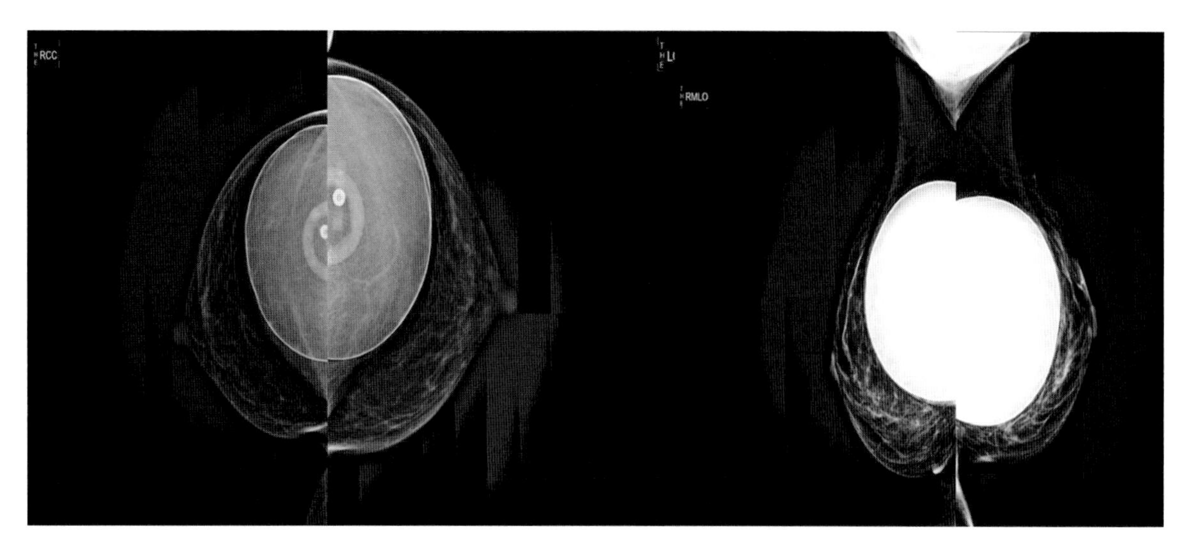

Fig. 4.21. Oblique and cranial caudal view of the right and left breast. Amplifier blocks become visible because of changes in image processing settings (Hologic Lorad Selenia DR)

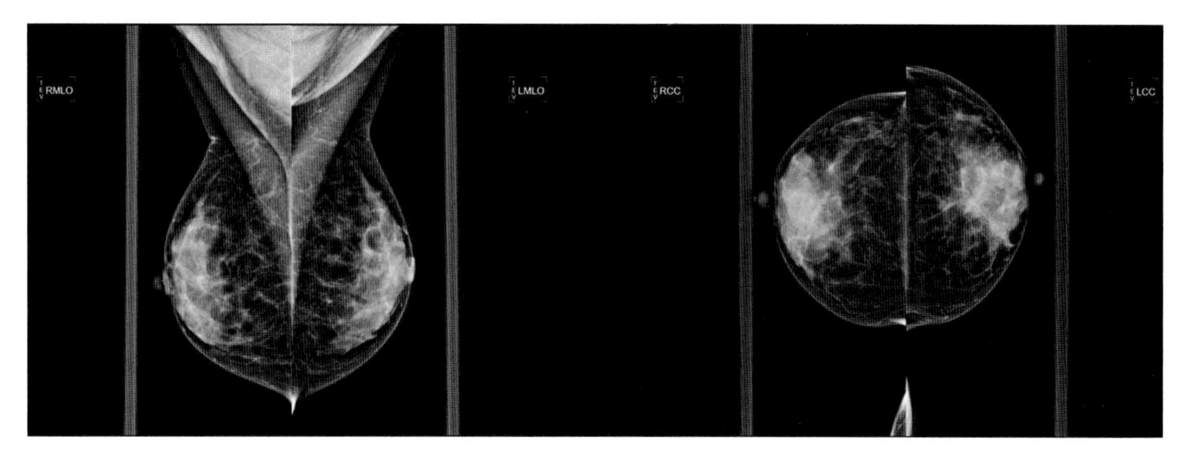

Fig. 4.22. Oblique and cranial caudal view of the right and left breast: presence of white bars in the background of the image, which represent the compression paddle (Hologic Lorad Selenia)

Incorrect interpretation of the DICOM header values or bugs in the viewing software could also introduce problems. Radiologists should be aware of this fact and always critically check the quality of their images after setup or after software upgrades.

Some of the structures seen outside the breast image, i.e., in the black background of the image, are difficult to categorize. By analogy with the image quality of FSM, the background of the digital mammogram must be also equally black. Some of the disturbing white structures, like vertical or horizontal lines at the borders of the images, can be easily omitted with adequate processing. Edges of the compression plate should not be visible in the images (Figs. 4.22 and 4.23a, b).

4.3 Conclusion

In QA procedures of the full digital imaging chain, the constant evaluation of clinical data for artifacts is crucial. Clinical images without artifacts are a sign of good quality care. Some of the mentioned artifacts could cause difficulties for the diagnostic evaluation of mammograms. It is important that the radiologist recognizes them and takes proper action, as he is responsible for the ultimate patient care in the radiology department: for certain artifacts, he must learn that an additional image is much better than the acceptance of a suboptimal examination.

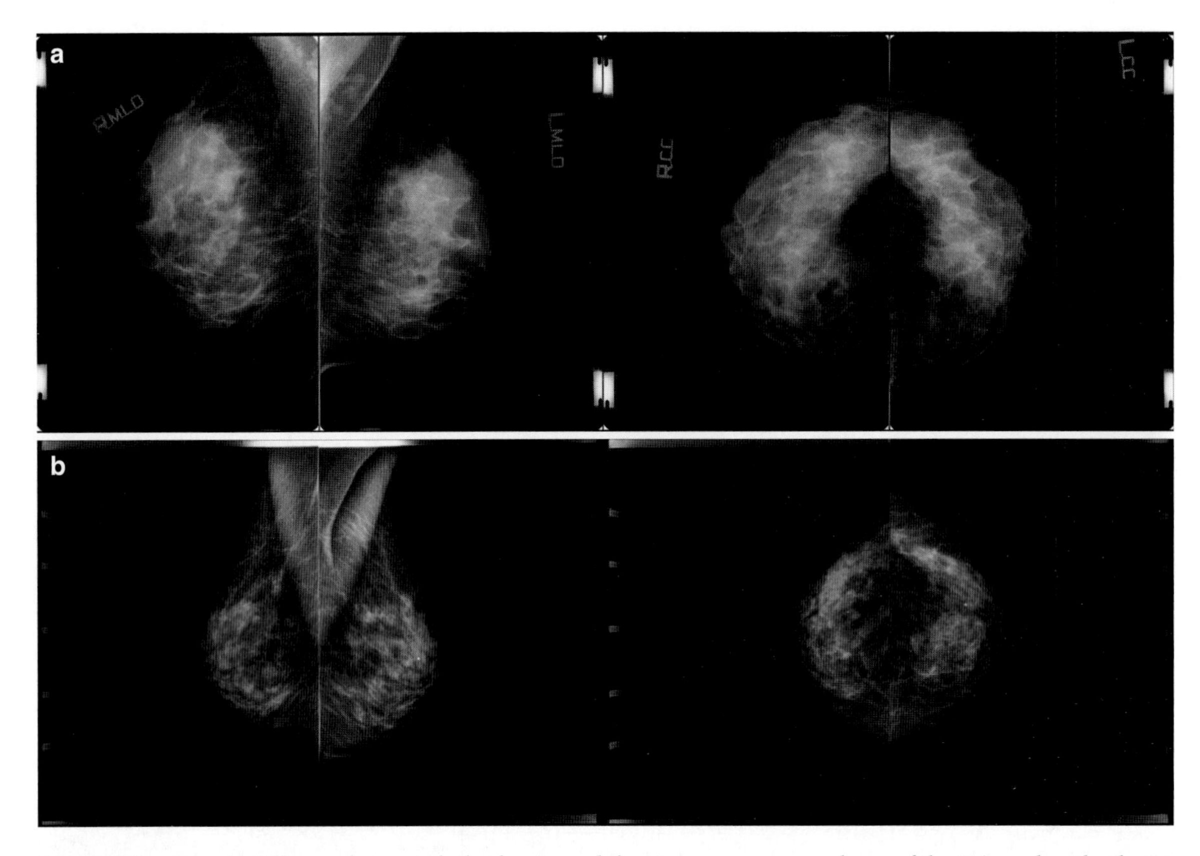

Fig. 4.23. Because of a difference between the bucky size and the IP size, an unexposed part of the IP is read out by the CR reader. (**a**) Fuji Profect CR, (**b**) Agfa CR85

The majority of these artifacts would also show up on images of homogeneous slabs of PMMA. This supports the idea in QC protocols to perform stability tests on a systematic basis. One of the challenges for the radiographer or physicist in charge of the quality supervision of these QC procedures is to link technical artifacts with the clinical relevance. Collecting, classifying, and reporting artifacts in clinical images is a time-consuming and tedious work, but will ultimately lead to a classification into unimportant artifacts and artifacts on which to react (eventually with a replacement of hardware or software components, and agreements to be negotiated during the purchase of a system). To simplify this process, vendors are encouraged to incorporate dedicated software tools for problem reporting in the radiological workstations or in the PACS. This would bring the task of maintaining quality nearer to the people who use the system on a day-by-day base. A good start here could also be to stress the importance and the appearance of these artifacts in the education of radiologists, technologists, and physicists (Van Ongeval et al. 2008).

We believe that a close collaboration between radiologists, technologists, and medical physicists is a must and is a necessary condition for QA. The European guidelines prescribe also a "long-term stability" evaluation. Present overview of artifacts shows that such a weekly (or daily) visual check of a homogeneous image is a relevant task and should therefore be organized (Jacobs et al. 2008). In the same document, technologists and radiologists get also the task to maintain high quality.

Initiatives to share artifacts between centers should be initiated. Databases of image artifacts would not only improve the knowledge, but also open the door toward an automated tracking or analysis.

References

Ayyala RS, Chorlton MA, Behrman RH, et al (2008) Digital mammographic artifacts on full-field systems: what are they and how do I fix them? Radiographics 28:1999–2008

Bassett LW (1995) Quality determinants of mammography: clinical image evaluation. In: Kopans DB, Mendelson EB (eds)

Syllabus: a categorical course in breast imaging. Radiological Society of North America, Oak Brook, IL, pp 57–67
Bloomquist AK, Yaffe MJ, Pisano ED, et al (2006a) Quality control for digital mammography in the ACRIN DMIST trial: part I. Med Phys 33(3):719–736
Bloomquist AK, Yaffe MJ, Mawdsley GE, et al (2006b) Lag and ghosting in a clinical flat-panel selenium digital mammography system. Med Phys 33(8):2998–3005
EUREF, Van Engen R, Young K, Bosmans H, et al (2006) Part B: digital mammography. In: European guidelines for quality assurance in breast cancer screening and diagnosis, 4th edn. European Commission, Luxembourg. Downloadable from www.euref.org
Hogge PJ, Palmer CH, Muller CC, et al (1999) Quality assurance in mammography artifact analysis. Radiographics 19:503–522
Jacobs J, Lemmens K, Nens J, et al (2008) One year of experience with remote quality assurance of digital mammography systems in the Flemish breast cancer screening program. Proceedings of the 9th International Workshop on Digital Mammography, Tucson Arizona
Marshall NW (2006) Retrospective analysis of a detector fault for a full field digital mammography system. Phys Med Biol 51(21):5655–5673
NEMA PS 3.14 (2000) Digital imaging and communications in medicine (DICOM) Part 14: grayscale standard display function. National Electrical Manufacturers Association, Rosslyn, VA
Saunders RS Jr, Baker JA, Delong DM, et al (2007) Does image quality matter? Impact of resolution and noise on mammographic task performance. Med Phys 34(10):3971–3981
Van Ongeval C, Van Steen A, Bosmans H (2008) Teaching syllabus for radiological aspects of breast cancer screening with digital mammography. Radiat Prot Dosimetry 129(1–3):191–194
Yaffe MJ, Bloomquist AK, Mawdsley GE, et al (2006) Quality control for digital mammography: Part II recommendations from the ACRIN DMIST trial. Med Phys 33(3):737–752

Image Processing

5

Nico Karssemeijer and Peter R. Snoeren

CONTENTS

KEY POINTS

Image processing is a crucial element of modern digital mammography. Optimizing mammogram presentation may lead to more efficient reading and improved diagnostic performance. Despite that the effects of image processing are often much larger than those of acquisition parameter settings, little is known about how image processing can be optimized. Experts agree that comparison of features in various mammographic views is very important. This issue must be addressed by processing. Variation of image presentation across views and subsequent mammograms should be minimized. The dynamic range of electronic displays is limited. Therefore, processing techniques should be designed to limit the dynamic range of mammograms. This can effectively be done by applying peripheral enhancement in the uncompressed tissue region near the projected skin–air interface. Adaptive contrast enhancement can be applied to enhance microcalcifications and dense tissue in the interior of the mammogram. Mammogram processing should be aimed at displaying all relevant information in good contrast simultaneously, as human interaction to manipulate contrast during reading is too time-consuming to be applied on a regular basis.

5.1 Introduction

The goal of mammography is to detect and diagnose breast cancer, a task which is generally performed by experienced and skilled radiologists. For optimal reader performance, mammograms have to be matched to the human visual system when they are

Nico Karssemeijer, PhD
Department of Radiology, Radboud University Nijmegen Medical Center, PO Box 9101, 6500HB Nijmegen, The Netherlands
Peter R. Snoeren, PhD
Department of Radiology, Radboud University Nijmegen Medical Center, PO Box 9101, 6500HB Nijmegen, The Netherlands

presented. In other words, characteristic features of cancer have to be displayed with optimal contrast to avoid being missed or misinterpreted. Within the limited possibilities of conventional screen-film mammography, this matching problem has received much attention in the past. Many innovations have been made in mammography to enhance visibility of cancers. Most notable in this respect is the gradual increase of contrast in the interior region of the breast at the cost of contrast in the periphery, where cancer seldom occurs. This change was only possible due to development of more accurate automatic exposure control (AEC) devices. With the latest generation of mammography films, the skinline and large parts of the periphery are hardly visible on mammography film alternators. With the introduction of digital mammography, a wide range of new possibilities has become available to enhance mammograms. An overview of digital mammogram processing techniques will be presented here.

Digital image processing is only one part of the mammographic imaging chain. Ideally, the design of a medical image processing method should be independent of image acquisition and display. For image display, this ideal can be achieved by ensuring that display devices conform to the DICOM display standard. Therefore, it is important that one is aware of the mechanisms used in the definition of this standard. Mappings to convert pixel values to luminance in display devices depend on parameter settings in the processed images, like window/level settings and values of interest lookup table (VOI LUT). Selection of appropriate values of these parameters is an issue that should be addressed by processing algorithms. On the other end of the chain is the acquisition device. Digital detectors in mammography devices differ and these differences have to be taken into account when processing algorithms are designed. Examples are variation in image resolution, gain, modulation transfer function, and noise characteristics. Fortunately, important acquisition parameters such as anode material, ltration, and kVp are provided in the DICOM header and can thus be used in processing algorithms. Despite many differences, there is a major advantage in the use of digital detectors: in the range of interest, pixel values are more or less proportional to X-ray exposure at the detector. This allows design of robust processing algorithms which can be applied to a variety of systems.

Digital mammography manufacturers have only just begun to explore the enormous benefit that digital processing and display may provide. An interesting pictorial essay of some mammographic processing techniques was given by Pisano et al. (2000a). Here, we will discuss some of the basic methods currently employed and more advanced methods that will likely become available in the next generations. Basic processing methods include grayscale transforms and adaptive contrast enhancement. A common dedicated mammogram processing method is peripheral enhancement, which has been adopted widely by manufacturers to overcome shortcomings of the dynamic range of digital displays.

5.2 Grayscale Transforms

Application of lookup tables (LUT) to change the grayscale of an image can be considered as the most elementary form of image processing, as it operates on individual pixels. Almost all digital imaging systems apply such transforms. Obviously, parts of the image with the highest diagnostic information content should be displayed with optimal contrast, while contrast may be reduced in parts which are less relevant. To achieve this, usually a nonlinear mapping of pixel values is required. The design of appropriate mappings has been facilitated by the DICOM standardization of displays. According to this standard, medical displays should be perceptually linear, which can be achieved by adhering to the DICOM Grayscale Standard Display Function. When a display is calibrated properly, similar differences in pixel values should be perceived as similar differences in luminance, regardless of the luminance level. This ensures that images in a clinical environment are presented in a predictable way, allowing optimization of image processing algorithms for a given diagnostic task.

Mammogram presentation has been perfected over the years in conventional mammography. Therefore, as a strategy to define a good grayscale transform for digital mammograms, one could aim at creating a film-like presentation. The transitions of signals in conventional mammography are well known. First, the response of a film-screen system to X-ray exposure E is determined by a characteristic curve, which expresses the relation between the logarithm of exposure and optical density D of the exposed film. Second, when viewed on a lightbox, the intensity of the transmitted light through the film is expressed by $I = I_0 \log_{10}(-D)$, with I_0 being the luminance of the lightbox. Using Weber's law a relation with perceptually linearized digital displays can be made. Within a wide range of intensities, just noticeable differences in the intensity correspond to equal differences in the optical density. Thus, by applying a characteristic curve with a shape similar to that of modern screen-film systems, the exposure representation of raw digital mammograms can be converted to a processed image which looks like a conventional mammogram when displayed on a DICOM calibrated display device. However, because higher optical densities represent lower intensity, after application of the characteristic curve, the image has to be inverted.

Examples of characteristic curves of mammographic screen-film systems are shown in Fig. 5.1. These curves can be well-modeled by

$$D(E) = D_{bf} + D_{max}\left[\frac{1}{1+e^{-g(\ln E - s)}}\right]^{q} \tag{5.1}$$

$$= D_{bf} + D_{max}\left[1 + (E/s)^{-g}\right]^{-q}$$

with X-ray exposure denoted by E, $D(E)$ being the optical density, D_{bf} being the base plus fog optical density of the unexposed film, and D_{max} being the optical density of the fully exposed film. The parameters g and s represent the gradient and speed of the film, respectively. The parameter q can be used to model an asymmetric shape of the curve. An example of a symmetric model ($q = 1$) fitted to the characteristic curve of a Kodak Min-R 2000 film is shown in Fig. 5.1.

To compute a proper grayscale transform for a mammogram using the model mentioned earlier, the parameter s has to be determined. This parameter has to be chosen, such that relevant information in the exposure domain maps to the steepest part of the characteristic curve. In conventional mammography systems, this parameter is fixed, while exposure is adjusted to the proper range by means of the AEC unit of the mammography system. The AEC shuts off the exposure when a limiting value is reached in a measurement field, which is located in a central location in the breast projection. In digital mammography, exposure may vary over a wide range, as restrictions to proper film exposure no longer exist. The issue of mapping the exposure values in a proper range of contrast can be addressed in a similar fashion though, by computing the parameter s from exposure values in the interior part of the breast. When the breast is segmented, s can be chosen proportional to the average pixel value in the interior part of the breast. To avoid segmentation, a sliding window can be used to scan the central part of the image. The location where the average pixel value in the sliding window has the highest value represents the densest part of the breast. This value can be used to adjusts. For robustness, the window should not be too small, in the order of several square cm. An example of prior and current digital mammograms of a patient acquired with a Lorad Selenia processed with this method is shown in Fig. 5.2. It can be seen that all images have excellent contrast and have a very similar appearance.

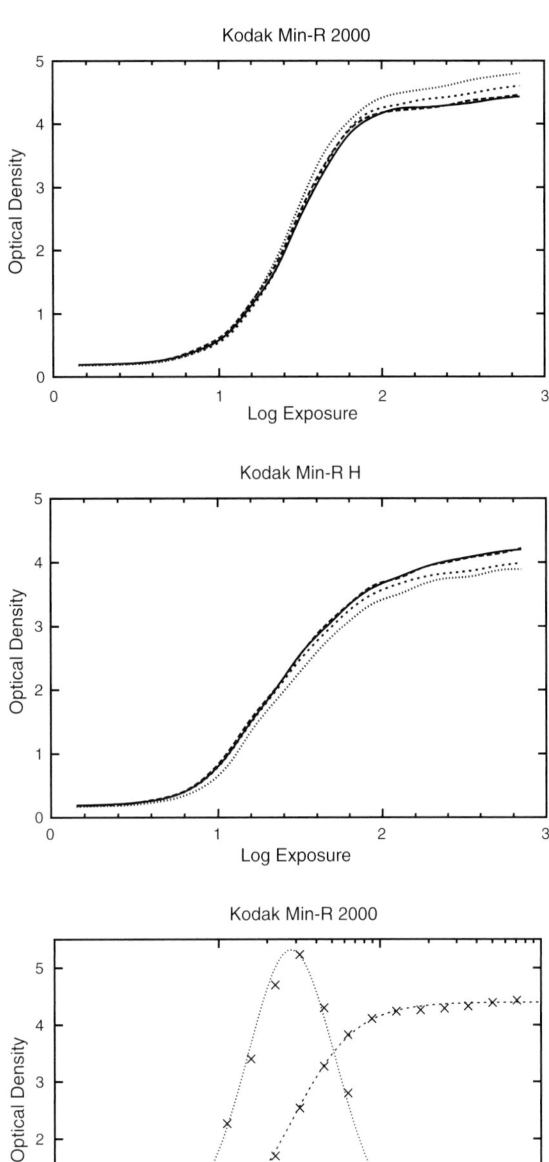

Fig. 5.1. Characteristic curves of a Kodak Min-R 2000 film and a Kodak Min-R H film. The *right figure* shows a model fit

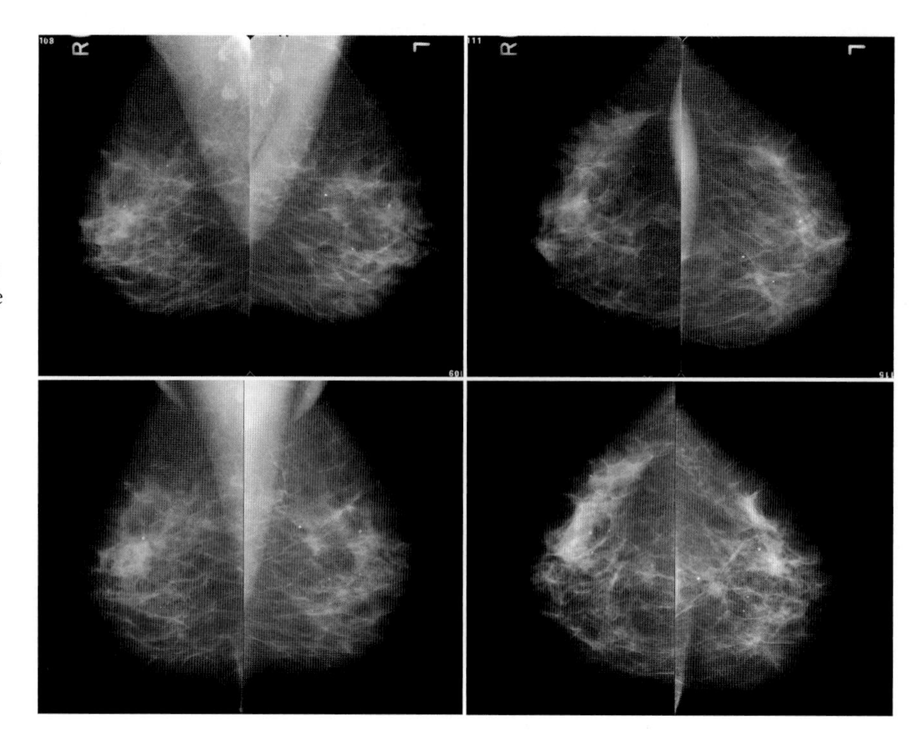

Fig. 5.2. Full field digital mammograms (Hologic Selenia) processed with a characteristic curve of a Kodak Min-R 2000 system to give them a film-like appearance. The two mammograms are prior (*top*) and current (*bottom*) mammograms of the same patient

Grayscale transforms are also be applied by the display system. DICOM has implemented several mechanisms for scale conversion of mammograms. Conversions are determined by values of tags in the DICOM header. Normally, a linear rescaling of pixel values within a range of interest is carried out based on the window and level settings. However, a nonlinear rescaling is also possible by defining a VOI LUT. Furthermore, a sigmoid function can be defined parametrically. Using these mechanisms, manufacturers may choose to maintain exposure-related pixel values and use DICOM functionality to define image presentation.

5.3 Spatial Enhancement

With digital imaging, a wide variety of methods have become available for spatial enhancement of images. In contrast to grayscale transforms, spatial enhancement techniques change pixel values based on spatial context. They can be used to sharpen edges, reduce noise, or to increase contrast of faint dense tissue regions. In this section, several general enhancement techniques are discussed, while more specific enhancement methods developed for mammography are discussed in the following sections. Image enhancement does not increase the information content in mammograms. Its aim is to increase contrast of relevant features in a mammogram, so that they can be detected and interpreted more easily. Because of the reduced dynamic range of softcopy display in comparison to film viewers, spatial contrast enhancement is seen by many as a requirement for proper presentation of digital mammograms.

5.3.1 Unsharp Masking

Unsharp masking is a technique for sharpening images by using a blurred mask of the original image. The technique was already in use in photographic processing and became very popular with digital imaging. Mathematically, the filter is expressed by

$$y_i' = y_i + f(y_i - s_i(\sigma)) \tag{5.2}$$

where y_i and y_i' are the original and processed pixel value at location i, the smoothed image value at i is $s_i(\sigma)$, and f is a function determining the amount of enhancement. The results are strongly dependent on f and on the scale of the smoothing kernel. Blurring is often performed by Gaussian smoothing. The scale of the smoothing function determines the frequency range in which contrast enhancement occurs. Use of a small scale limits the enhancement to smaller structures. The function f determines the amount of

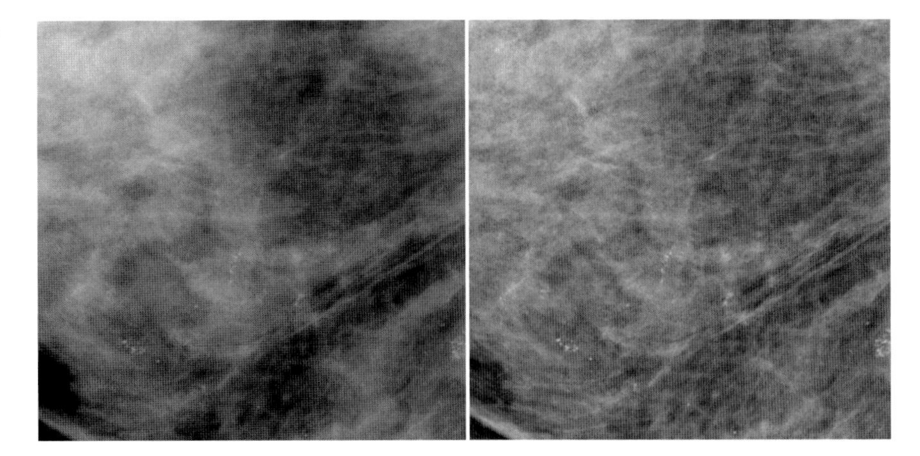

Fig. 5.3. Effect of unsharp masking on a mammogram with microcalcifications. On the *left*, the original image is shown. The right figure shows the processing result obtained with $\sigma = 1$ mm and f a sigmoid function with slope 1

enhancement. In the simplest case, a constant is used, causing a percentage of the difference image to be added to the original. However, this may render structures that already have high contrast too sharp or outside the pixel value range. To prevent this, a sigmoid function can be used to limit contrast enhancement to a fixed maximum value.

Unsharp masking is most often used to enhance high frequencies in an image. In digital mammography, the technique can be used to enhance microcalcifications. Figure 5.3 shows an example in which the visibility of microcalcification is clearly improved by processing. Blurring was performed by Gaussian smoothing with $\sigma = 1$ mm. When CRT displays are used, enhancement of high frequencies can compensate for unsharpness of the display device. A similar effect may be obtained by deconvolution, but in that case, characteristics of the display device have to be known. As most mammographic workstations are nowadays equipped with LCD displays, there is less need for compensating display unsharpness. Unsharp masking may also be advantageous to enhance digitized films, which often have reduced sharpness due to the digitizer characteristics. A disadvantage of unsharp masking is that it increases visibility of noise.

One can also use unsharp masking to enhance contrast of larger structures by using a larger scale for blurring. This may be advantageous if the dynamic range of images is large. Without processing, mapping of the full dynamic range to the available gray levels results in low contrast of image structures. This is shown in Fig. 5.4, where the same image is processed with different parameter settings. Contrast in the original image is low, while in the processed images with larger values, contrast is enhanced. Note that unsharp masking also gives rise to artifacts. These may especially be observed at sharp boundaries of bright structures. Enhancement causes darker and brighter rims at both sides of the boundary. This "ringing" effect is often visible when adaptive contrast enhancement techniques are applied.

5.3.2 Adaptive Histogram Equalization

Histogram equalization is a technique to compute a grayscale transform in such a way that in the processed image all gray values occur at equal frequency. The method is easy to implement and fast. However, in general, it does not produce an acceptable image, as it does not take into account that the information content in an image may depend on the signal level. When histogram equalization is applied to a mammogram, it redistributes gray levels in a way that expands contrast of the background at the cost of contrast in the tissue area. This can be easily understood: If we take a typical mammogram with one-third of the image representing the projected breast, the breast tissue maps to only one-third of the gray value range after processing. To overcome this limitation, improvements have been proposed (Pizer et al. 1987).

In adaptive histogram equalization, a different grayscale transform is computed at each location in the image, based on a local neighborhood, and the pixel value at that location is mapped accordingly. The local neighborhood is usually chosen as a square tile centered at the pixel to be processed. The diameter of the tile is an important parameter in the algorithm. When it is too small, the method becomes too sensitive to local variations, and when it is too large, limitations of the nonadaptive technique start to play a role. A typical

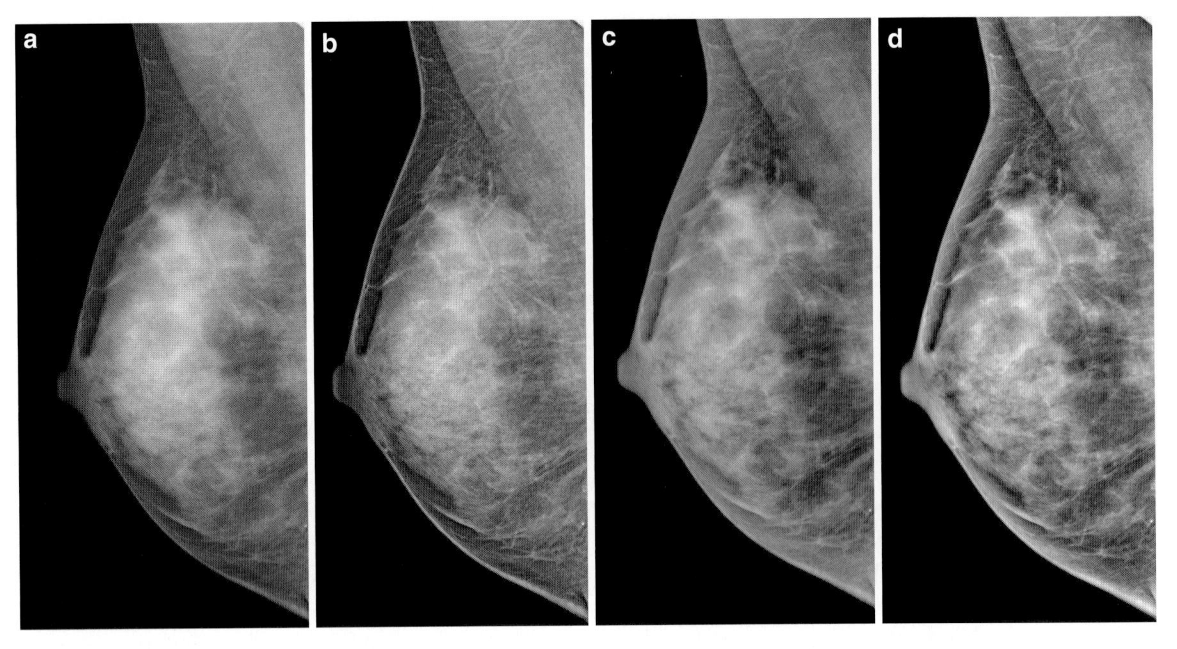

Fig. 5.4. Unsharp masking of a mammogram obtained with a GE SENOGRAPHE 2000D. The original image (**a**) is processed using the following settings: σ = 1 mm, slope = 1 (**b**) σ = 1 cm, slope = 1 (**c**) σ = 5 mm, slope = 4 (**d**)

value for the diameter is 1/16 of the original size of the image. Obviously, computing the transform at every pixel is computationally intensive. Optimizations have been proposed in which the transform is computed in a limited number of overlapping tiles, whereby processing of pixels not centered in tiles is performed by interpolation of the neighboring mappings.

By changing the slope of the transform that converts pixel values from the original to the processed image, contrast is increased. In this way, both signal and noise are enhanced proportionally. This may not be desirable, because when the enhancement of noise becomes too strong it may affect performance of the radiologists. With adaptive histogram equalization, there is a risk of increasing noise to an unacceptable level in image regions that have little signal variation, e.g., homogeneously dense tissue areas or background. To reduce contrast amplification in such areas, contrast limited adaptive histogram equalization (CLAHE) has been proposed. It can be easily shown that the slope of the transform computed by histogram equalization is proportional to the height of the histogram. Thus, by clipping and renormalizing the histogram before computing the transform, the slope can be limited. This is what CLAHE does. An additional parameter is introduced with which the maximum contrast enhancement can be adjusted. An example of CLAHE processing of a mammogram is shown in Fig. 5.5.

5.3.3 Multiscale Image Enhancement

With multiscale image processing techniques, it is possible to tune contrast enhancement to certain frequency bands. In this way, features occurring at different scales can be enhanced in a different way. For instance, one could aim at enhancing microcalcifications and masses in a range of scales, while suppressing other structures. A common multiscale processing technique is wavelet processing. Application in mammography was proposed by LAINE et al. (1994). In this method, multiscale edges identified within distinct levels of transform space provide local support for image enhancement. Mammograms are reconstructed from wavelet coefficients modified at one or more levels by local and global nonlinear operators. In each case, edges and gain parameters are identified adaptively by a measure of energy within each level of scale-space. The authors demonstrate that wavelet processing can reveal features that are barely seen in unprocessed traditional mammograms. However, the clinical benefit of displaying such features has not been demonstrated. It may be confusing for the readers to be presented with images that deviate strongly from the images they are trained with. More recently, HEINLEIN et al. (2003) also developed a mammogram processing method based on wavelets. The main novelty is the

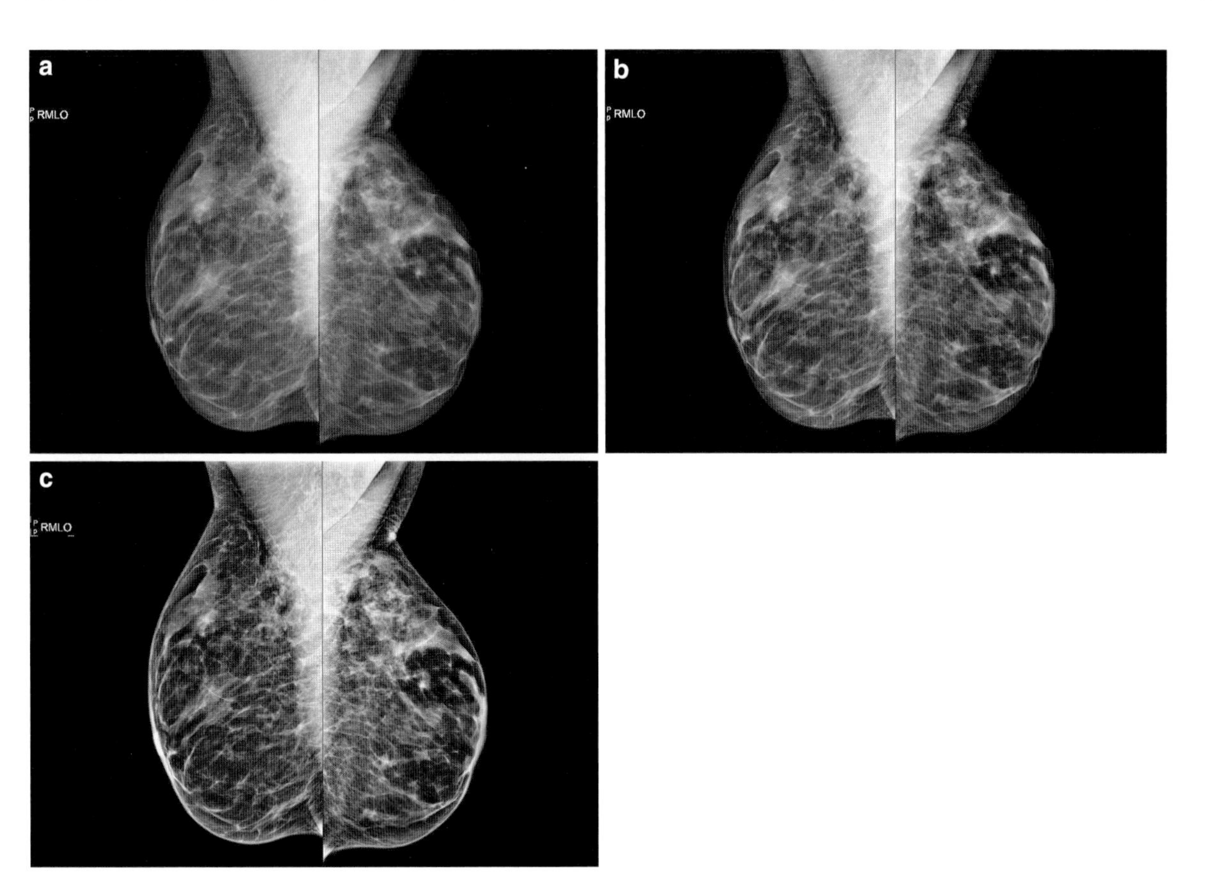

Fig. 5.5. A raw digital mammogram from a Lorad Selenia processed using nonspatial processing and peripheral enhancement (**a**). Figure (**b**) shows results obtained by applying contrast-limited adaptive histogram, equalization (CLAHE). The clinical image processed by the manufacturer, using a proprietary algorithm, is shown in (**c**)

application of a continuous wavelet transform. Furthermore, a model-based approach is used to make the method more specific for microcalcifications.

5.3.4 Peripheral Enhancement

Peripheral enhancement is a dedicated image processing technique developed for mammograms. It is used to improve the visibility of the peripheral uncompressed region of the projected breast, where tissue thickness is smaller than in the interior part of the mammogram. The technique is also referred to as peripheral equalization or thickness correction. In peripheral enhancement methods, the darkening due to decreased tissue thickness in the peripheral region is estimated from the mammogram and thereafter compensated for by a smoothly varying correction function. After correction, fatty tissues in the interior and peripheral regions have similar gray values. With peripheral enhancement, the dynamic range of the mammogram greatly reduces, and as a consequence, less manual adjustments of contrast settings are required to view details close to the skinline. This benefits workflow. In the following paragraph, we will discuss filter-based peripheral enhancement techniques and a parametric method in which the three-dimensional breast outline is modeled.

Peripheral enhancement was first developed as a preprocessing stage in computer aided detection (CAD) systems. Byng et al. (1997) were the first to propose the use of this technique for enhancement of mammogram display. The method that they describe is a nonparametric filter-based method. Filtering is used to obtain a blurred version of the mammogram representing tissue thickness. This approach can be used because breast thickness variations are smoother than tissue density variations. Thickness equalization is only applied in the periphery of the breast, which is simply determined by a threshold T representing gray values at the border of compressed and uncompressed part of the breast. Denoting s_i as the pixel in the

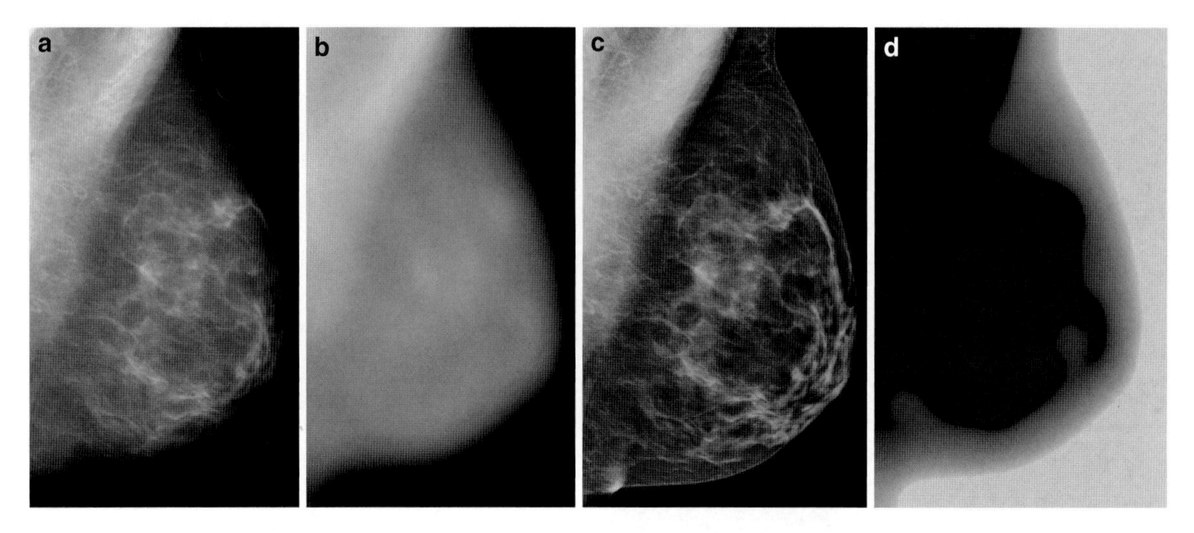

Fig. 5.6. Example of thickness correction with anisotropic smoothing and dense tissue interpolation. The original mammogram (GE FFDM) without enhancement is shown in (**a**). After segmentation and removal of dense tissue regions, the image is blurred by anisotropic diffusion (**b**). The corrected image (**c**) is obtained by adding an image representing the tissue thickness difference derived from the blurred image (**d**)

blurred image at location i, the equalized image is obtained by multiplying the pixel values in the periphery by a correction factor T/s_i. Only pixels for which $s_i < T$ are processed, which ensures continuity. In the method by Byng, a new threshold is determined in each image row by taking the average of a small region around the border point. Their method was evaluated with digitized screen-film mammograms, but is also applicable to full field digital mammograms.

Snoeren and Karssemeijer (2005) also used a blurred version of the mammogram to correct for thickness differences, but instead of using isotropic smoothing of the original mammogram, they used an anisotropic smoothing.

The processed image revealed details in the region behind the nipple which were not perceivable in the original. The bright area near the skin in the lower part is a skinfold. The interior part of the mammogram remained unchanged, but can be displayed in higher contrast after the correction.

Before smoothing, dense tissue areas are removed and interpolated using surrounding fatty tissue values (see Fig. 5.6). The idea is that by removing dense tissue, the image better reflects thickness differences in the mammogram. Linear interpolation was performed along the lines running equidistant to the skin edge. The reason is that thickness variations are much stronger in the direction perpendicular to the skin line. Gray values of fatty tissue on lines parallel to the skin are usually small. The segmentation of fatty and dense tissue itself is obtained in an iterative process in which a current equalized image is automatically thresholded by Otsu's thresholding method. Hence, when the thickness correction becomes better in subsequent iterations, the segmentation becomes better too. Anisotropic diffusion is used to smooth the mammogram in the direction parallel to the skin edge. This leads to more accurate estimates of the thickness profile and reduces artifacts. The correction is performed by adding a correction term to pixel values in the peripheral zone. Using a threshold T again on the blurred image, pixels yi representing higher exposure are replaced by $y_i' = y_i - s_i + T$, with s_i being the pixel value in the blurred image. The method is modality independent. With full field digital mammograms, it should be applied after a logarithmic transform of the pixel values, to avoid alteration of contrast. An example is shown in Fig. 5.6.

As a last technique, we describe a parametric method by Snoeren and Karssemeijer (2004) which is only suitable for unprocessed digital mammograms with a linear relationship between exposure and gray value. Instead of using a filtered image for correction, as in the previous two examples, a geometric model of the three-dimensional shape of the breast is used (see Fig. 5.7). The interior region is modeled by two nonparallel planes, requiring three degrees of freedom, one for the onset and two for the slopes. The exterior region is modeled by a band of semi-circles. This requires no additional degrees of freedom: The semi-circles are completely determined by the breast outline and the interior model. Given the parameters of the

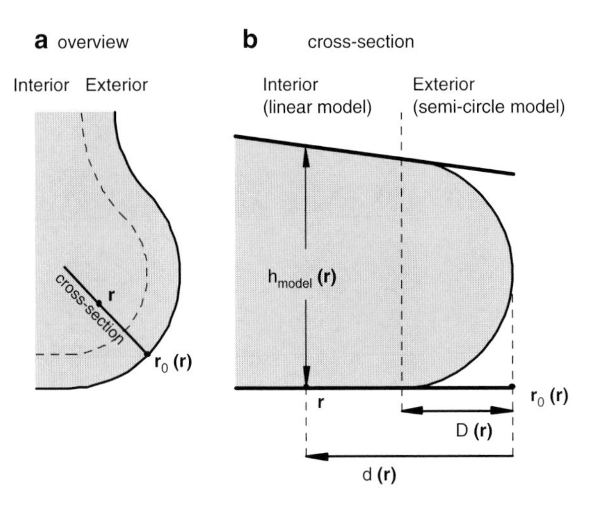

Fig. 5.7. Schematic representation of the interior and exterior part of the model for tissue thickness. The interior model (valid for $d(\mathbf{r}) \geq D(\mathbf{r})$): $h_{model}(\mathbf{r}) = a + \mathbf{a}\cdot\mathbf{r}$. The exterior model (valid for $d(\mathbf{r}) \leq D(\mathbf{r})$): $h_{model}(\mathbf{r}) = 2\sqrt{D^2(\mathbf{r}) - (D(\mathbf{r}) - d(\mathbf{r}))^2}$. Hereby, $\mathbf{r}_0(\mathbf{r})$ is the point on the skin line closest to $\mathbf{r}$; $D(\mathbf{r})$ is the distance of $\mathbf{r}_0$ to the interior/exterior border; $d(\mathbf{r})$ is the distance of $\mathbf{r}_0$ to $\mathbf{r}$. From SNOEREN and KARSSEMEIJER 2004 (©2004 IEEE)

geometric model and assuming a linear relationship between tissue thickness and log-exposure (Beer's law of attenuation), one can model the gray values of a breast that only consists of fatty tissue. Therefore, after fat/dense segmentation of the mammogram the model can be fitted to the "fatty" pixels in the unprocessed mammogram. The corrected image is obtained by adding a fatty tissue component in the periphery which fills in the air gap between the fitted planes and the breast. An example is given in Fig. 5.8. The case is challenging because the large cysts in the periphery may easily be distorted. The example shows that excellent results can be obtained using parametric model-based methods. It is noted that the method critically depends on accurate segmentation of the breast.

5.4 Matching Current and Prior Mammograms

In mammography, the comparison of images obtained in subsequent examinations of a patient is an important element in diagnostic and screening procedures. These comparisons are made to detect interval changes, to monitor progression of a disease, or to estimate the effect of treatment. Studies have shown that the use of prior mammograms in breast cancer screening effectively reduces the number of false-positive referrals (THURFJELL et al. 2000; BURNSIDE et al. 2002; ROELOFS et al. 2007). This is due to the fact that the use of priors allows radiologists to distinguish lesions that grow from normal dense structures in the breast that somehow look suspicious.

To make it easier for the radiologists to detect and interpret mammographic changes, mammograms in temporal image pairs should be presented in a similar way. Unfortunately, this is not always possible, because there are many sources of variability that are hard to control. Some already existed in conventional mammography, like variation in positioning and compression. However, variations related to exposure and film-screen differences were limited in conventional mammography. With digital mammography, a major new source of variation has been introduced: the variability in image processing methods. As image processing is usually performed on acquisition systems of modality manufacturers, using proprietary software, there is currently no way to make images look comparable on mammographic workstations if they are generated by systems from different vendors. This is a serious drawback of digital mammography which should be resolved.

Problems with display of priors are often striking when priors are digitized films. Digitization of prior screening mammograms is currently practiced on a large scale to bridge the transition period when screening transforms to a digital workflow. A typical example from a digital screening program is shown in Fig. 5.9. It is noted that the display problem will not disappear as soon as the transition to digital is completed. Similar problems occur when mammograms from different vendors have to be compared, and digitized priors will remain archived for use in future screening rounds, as radiologists in screening often like to look back several screenings when judging a potential abnormality. Therefore, proper display of digitized priors will remain a highly relevant issue. There is no need to say that display as shown in Fig. 5.9 is far from ideal.

Excellent matching of the presentation of prior and current mammograms is possible as long as no spatial enhancement algorithms have been applied by the manufacturer. The fact that pixel values in unprocessed mammograms are proportional to exposure provides an ideal setting for processing mammograms in a similar way, even if they are acquired with very different detector systems. Case-based mammogram processing based on raw data has been studied by SNOEREN and KARSSEMEIJER (2007). The approach suggested is

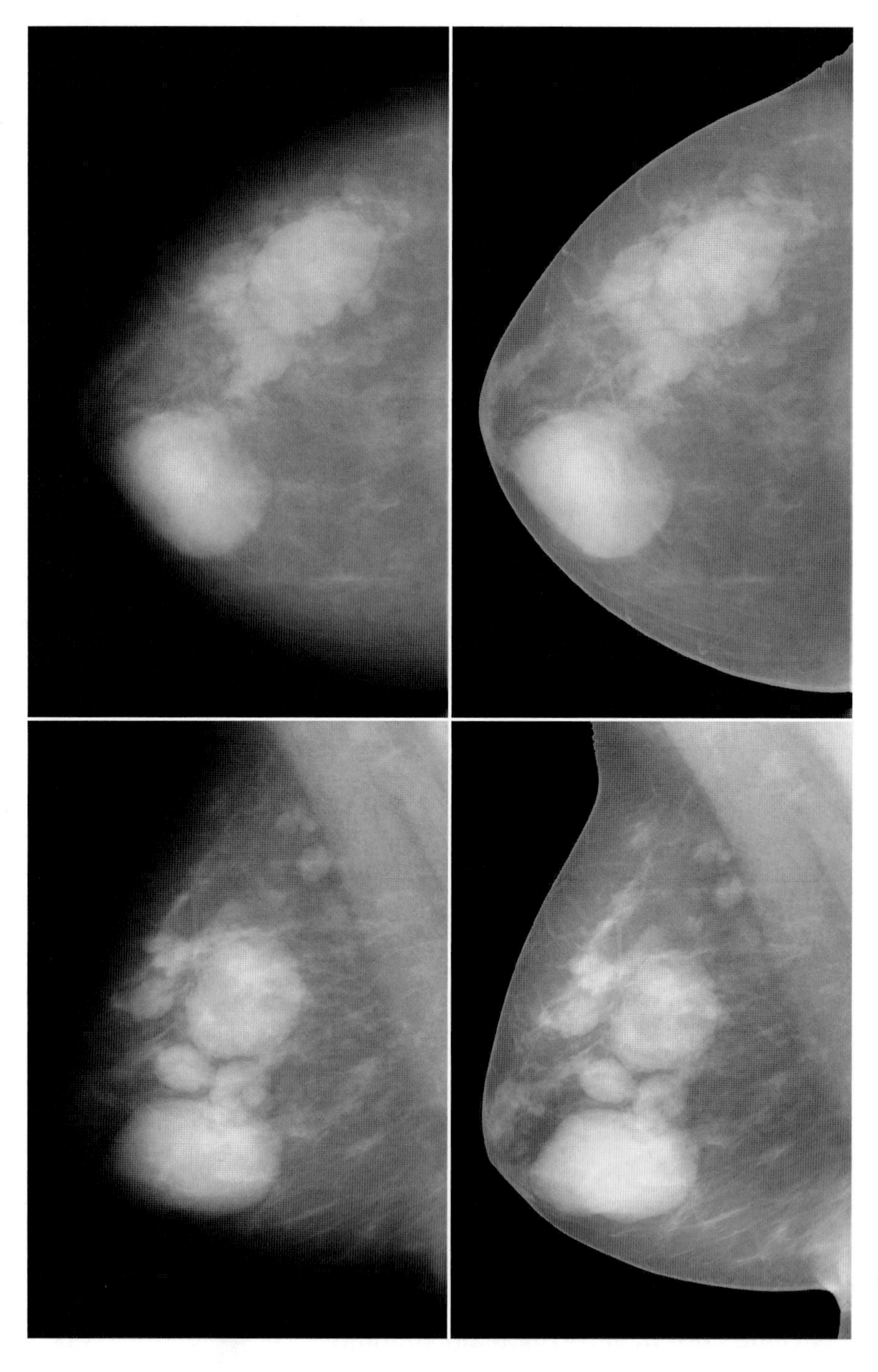

Fig. 5.8. Example of thickness correction with a geometric model. On the *left side*, cranio-caudal and medio-lateral views of the original mammogram are shown. On the *right side*, the thickness corrected images are depicted. The original mammogram was acquired with a GE Senograph 2000D. From Snoeren and Karssemeijer 2004 (© 2004 IEEE)

based on geometric image registration and subsequent derivation of a proper grayscale transform of all individual images. This transform is based on parametric histogram matching of overlapping regions in the registered images. Only transforms are allowed that are physically possible, given a model of the acquisition process. When digitized films are matched with full field digital mammograms, this model includes the characteristic curve of the film. Once images are transformed to the same domain, adaptive contrast enhancement can be applied. In this way, similarity in presentation of all views is maintained, as all images are processed by the same algorithm. An example is shown in Fig. 5.10. Also, the geometric registration has

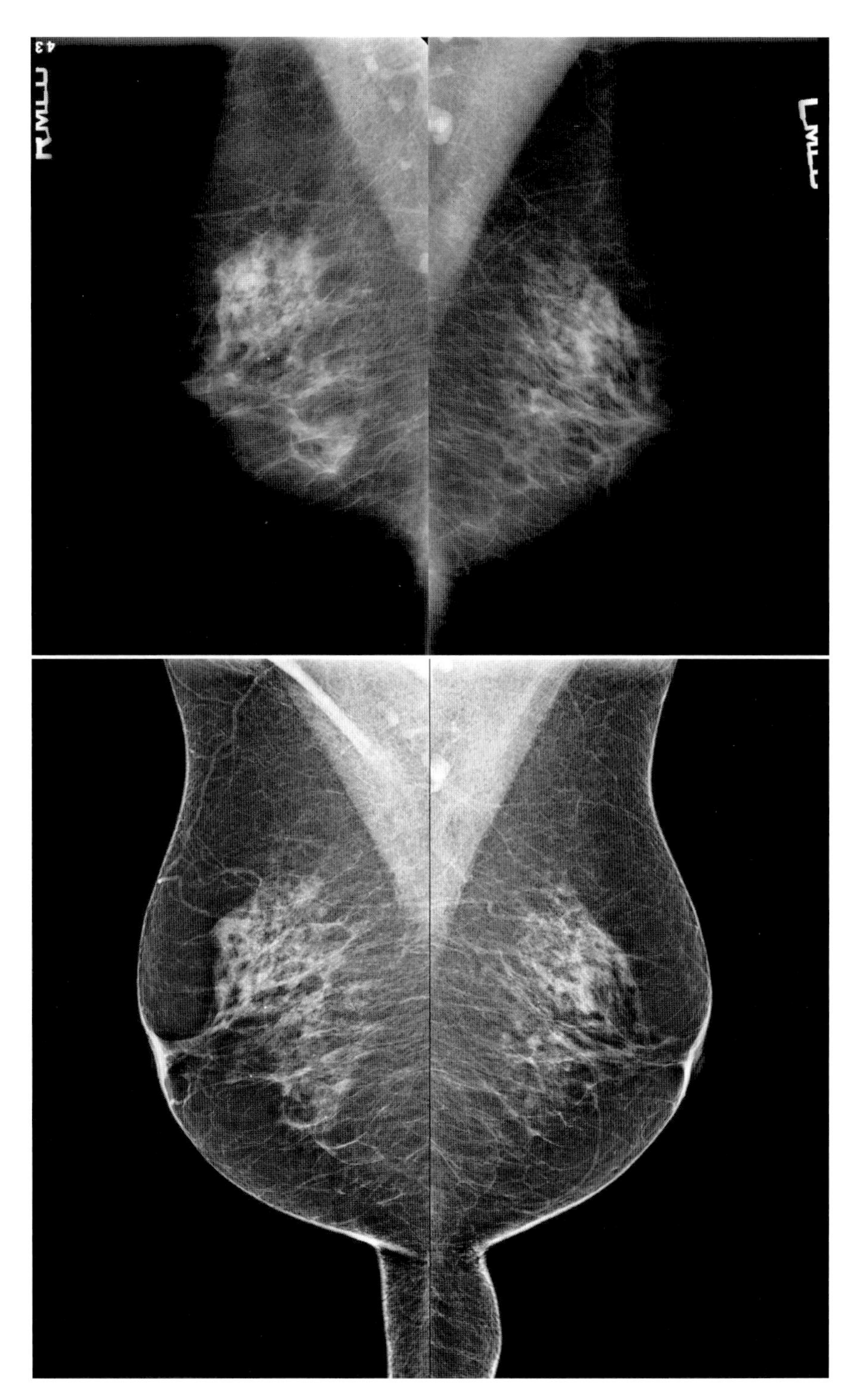

Fig. 5.9. A digital screening mammogram (*bottom*) compared with the previous screening mammogram which is digitized from l m. The FFDM image is acquired and processed by a Selenia system (Hologic)

advantages, as this ensures that corresponding mammographic structures are roughly located in the same space in the image matrix. When displaying mammograms electronically, toggling between current and prior mammograms on the same screen is a common technique. After registration structures appear in the same location on the screen, which makes comparison easier (Van Engeland et al. 2003).

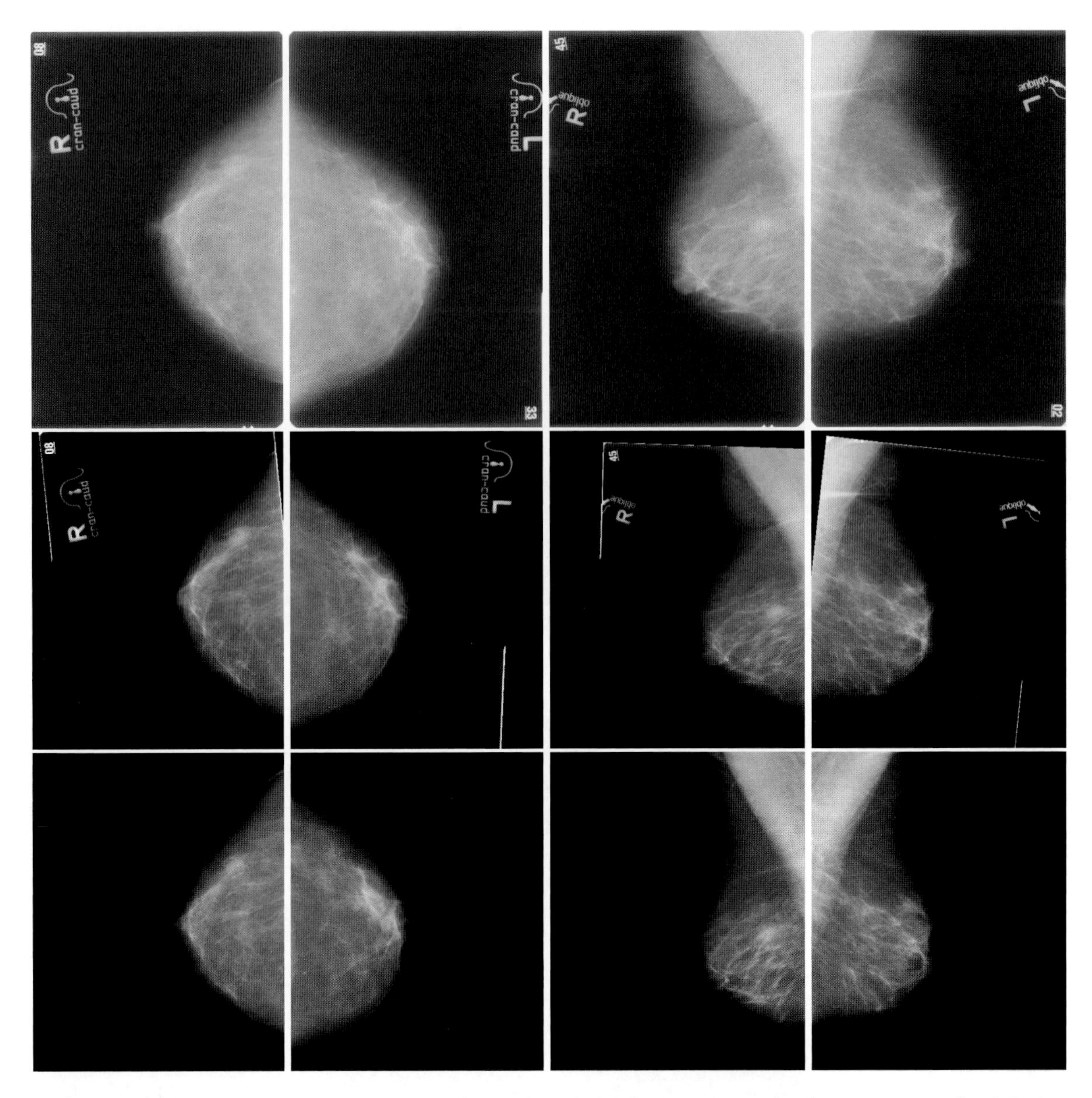

Fig. 5.10. A digitized prior screening mammogram (*top row*) is matched to a digital screening mammogram (*bottom*). The digitized prior views have first been registered geometrically to the unprocessed FFDM images. Subsequently, a model-based transform was computed for each view, which converts the digitized prior to the representation in which pixel values are proportional to exposure. After bringing them in the same space, any processing of the images will yield similarity in the presentation. Here, we applied only a lookup table as outlined in Sect. 5.2, without peripheral enhancement. The matched display of the prior is shown in the middle row

5.5 Physics-Based Methods

The physical processes involved in mammographic imaging are well understood. By modeling these, corrections methods may be designed to reduce variation of mammograms related to acquisition differences. An example is the scatter correction methods. The effect of scatter reduction in digital mammographic images was studied by Baydush et al. (2000). An iterative Bayesian estimation algorithm was formulated and used to process images of the American College of Radiologists (ACR) breast phantom acquired without a grid. The authors concluded that the technique can reduce scatter content effectively without introducing any adverse effects, such as grid line aliasing. Results suggest that the processing can increase contrast-to-noise ratio to values greater than that provided by a standard grid, which may potentially increase the visualization of subtle masses. This confirms the results of phantom measurements performed by VELDKAMP et al. (2003).

Physical modeling of the mammographic imaging chain was extensively studied by HIGHNAM and BRADY (1999). They came up with the idea of converting mammograms to a representation in which pixels represents the amount of so called "interesting" tissue (fibro-glandular tissue and cancerous tissue) projected in the pixel. Their normalization method is based on the assumption that the X-ray attenuation coefficients of fibro-glandular and cancerous tissue are nearly equal, but are quite different from that of fatty tissue. After normalization, a mammogram is corrected for scattered radiation and the dependency of image formation parameters like tube voltage, spectrum, and exposure time. In a later work by the authors, this representation is referred to as Standard Mammogram Form (SMF).

Conversion of mammograms to a standardized format has many advantages. It allows development of uniform image presentation methods that remove variation due to use of different image detectors and acquisition settings. In particular, temporal comparison of mammograms could be greatly improved by application of image normalization. Also, quantitative image analysis methods aimed at automated detection of lesions or measurement of breast density would benefit from standardization. Unfortunately, manufacturers do not embrace the idea and tend to move away from standardization by developing proprietary algorithms for image processing. As most clinics do not archive raw images, the relation between digital mammograms and the quantitative potential of exposure values in the raw data is lost.

In screen-film mammography, application of SMF was hampered by uncertainties in the image acquisition process. The method only became feasible when digital mammography was introduced. An effective procedure for standardization of digital mammograms was developed and validated by VAN ENGELAND et al. (2006). Assuming that pixel values are proportional to exposure, which is true for most digital detector systems, it can be derived that in good approximation, the thickness of dense tissue at a given location h_i can be written as

$$h_i = -\frac{1}{\mu_{d,eff} - \mu_{f,eff}} \ln \frac{\bar{y}_i}{\bar{y}_f}. \tag{5.3}$$

with $\mu_{f,\,eff}$ and $\mu_{d,\,eff}$ being the effective attenuation coefficients for fatty and dense tissue, respectively, $\bar{y}_i$ being the pixel value at i after thickness equalization, and $\bar{y}_f$ being a reference pixel value taken at a location with only fatty tissue. Note that there is no explicit dependency on breast thickness. The effect of breast thickness is included in the computation of the effective attenuation the coefficients, which also depends on other acquisition parameters. The method was validated for a series of cases by comparing it with MRI. By integrating dense tissue thickness over the breast, dense tissue volumes were obtained. These agreed well with dense tissue volumes measured in MRI data.

5.6 Evaluation of Mammogram Processing

Improved display of mammograms may lead to more efficient workflow and to more accurate detection of abnormalities. To determine the potential effect of processing in clinical practice, the performance of radiologists utilizing the processing methods has to be studied. In the literature, only a few examples of such studies may be found. PISANO et al. attempted to determine whether intensity windowing improves detection of simulated calcification in dense mammograms. Film images with no windowing applied were compared with film images with nine different window widths and levels applied. Using twenty students as observers, it was found that there was a significant variation in detection performance for clusters of calcifications, when the processing was varied (PISANO et al. 1997). It can be noted that, in practice, readers may apply intensity windowing interactively when reading mammograms on workstations. However, in practice it is too time-consuming to perform such operations on every mammogram.

Of the spatial enhancement methods, contrast-limited adaptive histogram equalization (CLAHE) has frequently been used in observer experiments. HEMMINGER et al. studied whether detection of simulated masses in dense mammograms could be improved with the technique and compared the effects of this processing method with histogram-based intensity windowing (HIW). The key variables in the experiments included the contrast levels of the mass relative to the background and the selected parameter settings for the image-processing method. Performance depended on the parameter settings of the algorithms used. The best HIW setting performed better than the best fixed-intensity window setting and better than no processing. Performance with the best CLAHE settings was not different from that with no processing. The authors concluded that CLAHE processing will probably not improve the detection of masses on clinical mammograms (Hemminger et al. 2001).

In a study by Pisano et al. (2000b), radiologists' preferences for digital mammographic display were investigated. Eight different image processing algorithms were evaluated using a series of twenty-eight images representing histologically proved masses or calcifications obtained with three clinically available digital mammographic units. Processing methods included histogram and mixture model-based intensity windowing, peripheral equalization, multiscale image contrast amplification, CLAHE, and unsharp masking. Twelve radiologists compared the processed digital images with screen-film mammograms obtained in the same patient for breast cancer screening and breast lesion diagnosis. Surprisingly, screen-film mammograms were preferred to most digital presentations, and none of the methods lead to a clear preference for the processed digital images. It is noted that this may have been due to inexperience of the radiologists with digital processing. Moreover, the comparison not only involved processing, because the comparison was made with screen-film mammograms. Also, acquisition differences may have played a role. In a later study, performance with digital mammography using different processing techniques was again compared with screen-film mammograms of the same patients (Cole et al. 2005). A total of 201 digital mammograms were used in combination with three processing methods: The manufacturers used default, multiscale image contrast amplification and a version of CLAHE. Three radiologists were involved in the experiments. It was found that for one manufacturer, the performance with digital mass cases was worse than screen-film for all digital presentations. The authors suggest that specific image processing algorithms may be necessary based on machine and lesion type.

A wavelet technique for spatial enhancement of microcalcifications in mammograms was evaluated by Kallergi et al. (1996). Digitized mammograms were used. Differences were observed between screen-film and unprocessed digitized mammography displayed on monitors. These differences were not significant when wavelet enhancement was included in the monitor display. Interobserver variation in the digitized reading was greater than that in film reading, but the wavelet enhancement reduced the difference. In a more recent study, the effect of wavelet processing on a mixture of mammographic findings was investigated (Kallergi et al. 2004). The study was designed as a localization response operating characteristic (LROC) experiment with 500 negative, benign, and cancer cases with masses and calcification clusters. Three observers reviewed the original and wavelet-enhanced images on a 5-Mpixel monitor, using a custom-made workstation user interface. Performance indexes were estimated for four different case combinations. It was found that wavelet enhancement improved the performance of all observers in all case combinations. The difference between enhanced and original performances was statistically significant. The authors argue that optimization of the softcopy quality is expected to require more advanced processing techniques than standard grayscale adjustments. Wavelet-based algorithms offer better softcopy quality than the originals, and a better starting point for additional manual grayscale adjustments or automated postprocessing.

Results of the studies summarized earlier indicate that positive effects of mammogram processing on diagnostic performance have not yet been clearly established. This may partly be due to the fact that most studies were performed with readers who were not yet familiar with softcopy reading and processed mammogram display. More research is needed to investigate the effects of various processing techniques.

References

Baydush AH, Floyd CE Jr (2000) Improved image quality in digital mammography with image processing. Med Phys 27(7):1503–1508

Burnside ES, Sickles EA, Sohlich RE et al (2002) Differential value of comparison with previous examinations in diagnostic versus screening mammography. AJR Am J Roentgenol 179(5):1173–1177

Byng JW, Critten JP, Yaffe MJ (1997) Thickness-equalization processing for mammographic images. Radiol 203:564–568

Cole EB, Pisano ED, Zeng D et al (2005) The effects of gray scale image processing on digital mammography interpretation performance. Acad Radiol 12(5):585–595

Heinlein P, Drexl J, Schneider W (2003) Integrated wavelets for enhancement of microcalcifications in digital mammography. IEEE Trans Med Imaging 22(3):402–413

Hemminger BM, Zong S, Muller KE et al (2001) Improving the detection of simulated masses in mammograms through two different image-processing techniques. Acad Radiol 8(9):845–855

Highnam R, Brady M (1999) Mammographic image analysis, 1st edn. Kluwer Academic Publishers, Dordrecht, the Netherlands

Kallergi M, Clarke LP, Qian W et al (1996) Interpretation of calcifications in screen/film, digitized, and wavelet-enhanced monitor-displayed mammograms: a receiver operating characteristic study. Acad Radiol 3(4):285–293

Kallergi M, Heine JJ, Berman CG et al (2004) Improved interpretation of digitized mammography with wavelet processing: a localization response operating characteristic study. AJR Am J Roentgenol 182(3):697–703

Laine AF, Schuler S, Fan J et al (1994) Mammographic feature enhancement by multiscale analysis. IEEE Trans Med Imaging 13(4):725–740

Pisano ED, Chandramouli J, Hemminger BM et al (1997) Does intensity windowing improve the detection of simulated calcifications in dense mammograms. J Digit Imaging 10 (2):79–84

Pisano ED, Cole EB, Hemminger BM et al (2000a) Image processing algorithms for digital mammography: a pictorial essay. Radiographics 20(5):1479–1491

Pisano ED, Cole EB, Major S et al (2000b) Radiologists' preferences for digital mammographic display. The international digital mammography development group. Radiology 216(3):820–830

Pizer SM, Johnston RE, Rogers DC et al (1987) Effective presentation of medical images on an electronic display station. Radiographics 7(6):1267–1274

Roelofs AA, Karssemeijer N, Wedekind N et al (2007) Importance of comparison of current and prior mammograms in breast cancer screening. Radiology 242(1):70–77

Snoeren PR, Karssemeijer N (2004) Thickness correction of mammographic images by means of a global parameter model of the compressed breast. IEEE Trans Med Imaging 23(7):799–806

Snoeren PR, Karssemeijer N (2005) Thickness correction of mammographic images by anisotropic filtering and interpolation of dense tissue. In: Fitzpatrick J, Reinhardt J (eds) SPIE medical imaging: image processing, vol. 5747, pp 1521–1527

Snoeren PR, Karssemeijer N (2007) Gray-scale and geometric registration of full-field digital and film-screen mammograms. Med Image Anal 11(2):146–156

Thurfjell MG, Vitak B, Azavedo E et al (2000) Effect on sensitivity and specificity of mammography screening with or without comparison of old mammograms. Acta Radiol 41(1):52–56

van Engeland S, Snoeren P, Karssemeijer N et al (2003) Optimized perception of lesion growth in mammograms using digital display. In: Chakraborty D, Krupinski E (eds) SPIE medical imaging: image perception. Observer Performance, and Technology Assessment, vol. 5034, pp 25–31

van Engeland S, Snoeren PR, Huisman H et al (2006) Volumetric breast density estimation from full-field digital mammograms. IEEE Trans Med Imaging 25(3):273–282

Veldkamp WJ, Thijssen MA, Karssemeijer N (2003) The value of scatter removal by a grid in full-field digital mammography. Med Phys 30(7):1712–1718

Computer-aided Detection and Diagnosis

6

ROBERT M. NISHIKAWA

CONTENTS

R. M. NISHIKAWA
Carl J. Vyborny Translational Laboratory for Breast Imaging Research, Department of Radiology and Committee on Medical Physics, The University of Chicago, 5841 S. Maryland Avenue, MC-2026, Chicago, IL 60637, USA

KEY POINTS

Computer-aided detection (CADe) and computer-aided diagnosis (CADx) are emerging technologies to help radiologists interpret medical images. In screening mammography, CADe can help radiologists avoid overlooking a cancer, while CADx can help radiologists decide whether a biopsy is warranted when reading a diagnostic mammogram. Even though there is much commonality in the techniques used in CADe and CADx algorithms, there are important differences in the input data and in the output of the algorithms. In particular, CADe outputs the location of potential cancers, while CADx outputs the likelihood that a known lesion is malignant. These differences affect the metrics used to evaluate their performance. Commercial CADe systems have been developed and clinical studies of CADe have indicated the ability to increase radiologists' sensitivity by approximately 10% with a comparable increase in the recall rate. Commercial CADx systems do not exist till date, but observer study results are very compelling. CADe and CADx schemes continue to evolve in terms of accuracy and user interface. It is expected that CADe and eventually CADx will play an increasingly important role in breast imaging in the future.

6.1 Introduction

Computer-aided diagnosis (CAD) has been defined as a diagnosis made by a radiologist who uses the output of a computer analysis of the images when making his or her interpretation. There are two main types of CAD systems. One is computer-aided detection

(CADe) and the other is computer-aided diagnosis (CADx). CADe schemes identify and mark suspicious areas in an image. The goal of CADe in mammography is to help radiologists avoid missing a cancer. CADx schemes help radiologists decide whether a woman should have a biopsy or not. CADx schemes often report the likelihood that a lesion is malignant, although there are other forms of output that will be discussed later. In simple terms, CADe schemes are used in screening mammography and CADx schemes are used in diagnostic mammography. In addition to improving radiologists' performance, a secondary goal of CAD is to reduce intra- and inter-variability of radiologists. Further, it is hoped that CAD can improve radiologists' productivity. There are other forms of CAD, such as estimating the risk of a woman to develop breast cancer (Huo et al. 2000, 2002a), but these will not be discussed in this chapter. Also, over the past 20 years, there have been several review articles and chapters written on CAD for mammography (Giger 1993, 2004b; Vyborny and Giger 1994; Karssemeijer and Hendriks 1997; Giger et al. 2000, 2008; Jiang 2002; Karssemeijer 2002; Nishikawa 2002, 2007; Roque and Andre 2002; Astley 2004a, b; Sampat et al. 2005; Hadjiiski et al. 2006a; Malich et al. 2006; Taylor 2007), and most have focused on the technical aspects of developing CAD algorithms and they supplement and overlap with some of the materials presented here.

6.2 Short Historical Overview

In 1955, Lee Lusted described the possibility of computers reading radiographs. He envisioned the possibility of computers automatically interpreting radiographs. The first published study using computers to analyze a mammogram was by Winsberg in 1967 (Winsberg et al. 1967). He reported about a technique to automatically identify breast lesions. By today' standards, the system was extremely crude. He used a facsimile machine to digitize xeromammograms with a pixel size of 0.14 mm and a pixel depth of 5 bits. Then, the image was subdivided into 64 square blocks, and four features related to the pixel values were extracted for each block. Next, the block values for the left and right breasts were compared with significant differences corresponding to the presence of a cancer. Winsberg then proposed that characteristics that define the increase in the density would be developed. After this study, there were sporadic reports in the literature (Ackerman and Gose 1972; Kimme et al. 1975; Wee et al. 1975; Hand et al. 1979; Spiesberger 1979; Semmlow et al. 1980). However, none of them were particularly convincing on the fact that an automated technique to replace or even complement radiologists was feasible. By today's standards, these studies used crude methods for film digitization, computer power was feeble, and knowledge of image processing, image analysis, and statistical classifiers were rudimentary. Further, the goal of these studies was automated reading by a computer, not an aid to a radiologist. This goal requires the CAD scheme to have performance comparable with a radiologist, which is difficult to achieve.

In the mid 1980s, at the University of Chicago, researchers in the Kurt Rossmann Laboratories were investigating digital imaging. While much of the research in the field at that time focused on digital detectors and PACS, the investigators in the Rossmann Labs were developing an understanding of how best to leverage the potential of digital imaging to improve radiologists' performance. This included practical research questions on how best to display a digital image (Loo et al. 1985; Chan et al. 1987b) and the basic perception studies on signal-to-noise ratio (SNR) metrics that correlate with the observer's performance (Loo et al. 1984; Giger and Doi 1985). From this research endeavor, it was decided that computers could be used to extract quantitative information from an image (Fujita et al. 1987) and also automatically detect lesions (Chan et al. 1987a; Giger et al. 1988). However, unlike past efforts in computerized analysis of medical images, the goal was to develop a tool to assist radiologists, not to device a stand alone system, which had been unsuccessful in the past. Hence, the concept of CAD was developed.

To test the viability of CAD, an observer study was conducted to see if an automated computer scheme could help radiologists detect clustered microcalcifications in mammograms. In their seminal paper, Chan et al. showed that a computer could improve radiologists' performance in detecting clustered microcalcifications in mammograms (Chan et al. 1990). That publication was the genesis of what we know today as CAD.

The publication of the study by Chan et al. prompted several groups to start developing other CAD schemes. Kegelmeyer et al. published an observer study that showed that radiologists could improve their sensitivity (at a fixed specificity) for the detection of spiculated masses in mammograms (Kegelmeyer et al. 1994). In 1988, Getty et al. showed for the first time that computers could aid radiologists in classifying breast lesions as benign or malignant (Getty et al. 1988). In an observer study, they showed that general radiologists, when aided by CADx, could perform at a level comparable with unaided expert mammographers.

In 1993, Susan Astley and Kevin Boyer organized a two-day session on CAD at the annual SPIE on Biomedical Image Processing and Biomedical Visualization (Acharya and Goldgof 1993). A book was subsequently published based on many of the talks given at that conference (Bowyer and Astley 1994). The following year, Alastair Gale, Susan Astley, David Dance, and Alistair Cairns organized a stand-alone conference: the International Workshop on Digital Mammography. This has become a biennial meeting focusing on digital mammography and CAD. The published conference proceedings provide an excellent snapshot of the state-of-the-art in CAD in mammography since 1994, in 2-year increments. The SPIE Medical Imaging conference held yearly in February is also an excellent source for the developments in CAD.

In 1993, a group of five men (Bob Wang, Bob Foley, Jimmy Roehrig, Julian Marshall, and Harold Rutherford) came to the University of Chicago and obtained licenses to CAD software technologies and the software code developed at the University. They returned to California and formed a company named R2 Technology. With further research and development and licenses for other CAD techniques, they developed the ImageChecker M1000, the world's first commercial CADe system. This system digitized screen-film mammograms, automatically detected suspicious regions, and displayed those detections to a radiologist. In a landmark study, Warren-Burhenne et al. (2000) used the ImageChecker system to show that the system had the potential to reduce radiologists' miss rate by 77%. Based partly on this study, in June 1998, the ImageChecker M1000 became the first CAD system to receive clearance from the United States Food and Drug Administration (FDA). Shortly there after, in 2001, Medicare approved reimbursement for CAD in mammography. Two other CADe companies were formed in the 1990s, CADx and iCAD, and they later merged into a single company called iCAD. There are now at least five companies actively developing CAD systems for mammography. It is estimated that approximately 50% of all mammograms read in the USA are done with the assistance of CADe.

In addition to the development of several clinical CADe systems, our knowledge of how to develop and evaluate CAD schemes has matured greatly. The effects of databases, scoring methodology on measured performance has been studied (Nishikawa et al. 1994; Nishikawa and Yarusso 1998; Kallergi 1999). Biases in training and testing and methods to reduce the biases have been elucidated (Chan et al. 1999b; Kupinski and Giger 1999; Sahiner et al. 2000; Gur et al. 2004c; Yousef et al. 2004, 2005, 2006; Wagner et al. 2007; Sahiner et al. 2008). While the methodology for developing and evaluating CADe schemes has matured, methodology for evaluating CADe scheme clinically is still evolving (Nishikawa and Pesce 2008).

6.3 Clinical Need for CAD in Mammography

6.3.1 Missed Cancers

The sensitivity of mammography is often quoted to be 85%. However, this number is obtained from screening mammography where a cancer that is undetected one year and that is found in a subsequent screening mammogram is not counted as a missed cancer in year one. Comparison of the sensitivity of mammography with breast MRI paints a different story. Granader et al. (2008) conducted a meta-analysis of screening women who are at high risk for breast cancer. MRI clearly out performed mammography in cancer detection (although at a higher false detection rate) and for this type of patient population, the sensitivity of mammography was only 37.5%.

The MRI data is specific to screening high-risk women. In general mass screening, a lower bound to the miss rate of mammography, can be found in the Digital Mammography Imaging Screening Trial (DMIST) (Pisano et al. 2005). In this study, 49, 528 women were screened with both a conventional screen-film mammogram (SFM) and a full-field digital mammogram (FFDM). Comparing cancers detected on SFM and FFDM showed that of all the cancers that were detected by either modality, only 73% were on SFM and 78% on FFDM. That is, mammography was found to miss at least 22% of cancers that are mammographically detectable. There were cancers that were missed by both the modalities, and hence, the overall result was that the sensitivity of conventional mammography and digital mammography were comparable (52% vs. 55%). Since the DMIST study involved a large number of screening exams, 33 sites and more than 100 radiologists, the results give a fairly accurate representation of the performance of mammography, being that 45% of women who have cancer were called normal on screening mammography.

In the retrospective analysis of missed cancers, Martin et al. found that in 30% of cases, the cancer was not visible on the mammogram even in retrospect (Martin et al. 1979). Approximately 7% were missed for technical deficiencies of the mammogram (e.g., improper patient positioning or improper X-ray exposure). In an additional 33% of the cases, the finding was subtle, but in 30%, there was an obvious cancer that was missed. It is believed that CAD will be helpful in reducing the number of missed cancers from the last two categories. One caveat of this study is that it was performed 30 years ago. Mammography has improved since that time, and the type of cancers missed in that study may not be representative of the types of cancer missed in screening performed in more recent times.

6.3.2 Low Positive Predictive Value for Biopsy Recommendations

Clinically, differentiating benign from malignant lesions is a difficult task. In the USA, the positive-predictive value (PPV) for diagnostic breast imaging is generally less than 50%. The PPV measures the percentage of all breast biopsies that are positive for cancer. Using data from the Breast Cancer Surveillance Consortium, Barlow et al. determined that the PPV based on 41,427 diagnostic mammograms was 21.8% (Barlow et al. 2004). Elmore et al., by examining the results from eight large mammography registries (containing the follow up information on more than 300,000 screening mammograms), found that the PPV ranged from 16.9 to 51.8% with a median value of 27.5% (Elmore et al. 2003). Thus, approximately three benign biopsies are performed for every biopsy of a malignant lesion. Unnecessary biopsies are both physically and emotionally traumatic to the patient, they are costly to the health care system, and add unnecessarily to the workload of radiologists, pathologists, and surgeons. Improving radiologists' PPV can have a substantial positive effect on patient care and the healthcare system.

6.3.3 Reader Variability

In their seminal paper, Beam et al. showed that there is a great variability among radiologists (Beam et al. 1996). They found that the sensitivity of radiologists reading screening mammograms could vary by as much as 45%. Miglioretti et al. found considerable variability in performance among radiologists reading diagnostic mammography (Miglioretti et al. 2007). In their study, which included 123 American radiologists, the sensitivity ranged from approximately 40–100% with false positive rates of approximately 1–11%.

A part of variation among the radiologists may be owing to the different radiologists having different thresholds for recalling (in screening) or recommending a biopsy (diagnostic mammography). However, both the Beam and Miglioretti studies estimated ROC curves for the radiologists. They both concluded that the radiologists had different ROC curves so that the variability is not owing to different thresholds which would have meant that the radiologists were on the same ROC curve. Therefore, the variability between radiologists is caused by the range in skills of different radiologists. In addition to the variability among the radiologists, it is well known that the same radiologist may not be consistent in their interpretation, which is another source of variability. That is, the same radiologist may make a different recommendation given the same patient at a different time. Variability between radiologists decreases the clinical effectiveness of diagnostic imaging.

Generic Description of CADe and CADx Schemes

6.4.1 Methodology

Although the output of CADe and CADx are different, generically they operate in a similar fashion. A generic flowchart of CADe and CADx is shown in Fig. 6.1. While the vast majority of CAD algorithms follow this approach, there are CADe schemes that attempt to directly analyze the image data without segmenting a lesion and extracting features (El-Naqa et al. 2002; Campanini et al. 2004).

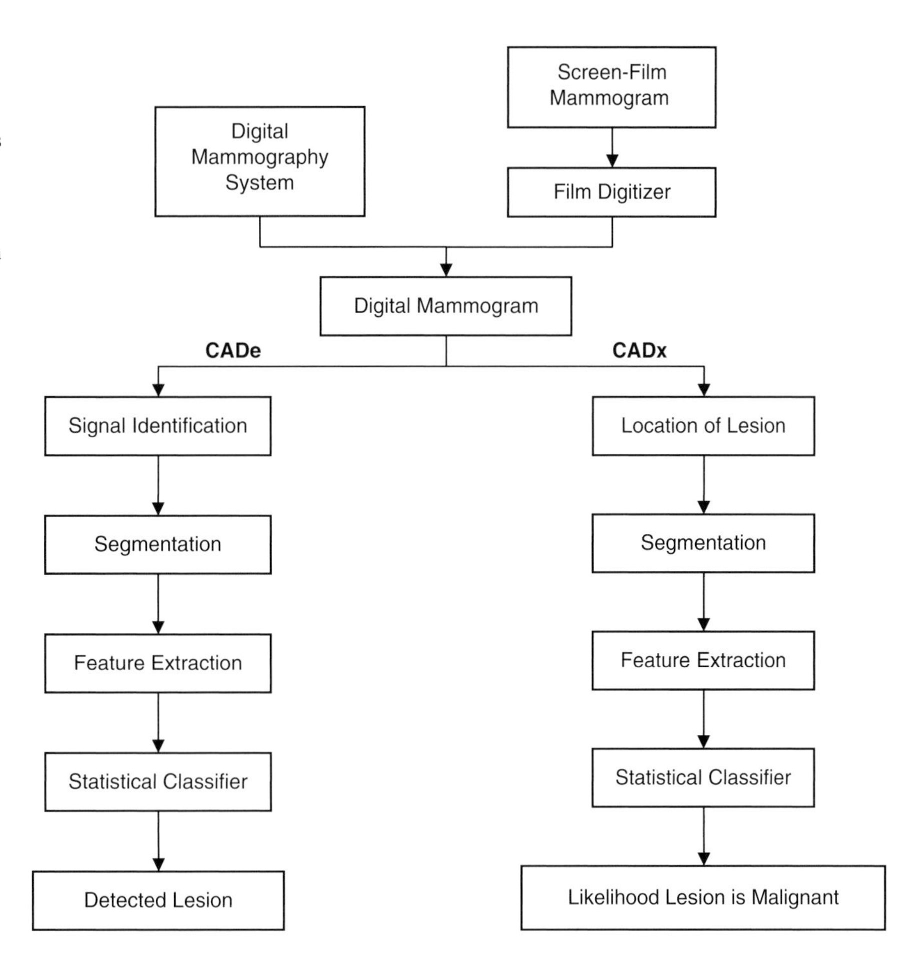

Fig. 6.1. Generic flowchart of a CADe and a CADx scheme. In CADe, the location of potential lesions is determined by the computer. In CADx, the lesion location is known, either by a CADe algorithm or manually indicated by a radiologist. The output of the two schemes also differs. CADe output gives the location of potential malignancies, while CADx gives the likelihood that the indicated lesion is malignant. Further the features that are extracted from the image may differ for CADe and CADx, but there can be overlap. The statistical classifier in CADe is used to distinguish actual lesions from false detections, while in CADx the classifier distinguishes benign from malignant lesions

Starting with a digital mammogram, which can be obtained by digitizing a screen-film mammogram or using a digital mammogram, a lesion or lesions (CADx) or potential lesions (CADe) are identified. For CADx schemes, the lesion is usually identified by the radiologist, although a CADe scheme could be used. For CADe scheme, there are many different methods. An intuitively straightforward method is to use some form of grey-level thresholding, as cancer usually appears brighter than its surrounding background in a mammogram. For example, taking the brightest 2% of the pixels in a mammogram can be effective in identifying potential microcalcifications (Chan et al. 1987a).

After a lesion or lesions have been identified, they are segmented from the image. This essentially involves identifying the margin or border of the lesion(s). For example, Kupinski and Giger computed the radial gradient in an area that contains a seed point for a detected lesion (Kupinski and Giger 1998). A pixel belongs to the lesion if the radial gradient is greater than a threshold value. In this way, all pixels that are a part of the lesion are identified. At this stage in a CADe scheme, there are usually a large number of potential or candidate lesions, perhaps, ten to hundreds.

Next, features are extracted from the image, which characterize the lesion and in some cases, its background. The goal here is to characterize the lesions so as to differentiate benign from malignant lesions (CADx) or actual lesions from falsely detected candidates (CADe). Many of the extracted features are based on features that radiologists use to distinguish benign from malignant lesions (e.g., how spiculated the lesion appears), but others are not (e.g., the use of co-occurrence matrices to characterize texture). The optimal set of features is not known, but different features are used for masses and clustered microcalcifications.

These features are then input to a statistical classifier, such as an artificial neural network. The classifier is trained to distinguish benign from malignant lesion (CADx) or true from false detected lesions (CADe). The output of the statistical classifier is a

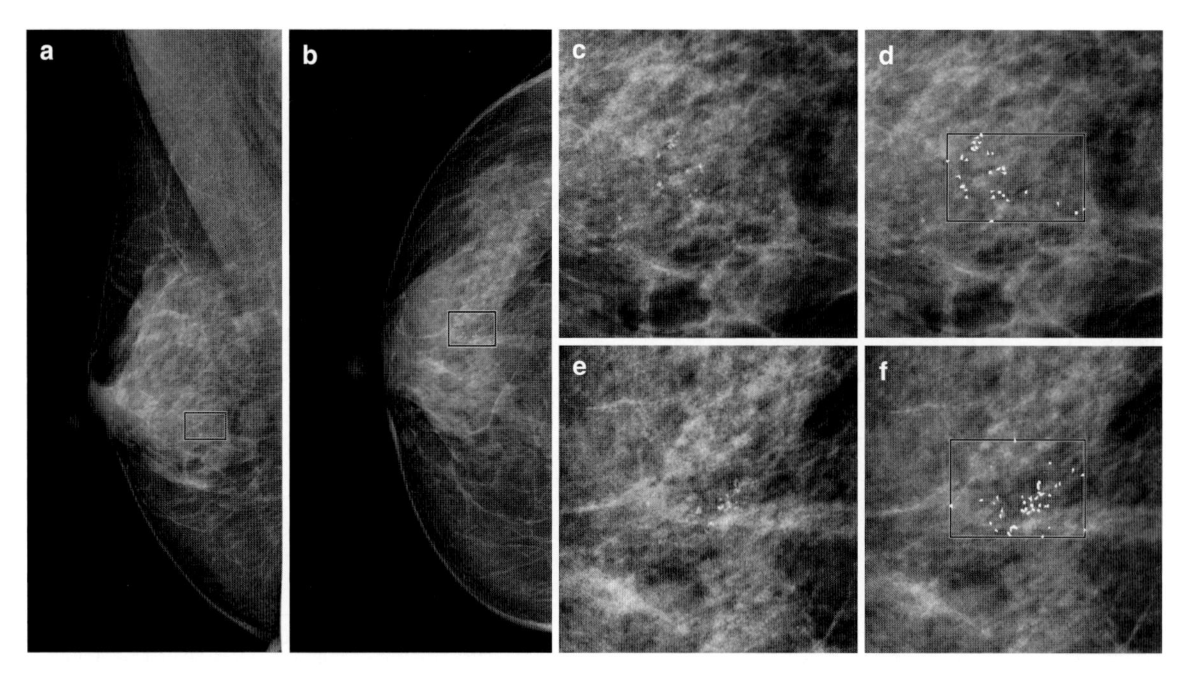

Fig. 6.2. Fifty-one-year-old female with microcalcification-associated, 10-mm high-grade DCIS detected during routine mammography screening. Right MLO (**a**) and CC (**b**) view with CAD detection marks (iCAD SecondLook V7.2) shown. Enlarged regions of interest of the suspicious microcalcifications in MLO (**c, d**) and CC (**e, f**) view, each shown as original image (**c, e**) and with individual CAD-detected microcalcifications marked as overlay (**e, f**)

number that is monotonically related to the likelihood that the lesion is malignant (CADx) or the lesion is an actual lesion (CADe). The performance of a CADe scheme does not vary greatly with the type of statistical classifier used (Wei et al. 2005).

The output of a CADe scheme is usually conveyed to the radiologist by annotating a copy of the image. That is, the symbols are placed on the image to indicate where the computer has identified a potential cluster of microcalcifications or a mass lesion (e.g., see Fig. 6.2). In CADe, a mass lesion refers to masses (circumscribed or spiculated), architectural distortions, asymmetries, and developing densities; that is, any non-calcific lesion. Different symbols are used to identify masses and clustered microcalcifications. For the analysis of film mammograms, the computer output can be a low quality version of the mammogram with symbols marking potential lesions printed on paper or displayed on low resolution monitors. The output images are not meant to be of diagnostic quality, but are to be used as "roadmap" by the radiologist. For mammograms acquired digitally and viewed on softcopy monitors, the output can be placed in an overlay on the image and turned off and on as needed. If the digital mammograms are printed on the film, then the methods for film mammograms can be used to display the CADe output.

The output for CADx results can be more complex. A simple method is to report the output of the CADx scheme to the radiologist, which is related to the likelihood that the lesion is malignant. This output value can be converted to be an exact likelihood of malignancy (Jiang et al. 1999), and further, it can be adjusted to account for the differences in the prevalence of cancer in the dataset used to train the statistical classifier and the radiologists' patient population (Horsch et al. 2008).

Two additional pieces of information have been proposed to either augment or replace the likelihood of malignancy number. Swett and Miller originally proposed that it may be useful to show radiologists some examples of lesions with known pathology that are similar to the lesion under investigation (Swett and Fisher 1987; Swett et al. 1989, 1998). Such a system has been developed independently by Sklansky (Sklansky et al. 2000) and Giger et al. (1988, 2003). Giger further proposed to show a graph that shows the probability of malignancy of the unknown lesion to all lesions in a known reference database (an online atlas). An example of the user interface is given in Fig. 6.3. The use of a reference library is also being developed for CADe applications, as will be discussed at the end of this chapter.

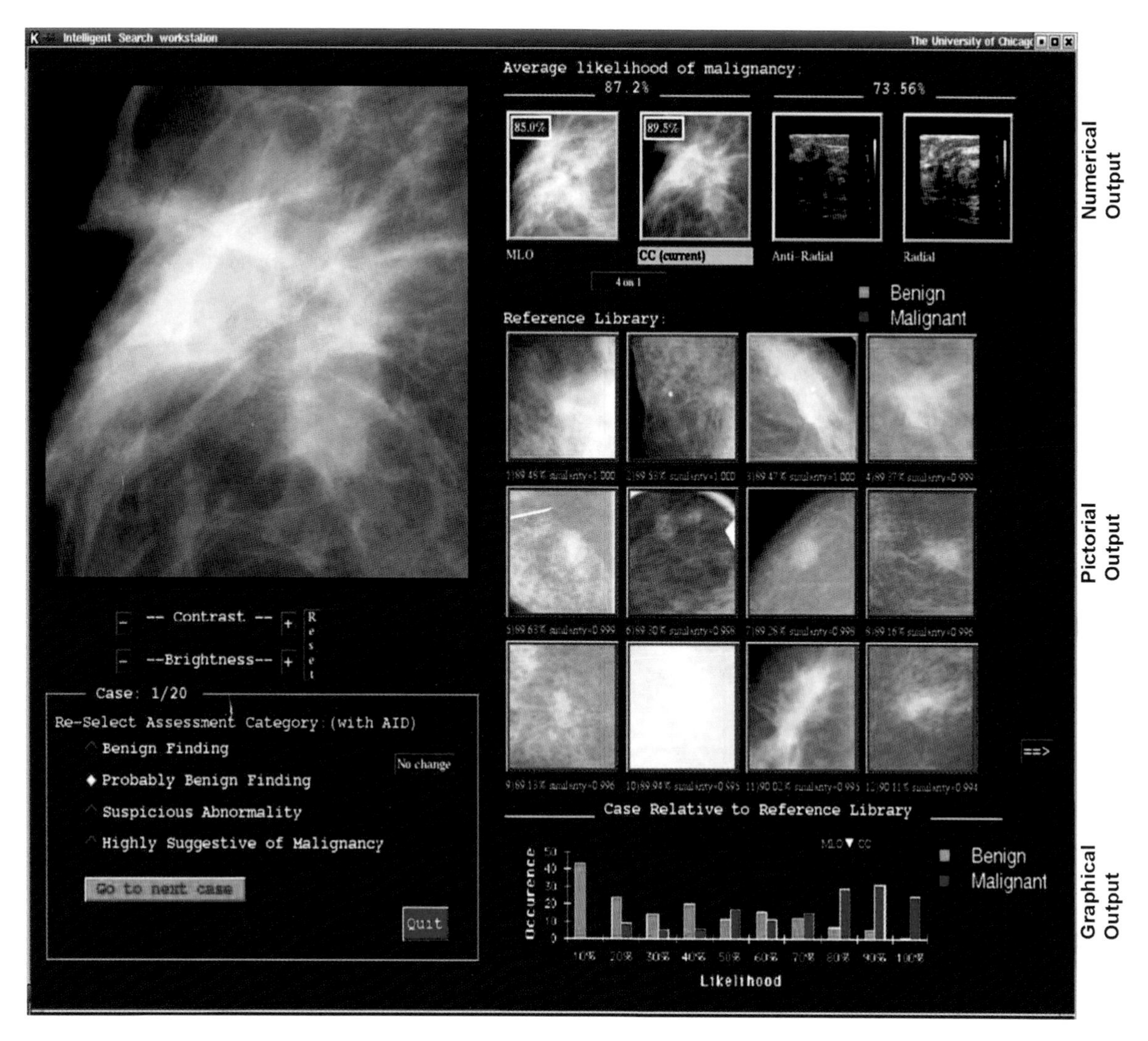

Fig. 6.3. An example of the user interface to implement an online reference library for CADx. Features of an unknown lesion, given in upper left corner of the computer screen, are extracted and are used to compute the likelihood that the unknown lesion is malignant. The likelihood is also used to search a reference database for lesions with either a similar likelihood of malignancy of similar feature values. In this particular interface, the user is also presented with a graphical representation of how the likelihood of malignancy is compared with other lesions in the reference library. Similar interfaces are being developed for CADe (taken from Giger 2004a. With permission from the Radiological Society of North America)

The difficulty with the use of reference library is the difficulty in finding lesions that appear similar to the unknown lesion. This is primarily for two reasons. First, there are many different features that can be used to characterize a lesion, and depending on the lesion, these characteristics are weighted differently when humans view the image (Nishikawa et al. 2004). An example is shown in Fig. 6.4. Second, radiologist often cannot agree on whether a pair of lesions looks similar. This is particularly true for judging microcalcifications. Without the "truth" of which pair of lesions looks similar, it is extremely difficult to develop a computerized technique to judge whether two lesions are similar. Several different groups are working on this problem (El-Naqa et al. 2003, 2004; Muramatsu et al. 2005; Zheng et al. 2006b; Tourassi et al. 2007).

6.4.2 Required Pixel Size

Whether a digital image is obtained by digitizing a film mammogram or it is acquired on a full-field

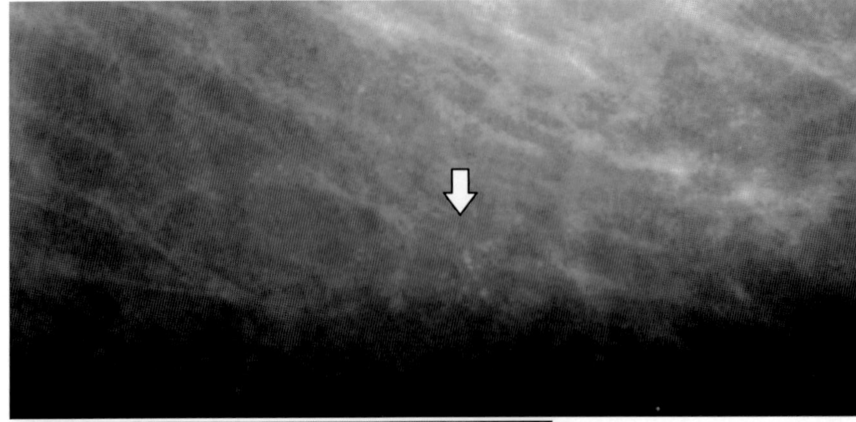

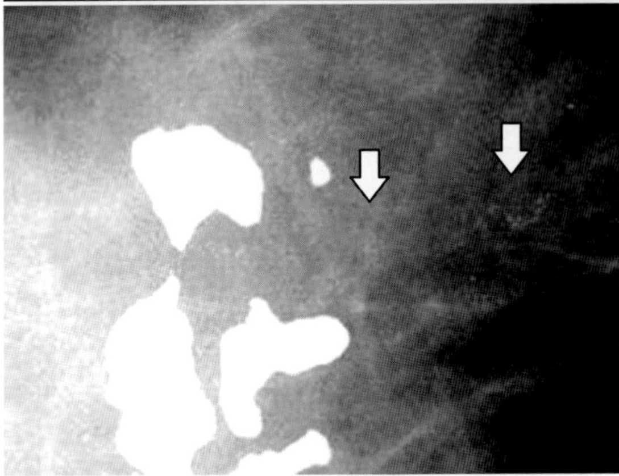

Fig. 6.4. An example of the difficulty in determining similar images. For this cluster of calcifications, the large adjacent macrocalcifications can influence how a radiologist perceives this cluster. Some radiologists may choose to ignore them and consider the microcalcifications similar, while others will consider the two clusters dissimilar because only one contains macrocalcifications. The arrows indicate the location of clustered calcifications (taken from Nishikawa et al. 2004. With permission from SPIE)

digital mammography (FFDM) system, the quality of the digital image will affect the performance of the CAD scheme. Perhaps the most important factor is the pixel size. Consistent with the pixel size in full digital mammograms, film mammograms are digitized for CAD with pixels ranging from 50 up to 100 μm, based on the published literature. It should be noted that all commercial CADe systems digitize the image at 50 μm. As it is generally believed that pixel size is only important for microcalcifications and not for masses, all published studies only examine microcalcifications.

Chan et al. published two studies on the effect of pixel size on CADe and CADx schemes. They found that as the pixel size decreased from 140–105 to 70–35 μm, the performance of their CADe scheme improved, with a statistically significant improvement between 35 and 70 μm and between 70 and 105 μm (Chan et al. 1994). For their CADx scheme, with pixel sizes of 35, 70, 105, and 140 μm, there were no statistically significant differences between any sizes (Chan et al. 1996). In their study, they employed a fairly comprehensive set of texture features (using co-occurrence matrices) and morphological features. For pixel sizes of 35, 70, 105, and 140 μm, there were no statistically significant differences between any sizes.

This result was also observed by Gavrielides et al. (1997), who conducted a smaller study by comparing the performance of their own CADx scheme as a function of the pixel size of the image. They examined 30, 60, and 90-μm pixels. For manual segmentation and considering only cases with actual calcifications (as opposed to simulated clusters), their classifier obtained performances of 48, 76, and 80%, for 30, 60, and 90-μm pixels, respectively. Again, a smaller pixel did not improve performance of CADx schemes.

These studies indicate that for a CADe scheme to distinguish actual microcalcifications from false positives, high spatial frequency information is needed. False positives can be caused by film artifacts, which appear with sharp edges on a screen-film mammogram, and therefore, can be distinguished from the actual microcalcifications that do not appear as sharp. When the film is digitized, if the spatial resolution is not sufficiently high, the artifact will become blurred and will look like an actual microcalcification.

6.4.3 Full-Field Digital Mammography

The transition from screen-film mammography to FFDM has several advantages for CAD and one potential drawback. These are summarized here, but a detailed discussion has been presented previously (NISHIKAWA 2003). The major drawback is that CAD algorithms written for screen-film mammography may not work on FFDM images. Early studies, however, indicate that only small adjustments such as retraining classifiers and preprocessing of the images may be sufficient.

The main advantage of applying CADe to FFDM images is that whole process can be streamlined. There is no need to digitize the image, so the cost of the system can potentially be decreased, because a film digitizer is not needed. Further, the radiology technologist no longer has to spend time digitizing the films, and thus, his or her effort in implementing CADe is reduced. The display of the CADe results is simplified because they can be displayed as an overlay on the digital image, assuming that softcopy reading is used. The archiving of the CADe output is also simplified because they can be stored as a structured report in the DICOM header.

6.5 Evaluation Methods for CADe and CADx

CADe and CADx schemes are evaluated using different techniques. CADx is basically a binary problem: a lesion is either benign or malignant. This type of problem is well served by receiver-operating characteristic (ROC) analysis (METZ 1978, 1996, 2000). CADe schemes, on the other hand, have many different possible outcomes, from no lesion detected to possibly ten or more. This type of problem is not amenable to ROC analysis, and free-response ROC (FROC) analysis is often employed.

It is important to note that the cases used to evaluate a CAD scheme and for the CADe schemes, the use of scoring criteria can greatly influence the measured performance (NISHIKAWA et al. 1994). This is problematic when comparing the published results from different CAD schemes that were evaluated using different databases. The differences in the difficulty of the cases used can easily overwhelm the actual differences of the two schemes.

6.5.1 Evaluation of CADe Schemes

FROC curves are a plot of the sensitivity and the number of false detections per image (see Fig 6.5). As a typical screening exam consists of two views of each breast, the number of false detections per case can also be reported. Analyses of CADe schemes often consider sensitivity at the breast level and at the image level. The breast level measure is probably more clinically relevant, because as long as the radiologist finds the cancer in at least one of the two views, the woman will be recalled. The image level measure is probably more representative of the

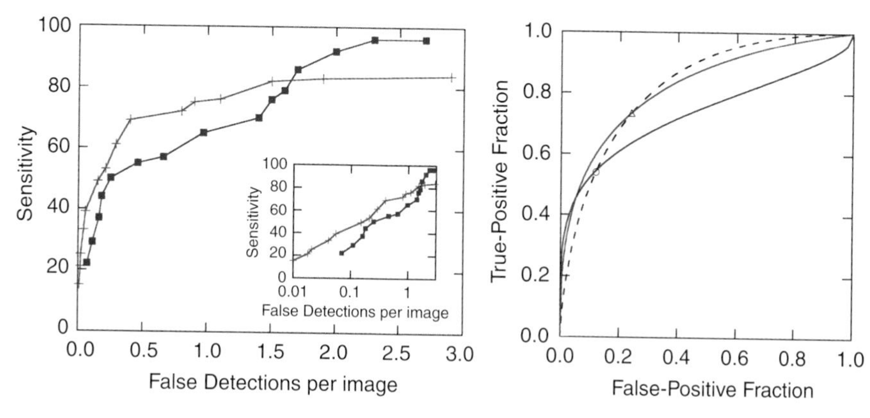

Fig. 6.5. A comparison of FROC and ROC curves. For ROC curves the x-axis is fixed to lie between 0 and 1.0. For FROC curves, values can be greater than 0 with a theoretical limit imposed by the finite size of the image. In each graph, the two points given by the circle and triangle represent the performance of two hypothetical CADe or CADx schemes. If the two points lie on the same curve (*broken line*), then the two schemes have the same overall performance. If the two points lie on different curves (*solid lines*), then the curve closest to the upper left corner of the graph has the best performance. Shown as an inset are the FROC curves plotted on a linear-log scale illustrating fairly linear curves

actual performance of a CADe scheme – it treats image independently.

CADe schemes are often evaluated by "stand-alone performance". That is, the CADe scheme performs by itself, ignoring whether there is any benefit to the radiologist. While this metric is important, it does not indicate directly whether the CADe scheme will help radiologists find missed cancers. When used as a "second reader", the CADe scheme needs to have high sensitivity or the radiologist will not have confidence that CADe scheme is capable of detecting cancer and he or she will not use CADe optimally. Further, even if the CADe scheme has high sensitivity, if the CADe scheme misses the same cancers that radiologist misses, then the CADe scheme will not reduce the radiologist' miss rate. Observer studies are often conducted to directly measure the potential for CADe to improve radiologists' performance.

An FROC curve shows all sensitivity-false detection rate pairs possible for a CADe scheme; when a CADe scheme is used clinically, a single operating point on the curve must be chosen. There is no clear guideline on how to select the point – a point higher on the curve will have higher sensitivity, but also more false detections, while a lower point will have fewer false detections as well as a lower sensitivity. Some clinical systems have an option to operate at different points on the FROC curve, allowing the radiologists the choice of having the CADe to have higher sensitivity or lower number of false detections.

To compare two different FROC curves is difficult because the upper limit to the number of false detections per image is undefined. This means that there is no way to define the area under an FROC curve, like it is done for ROC curves. Currently, the most common method for comparing FROC curves is to use jack-knife FROC (JAFROC) (Chakraborty and Berbaum 2004). Another option is to compare the sensitivity values at a fixed false detection rate or to compare the false detection rates at a fixed sensitivity. This, however, has less statistical significance than comparing the whole curve using JAFROC. Free software for JAFROC is available from Dev Chakraborty at: http://www.devchakraborty.com.

6.5.2 Evaluation of CADx Schemes

ROC analysis is often used to evaluate the medical imaging technology, because it avoids the problem of the tradeoff between sensitivity and specificity. That is, radiologists can change their sensitivity by reading more aggressively. However, this increase in sensitivity is necessarily accompanied by a decrease in specificity (when the radiologist lowers his or her threshold for calling a cancer). Unless it is known whether the increase in sensitivity offsets the decrease in specificity, it may not be clear whether the radiologist changing his or her threshold is an improvement or not. ROC curves plot the sensitivity or true-positive fraction vs. one minus the specificity or the false-positive fraction (see Fig. 6.5).

For a CADx scheme, pairs can be generated by applying a series of threshold to output of the CADx scheme, which spans from the lowest output value possible to the highest output value possible. Then, at each threshold, the number of cases with malignant lesions where the output is greater than the threshold is determined to get the number of malignant lesions classified as malignant, and the number of cases with a benign lesion above the threshold is counted to get the number of benign lesions classified as malignant (i.e., the number of false positives). By knowing the total number of malignant and benign lesions, the true-positive fraction (TPF) and the false-positive fraction (FPF) can be computed and plotted on a ROC graph. Statistical tools are then used to fit a curve to the points, and the area under the curve (AUC) is used as a summary index of the performance of the CADx scheme. Statistical methods are also available to compare AUC values from different CADx schemes to compare two curves. This is used by developers of CADx schemes to determine a new scheme having higher performance than a previous scheme. Free ROC software, developed by Charles Metz and Lorenzo Pesce, is available at http://xray.bsd.uchicago.edu/krl/roc_soft.htm.

In an observer study, pairs of sensitivity and specificity are generated by requiring the radiologist to use a monotonic scale related to the likelihood that a lesion is malignant. Then the thresholds are applied to the ratings, as with the output of the CADx (as described earlier), to generate an ROC curve for the radiologist. The Metz ROC software can fit a curve to a set of TPF–FPF pairs and perform sophisticated statistical analyses that among other calculations, allow one to compare two different ROC curves. The analyses can account for variations due to the differences in cases and differences in readers, allowing one to generalize the results of the experiment to a general set of readers and cases (that are

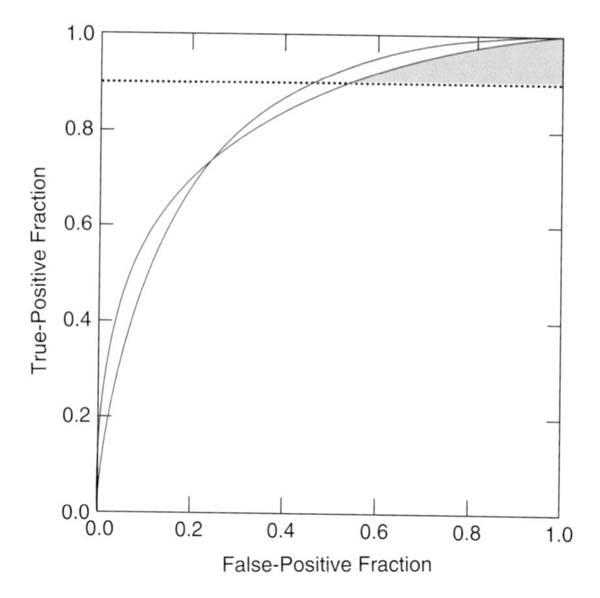

Fig. 6.6. The graph on the left shows two ROC curves that have equal area under the curve (AUC), but the curves cross. One system is superior if high sensitivity is desirable, while the other system would be preferred for high specificity. On the right is an illustration of partial area under the ROC (pAUC), where it is assumed that high sensitivity is desirable. The shaded area divided by the area above the 0.9 value line is $_{0.9}$AUC, the area under the ROC curve above a sensitivity of 90%. When high sensitivity is desired, $_{0.9}$AUC shows the advantage for the *blue curve*, when compared with using AUC for the whole curve

similar to the ones used in the study) (Dorfman et al. 1992; Beiden et al. 2000; Yousef et al. 2006; Wagner et al. 2007).

By computing AUC, it is assumed that all points on the ROC curve have equal utility (i.e., produce an equal tradeoff in benefit and cost). There are certain situations where it is known, at least qualitatively, that all points on the ROC curve do not have equal utility. For example, in diagnostic mammography, radiologists operate clinically such that missing a cancer is worst than biopsying a lesion that is benign. That is, radiologists operate at a relatively high sensitivity and relatively low specificity. In such situations, a partial area under the ROC curve (pAUC) is a more appropriate summary metric (McClish 1989; Jiang et al. 1996). To compute pAUC, the area under the ROC curve between a sensitivity threshold (e.g., 90%) and 1.0 is divided by the same partial area for a perfect ROC curve (which is just one-sensitivity threshold). The normalization causes the pAUC to have values between 0 and 1.0. For random guessing, the AUC is 0.5, but the pAUC will be 0.5*(one-sensitivity threshold)^2. The pAUC is also useful for comparing two ROC curves that cross. That is, crossing curves can have the same AUC, but one will have higher pAUC (see Fig. 6.6).

6.6 Observer Studies for CADe and CADx

Observer or reader studies are experiments meant to mimic the clinical reading situation. The goal is to control the parameters of the experiment to generate data that can be extrapolated to predict how a technology will operate clinically for a general population of patients and a general population of readers.

There have been six observer studies for CADe, which are summarized in Table 6.1. These include two earliest studies that were smaller and did not simulate the actual clinical reading conditions, as well as the more recent studies. For example, in the study by Chan et al. (1990), the radiologists read exams that consisted of a single image and the only type of lesions were clustered microcalcifications. In the study by Kegelmeyer et al. (1994), the CADe scheme had 100% sensitivity and only spiculated cancers and normal cases were included. Nevertheless, these two studies were instrumental in igniting the field.

The study by Taylor et al. (2005) had two parts. In the first part, 50 readers (30 consultant radiologists, 5 breast clinicians, and 15 trained radiographers) read 180 cases that contained 60 cancer cases (40 consecutive screen-detected cancers and 20 interval cancers). They found no change in either sensitivity or specificity when the readers used CADe. Subsequently, they performed a second study, where they enriched the case set with clinically missed cancer cases. In total, they had 120 cases, 40 that contained a clinically missed cancer and 4 other cancer cases. They used a subset of 35 readers from the first study (18 consultant radiologists, 2 breast clinicians, and 15 trained radiographers). Again, they did not find a statistically significant change in either sensitivity or specificity.

The study by Astley et al. is by far the most comprehensive study to date. It is probably the largest observer study ever conducted. The strengths of their study were that there was an extensive training period, a large number of cases, and experienced readers. Unlike the other observer studies, in this study, each case was read just once by one of the eight

Table 6.1. Summary of CADe observer studies. Important details from the different studies are given in the text

Study lead author	CADe stand-alone performance	Total no. of cases (no. of cancers)	Number of readers	Readers unaided	Readers aided	p-value
Chan et al. (1990)	87% sens., 3 false cluster/image	60 (30)	15	AUC = 0.94	AUC = 0.97	<0.001
KEGELMEYER et al. (1994)	100% sens. 0.28 false detections per image	85 (36)	4	Sens = 81% Spec = 96%	Sens. = 93% Spec = 97%	0.027 0.49
Taylor Study 1 (TAYLOR et al. 2005)	R2 ImageChecker version 2.2 (75% sens., 1.9 false detections per image)	180 (60)	50	Sens = 0.78; 95% CI: (0.76–0.80) Spec = 0.84 (0.81–0.86)	Sens = 0.78; 95% CI: (0.76–0.80) Spec = 0.84 (0.81–0.87)	Not statistically significant
Taylor Study 2 (TAYLOR et al. 2005)	Same as study 1 above	120 (44)	35	Sens = 0.77; 95% CI: (0.73–0.81) Spec = 0.85 (0.81–0.87)	Sens = 0.80; 95% CI: (0.76–0.84) Spec = 0.86 (0.84–0.88)	Not statistically significant
GILBERT et al. (2006)	R2 ImageChecker M1000 version 5.0	10,096 (315)	8	[a]Sens = 32.7% [a]RR = 6.5	Sens = 40.0% RR = 8.6	0.00 <0.001

[a]Double reading without CAD taken from original clinical readings

radiologists using CADe. Further, this study compared single reading using CADe with independent double reading, with the double reading results being taken from the clinical reading of the cases when they were originally acquired and read. This study clearly demonstrated that CADe can help radiologists detect more cancers – the difference in sensitivity was 25% ((40–32)/32). However, this was done at a cost of increase in the number of recalls of 32% ((8.6–6.5)/6.5).

The results from the CADe observer studies are equivocal. The early studies showed a definite advantage in using CADe, but the more recent and clinically realistic studies have been less clear. The studies by Taylor et al. showed no net benefit from using CADe, while the Gilbert study showed a 25% increase in sensitivity, but a 32% increase in the recall rate. As we shall see in the next section, the clinical evaluations of CADe presented a clearer picture.

There have been nine observer studies for CADx and a summary of the results are given in Table 6.2. These studies give strong evidence that CADx can improve radiologists' ability to make accurate biopsy recommendations. In particular, when compared with the results of CADe observer studies, these CADx studies present a coherent picture that CADx has the potential to be clinically beneficial.

6.7 Clinical Studies for CADe

6.7 1 Methodology

There have been 13 clinical studies of CADe to date (FREER and ULISSEY 2001; GUR et al. 2004b; HELVIE et al. 2004; BIRDWELL et al. 2005; CUPPLES et al. 2005; KHOO et al. 2005; DEAN and ILVENTO 2006; KO et al. 2006; MORTON et al. 2006; FENTON et al. 2007; GEORGIAN-SMITH et al. 2007; GILBERT et al. 2008; GROMET 2008), with multiple comments on them (ELMORE and CARNEY 2004; FIEG et al. 2004; GUR et al. 2004a; GUR and SUMKIN 2006; BREM 2007a, b; HALL 2007; NISHIKAWA et al. 2007). One study examined the use of CADe as an adjunct to double reading (KHOO et al. 2005). They found no benefit to using CADe with double reading. Three studies compared single reading using CADe with double reading (GEORGIAN-SMITH et al. 2007; GILBERT et al. 2008; GROMET 2008). These studies, in general, found the two reading conditions comparable. The remaining studies compared aided with unaided reading, and the results from these studies have been mixed.

Two different methods have been employed to conduct the clinical evaluations of CADe. The first is

Table 6.2. Summary of CADx observer studies

Study	Type of lesion	CADx stand-alone performance (AUC)	Number of cases (no. of cancers)	Number of readers	Readers' unaided AUC	Readers' aided AUC	*p*-value
Getty et al. (1988)	All	0.86	118 (58)	6	0.83	0.88	0.02
Jiang et al. (1999)	Calc	0.80	104 (46)	10	0.61	0.75	<0.0001
Chan et al. (1999a)	Mass	0.92	76 (39)	6	0.92	0.96	0.026
Leichter et al. (2000)	Mass	0.95	40 (18)	1	0.66	0.81	<0.001
Sklansky et al. (2000)	Calc	0.75	80 (23)	4	0.69	0.82	0.003
Huo et al. (2002b)	Mass	0.90	110 (50)	12	0.93	0.96	<0.001
Hadjiiski et al. (2004)	Mass	0.87	253 (138)	10	0.79	0.84	0.005
Hadjiiski et al. (2006b)	Mass	0.90	90 (47)	8	0.83	0.87	<0.05
Horsch et al. 2006	Mass	0.81, 0.93[a]	97	10	0.87	0.92	<0.001

[a]CADx performance for mammograms, CADx performance for sonograms

a sequential or cross-sectional method (Freer and Ulissey 2001; Helvie et al. 2004; Birdwell et al. 2005; Dean and Ilvento 2006; Ko et al. 2006; Morton et al. 2006; Georgian-Smith et al. 2007). Here, the radiologist reads the case before looking at the CADe output and records an interpretation. Then, the CADe output is reviewed and the radiologist can modify his or her interpretation. The increase in the number of cancers detected and the number of women recalled when CADe is used is recorded, and compared with unaided values. The second method is a longitudinal method, which used historical controls (Gur et al. 2004b; Cupples et al. 2005; Fenton et al. 2007; Gromet 2008). In this method, the cancer detection rate and the recall rate from two independent time periods were compared: one when CADe was not used and the other, when CADe was used.

In general, the studies that employed historical controls showed very little increase in the cancer detection rate, while the cross-sectional studies showed approximately 10% increase in the number of cancers detected when CADe was used. [The exceptions to this general result are smaller studies with larger uncertainty (Cupples et al. 2005, Georgian-Smith et al. 2007).] The reason for the difference between the study types is quite subtle (Nishikawa and Pesce 2008). When CADe is introduced, it changes the underlying cancer prevalence in the screened population. CADe helps the radiologist to detect cancers that they would have otherwise overlooked. If the radiologist misses the cancer, it is still available in "pool" of cancers in the screened population. If the cancer is caught because the radiologist used CADe, then a fraction of the cancers are removed from the "pool," decreasing the cancer prevalence in the screened population. If the cancer prevalence is lower, then one would expect the cancer detection rate to be lower, assuming that radiologists' skill level has not changed (a fundamental assumption of the longitudinal method). Therefore, the change in the cancer detection rate will be smaller than it would, if the cancer prevalence did not change.

Finally, it is important to note that the goal of CADe is not to find more cancer per se, but to find cancer earlier when it is potentially more treatable. While the longitudinal method does not measure this, the cross-sectional method does. However, the cross-sectional method does have a potentially large bias. As the radiologist renders two decisions and the second decision is always with CADe decision, two different scenarios can arise. The radiologist may read less thoroughly without CADe than they would, if there was not a second chance to make an interpretation (Sumkin and Gur 2006). This would tend to increase the number of cancers detected when CADe is used. However, the contrary can also possibly occur. The radiologist may read extra carefully so as not to miss any cancers (Birdwell et al. 2005). It is not possible to determine if either of these scenarios occurred in a cross-sectional study, and if one did occur, how large an effect it had is difficult to determine.

Given that the goal of CADe is to detect cancers at an earlier time point, a better endpoint maybe the size or stage of the cancer. If CADe is effective in detecting cancers earlier, then the size of the cancers detected with CADe should be smaller and the number of advanced or late stage cancers should also decrease.

6.7.2 Recall Rate

Recall rates increase when CADe is used. Theoretically, this occurs because radiologists not only overlook cancers in mammogram, but they also overlook benign lesions that appear as malignant and pseudolesions - overlapping normal tissue - that appear malignant. If the computer detects these malignant-appearing non-cancers along with actual cancers, then the radiologist may detect not only more cancers, but also more benign and pseudolesions. In theory, the increase in cancer detection rate should be comparable with the increase in the recall rate under three conditions. First, the radiologist is equally likely to overlook a cancer as a malignant-appearing benign or pseudolesion. Second, the computer is equally likely to detect an overlooked cancer as an overlooked malignant-appearing benign or pseudolesion. Third, the radiologist evaluates computer-detected lesions with the same objectivity as lesion found unaided. That is, the radiologist does not give an "extra level of suspicion" to CADe-detected lesions.

Most of the clinical studies to date show a comparable increase in the cancer detection rate and recall rate. However, in some studies, the increase in recall rate is much greater than the increase in the cancer detection rate (Fenton et al. 2007). This may be attributed to the time it takes for the radiologists to learn how to use CADe most efficiently (Hall 2007). In one study (Dean and Ilvento 2006), the radiologists' recall rate initially increased by over 100% and did not decrease to a reasonable value (which is approximately a 10% increase) until after 22 months of use. This indicates that it takes almost 2 years for some radiologists to be able to judge CADe-detected lesions in the same manner as they judge lesions that they detect without the aid (condition 3 in the preceding paragraph).

Table 6.3. Comparison of positive predictive value (PPV) of CADe and radiologists reading screening mammograms (taken from Bick and Diekmann (2007). With permission)

Average number of CADe marks per normal case	Positive predictive value (ppv)[a]	Number of positive CADe marks/ abnormal readings to detect one cancer
5	0.001	1,000
1[b]	0.005	200
0.1	0.05	20
Radiologists[c]	0.1–0.5	2–10

[a]Assuming a cancer detection rate of 5 per 1,000 screening exams
[b]Performance of current commercial CAD systems
[c]Based on a range of radiologists' recall rates between 1 and 5%

One of the reasons why radiologists may have increased recall rate when first using CADe is that false detection rate of CADe is high, especially when compared with the number of correct marks. Given that the cancer prevalence in a screening population is 5 per 1,000 women screened, at most there will be ten correct CADe marks and 1,000 false marks. As illustrated in Table 6.3, the false detection of current CADe schemes would have to decrease by at least a factor of 10 to be comparable with the false detection rate by radiologists.

6.7.3 Comparison with Double Reading

CADe is often promoted as means of providing a "second opinion". This implies that double reading could be performed with a single radiologist. There are several different ways of implementing double reading, such having a third radiologist, when the first two disagree, or having the two radiologists discuss the case when they disagree. However, CADe most closely mimics independent double reading, where two radiologists independently read the case and a patient is recalled if either radiologist recommends a recall.

Taylor and Potts performed a meta-analysis and found that independent double reading increases the number of cancers detected by 13%, whereas the recall rate increases by 31%. These numbers are comparable with what is found by the clinical CADe studies that used the sequential or cross-sectional method (9% increase in the number of cancers detected), with CADe having a lower increase in the recall rate (10%), when compared with the independent double reading. In their paper, Taylor and Potts compared all the CADe studies with all forms of double reading to draw their general conclusions. There are several issues to consider that make such general conclusions doubtful. First, all CADe studies were performed in the USA, while nearly all the double reading studies were performed in Europe. There are differences in the way screening is implemented and in the performance benchmark that makes the interpretation of such a comparison problematic. Further, as discussed in a previous section, the two different methods used to measure the effectiveness of CADe were not equivalent, but were treated as such in the Taylor and Potts analyses.

Although there is ambiguity in the published literature, we believe that CADe is atleast as effective as independent double reading. In spite of some of the early overly enthusiastic expectations for CADe, this somewhat modest performance is, in retrospect, not unexpected.

6.8 Current Research in CADe and CADx

Although CADe systems have been in use clinically for more than 10 years, there is still much needed research ongoing to improve the performance of the CADe schemes, to examine new uses of CADe, to establish CADx as a useful clinical tool, and application of CADe and CADx to new breast imaging modalities. These will be briefly discussed here.

6.8.1 Improving CADe Scheme Performance

There is a general consensus that CADe for clustered calcifications is more effective clinically than CADe for masses. In fact, there is a discontinuity in radiologists' satisfaction in using CADe for calcifications vs. masses. This is partly because radiologists have a harder time detecting masses than calcifications. In general, there is very little in a mammogram that looks like a cluster of calcifications other than a cluster of calcifications. On the other hand, superposition of breast tissue can often mimic a cancer or partially or completely obscure a cancer.

When evaluating a suspicious area on the mammogram that may contain a mass, radiologists make comparisons. They compare the craniocaudal view with the mediolateral oblique view; they compare the images of the right breast with the left breast; and finally, they compare the present exam with past exam(s). For a human, these comparisons are not too difficult to perform. For a computer, on the other hand, this is a complex task.

The appearance of the internal structure of the breast on a mammogram can change with the mammographic technique (type of X-ray detector, kVp, anode material, filter material, X-ray exposure, etc.), the amount of compression, how the breast is positioned by the technologist, and the way in which the mammograms are displayed (film vs. monitors and the type of image processing). While radiologists are fairly robust at accounting for these potential sources of variation in the appearance of the mammogram, computer programs are not. As the change in the appearance of the mammogram is extremely variable, it is very difficult to compare the regions on different mammograms to know whether they are images of the same breast tissue. Nevertheless, there are several different approaches being developed.

6.8.1.1 Temporal Comparison (Comparison with Prior)

The goal here is to determine whether a lesion present in the current mammogram is present in the previous mammogram, and if so, has the lesion changed in appearance. Clinically, a stable lesion is an indication that the lesion is benign. If the lesion is new or has changed in size or shape, it is an indication that the lesion maybe growing and maybe malignant. Detecting a change can be difficult for a radiologist and is often a tedious and time-consuming process. The challenge here for CAD is to determine the location of the lesion in the temporal image pair. Once the lesion is properly identified, computers can be very accurate in extracting quantitative information about the lesion, making the determination of even subtle changes possible (Zheng et al. 2002, 2003; Hadjiiski et al. 2006b; Timp and Karssemeijer 2006; Qian et al. 2007; Timp et al. 2007; Wei et al. 2007).

6.8.1.2 Spatial Comparison (Different Views)

The goal here is to determine whether a lesion visualized in one view is visible in either the contralateral breast or the other view of the same breast. If a similar lesion is present in the same location in the contralateral breast, it is an indication that the lesion is either benign or normal tissue. If the lesion is present in the other view of the same breast, this is an indication that the lesion is real, as opposed to a superposition of normal tissue mimicking a cancer.

Most approaches rely on geometry. That is, given the location in one view, there is a band, based on the distance from the nipple, in the contralateral view where the lesion should appear. All possible combinations of lesion pairs are found and the most likely pairing is selected. This is then determined based on the fact that, while the background appearance of the breast may change, the characteristics of a lesion is more stable. Therefore, features are extracted for lesions in different views and these are used in a statistical classifier to predict the likelihood that the two lesions that are being compared represent the same lesion seen in two different views (as opposed to two different lesions) (Paquerault et al. 2002; Zheng et al. 2006a; Van Engeland and Karssemeijer 2007; Yuan et al. 2008; Velikova et al. 2009).

One of the difficulties in this approach is that a lesion is not always detected or even visible in both views. It is necessary to include this possibility when designing the CADe scheme, so as not to eliminate cancers that are visible in only one view and are detected by the single-scheme method.

6.8.2 CADe as Pre-screen

As the performance of CADe schemes has improved, the number of false detections has decreased while the sensitivity has improved or remained the same. The performance of clinical CADe systems is now at the point where it may be possible to use CADe as a pre-screener of screening mammograms. That is, before the cases are reviewed by a radiologist, a CADe scheme is run and cases without any detections are considered normal and not read by the radiologist. The goal here is to reduce the radiologists' workload. If the sensitivity of the CADe system is high, the number of false negatives (FN) will be low. Therefore the negative predictive value (NPV) will be high, as:

$$NPV = TN/(TN + FN) \tag{6.1}$$

where TN is the number of true negatives. In fact, the actual NPV will be slightly higher, as it is likely that even if the CADe does not correctly locate the cancer, there is a chance that there will be at least one false positive detection somewhere in the four views, and hence, the case would not pass the pre-screen stage.

With further improvements in CADe performance, CADe as a pre-screener may be possible. However, it remains to be seen if such a system is actually beneficial. Typically, the CADe scheme will pre-screen out "easy" cases, most likely mammograms of fatty breasts, which are relatively quick and easy for the radiologists to read, because there are no places for cancers to "hide" as there are in women with dense breasts. By removing all the "easy" cases, radiologists may become more exhausted, as every case will be "difficult". Further, the apparent false positive rate of the CADe scheme will be higher, because if the pre-screened cases had no false detections, then the average number of false detections in the remaining cases will be higher than the average for all cases. This could potentially cause more cancers to be missed or make the radiologists less efficient and effective in using CADe.

6.8.3 Concurrent Reading with CADe

Currently, CADe is used as a second reader, where the radiologist first reads the images without CADe and then re-reads with the knowledge of the CADe findings. An alternative paradigm is to have CADe as the first reader and have the radiologist verify (i.e., accept or reject) each CADe prompt. For this to be effective, the sensitivity of the CADe system must be very high. For mass detection, which is a very difficult task, the sensitivity is not high enough. However, the sensitivity of clustered microcalcification detection is approximately 98%. At this level of performance, it is conceivable that CADe could be the first reader, which is sometimes referred to as concurrent reading.

The use of concurrent reading for microcalcification clusters has many appealing advantages. Perhaps, the most tedious and time-consuming aspect of reading screening mammograms is the search for microcalcifications, which often involves scanning the film with a magnifying glass, or for a digital image, zooming the image to full resolution and panning the image, as the full image often does not fit onto the softcopy monitor at full resolution. CADe may relieve the radiologist from performing this task, as CADe finds clustered microcalcifications at a very high rate.

There have been no studies to date to determine whether the performance of radiologists remains as high as when they read using CADe as a second reader, when searching for microcalcifications. There is at least one study published that suggests that CADe can miss amorphous calcifications (Soo et al. 2005). Therefore, it is important to carry out carefully designed and conducted studies.

6.8.4 Interactive CADe

Karssemeijer et al. (2008) proposed a new implementation for CADe. Instead of presenting all the CADe detections to the radiologist, the radiologist can mark suspicious area on a softcopy display and if the area was identified by the computer, the computer's likelihood that the area contains a cancer is given. This information can be used by the radiologist in two ways. If the CADe scheme has a high negative predictive value, then if the area selected by the radiologist has a low likelihood of malignancy or if the computer did not select the region as suspicious, then the radiologist may have selected a region that does not contain an actual cancer. If the CADe scheme has high PPV and if the region selected by the radiologist has a high likelihood of malignancy, then there is a high probability that the selected region contains a cancer. Karssemeijer achieved high PPV by operating the CADe scheme at a low false detection rate of 0.1 per image. They have shown previously that at this false detection rate, the performance of the CADe scheme was comparable with that of the radiologists, if the analysis was restricted to detections that were considered suspicious by the radiologists (Karssemeijer et al. 2006). They found preliminary evidence that users were able to increase their sensitivity by using this interactive CADe system (Karssemeijer et al. 2008). When a radiologist queried the CADe system about a lesion he or she was uncertain about, if the CADe returned a high likelihood that the lesion was malignant, then it increased the chances that the radiologist would recall the patient, thereby increasing the sensitivity. However, they did not report the effect on recall rate.

Tourassi et al. (2007) and Zheng et al. (Zheng et al. 2006b, 2007; Park et al. 2007) independently proposed a different interactive system, where instead of the radiologist receiving the likelihood of the lesion in question being malignant, the lesions that are similar to the lesion in question are displayed along with the pathology of each lesion. The radiologist can then review the retrieved lesions and decide whether the questionable lesion needs to undergo diagnostic imaging (i.e., recall the woman). This is similar to the methods developed for CADx described in a previous section. The approach of Zheng et al. was to extract features of the lesion in question and compare them with the features of lesions in a reference library containing 3,000 regions with known pathology. The features are used to select similar-looking lesions from the reference library (Zheng et al. 2007). The approach by Tourassi et al. employed a featureless algorithm that examines the statistical relationships between the corresponding image pixels to make inferences about lesion similarity (Tourassi et al. 2007).

6.8.5 CADx Multimodality

When a woman is recalled because of an abnormal screening mammogram, or there is a palpable lump or a suspicious physical finding, she is given a diagnostic exam. Here, specialized X-ray views are taken and these are often supplemented with ultrasound and sometimes, MRI. As these three modalities capture different information about the breast, it can be difficult to determine the correspondence between the modalities. That is, if there is more than one lesion in any of the modalities, it may not be trivial to determine which of the multiple lesions correspond to the same lesion on different modalities.

Further, a radiologist reviews all the available images to decide whether to recommend that the woman should have a biopsy, and hence, it is necessary for CADx to be available for all imaging modalities in use. Developing CADx for ultrasound and MRI is a very active area of research.

One of the challenges of multimodality CADx is to determine how to combine the analyses from the different modalities. Presenting CADx output for each modality is feasible, and there may be higher utility in combining the multiple analyses into a single analysis presenting a single output to the radiologist. Research to determine the best approach is currently under way (Drukker et al. 2005; Horsch et al. 2006; Jesneck et al. 2007).

6.8.6 CADe and CADx for Tomosynthesis and Breast CT

Like other modalities, X-ray imaging of the breast is becoming three-dimensional (3D). Breast tomosynthesis and breast CT are emerging technologies that have a large potential to improve screening and diagnostic mammography. One of the potential drawbacks of these technologies is that a large volume of images is produced per breast. On the other hand, a screening mammogram produces two images of each breast, while a tomosynthesis exam could produce more than 100 images of each breast. Given the large amount of data that the radiologist needs to review, CAD may play an important role in assisting radiologists, so that they do not overlook a cancer. Several research groups are developing CADe and CADx schemes for these new modalities (Reiser et al. 2004, 2006; Chan et al. 2005, 2008; Singh et al. 2008); however, at the time when this chapter was written, there were no publications on CAD for breast CT.

One novel aspect of CADe development for tomosynthesis is that the CADe algorithm can be applied to either the reconstructed slices or the projection images (the images that are used to reconstruct the slices). When operating on the projection images, a CADe algorithm analyzes each projection independently, and the detections from projection are back-projected in to a virtual breast volume. Actual lesions are more likely to be detected in many slices, while false detections due to superposition of normal breast tissue may not. This knowledge can be used to reduce the false detection rate.

There are theoretical advantages to analyzing the projections including fewer images to analyze, fewer artifacts, and direct translation of CAD algorithms developed for conventional mammography, not dependent on the reconstruction algorithm (Reiser et al. 2006). The main disadvantage is that each projection image is noisier than a reconstructed slice, due to increase in the X-ray quantum noise and possible detector noise. This increase in the stochastic noise may not be a problem for mass detection, where the main source of noise is from the anatomical structure (Burgess et al. 2001), but may probably limit the performance of microcalcification detection schemes, because for small objects, the X-ray quantum noise is the dominant noise source.

6.9 Financial Disclosure

Robert Nishikawa is a shareholder in Hologic Inc. He and the University receive royalties from Hologic Inc. He is also a consultant to Siemens Medical Solutions and Carestream Health.

References

Acharya RS, Goldgof DB (1993) Biomedical image processing and biomedical visualization. In: Proceedings of SPIE, pp 442–553, 690–871

Ackerman LV, Gose EE (1972) Breast lesion classification by computer and xeroradiography. Cancer 30:1025–1035

Astley SM (2004a) Computer-aided detection for screening mammography. Acad Radiol 11:1139–1143

Astley SM (2004b) Computer-based detection and prompting of mammographic abnormalities. Brit J Radiol 77 Spec No 2:S194–200

Barlow WE, Chi C, Carney PA, et al (2004) Accuracy of screening mammography interpretation by characteristics of radiologists. J Natl Cancer Inst 96:1840–1850

Beam CA, Layde PM, Sullivan DC (1996) Variability in the interpretation of screening mammograms by US radiologists. Findings from a national sample. Arch Intern Med 156:209–213

Beiden SV, Wagner RF, Campbell G (2000) Components-of-variance models and multiple-bootstrap experiments: an alternative method for random-effects, receiver operating characteristic analysis. Acad Radiol 7:341–349

Bick U, Diekmann F (2007) Digital mammography: what we know and what we don't know. Eur J Radiol 17:1931–1942

Birdwell RL, Bandodkar P, Ikeda DM (2005) Computer-aided detection with screening mammography in a university hospital setting. Radiology 236:451–457

Bowyer KW, Astley S (1994) State of the art in digital mammographic image analysis. World Scientific, London

Brem RF (2007a) Blinded comparison of computer-aided detection with human second reading in screening mammography: the importance of the question and the critical numbers game. AJR Am J Roentgenol 189:1142–1144

Brem RF (2007b) Clinical versus research approach to breast cancer detection with CAD: where are we now? AJR Am J Roentgenol 188:234–235

Burgess AE, Jacobson FL, Judy PF (2001) Human observer detection experiments with mammograms and power-law noise. Med Phys 28:419–437

Campanini R, Dongiovanni D, Iampieri E, et al (2004) A novel featureless approach to mass detection in digital mammograms based on support vector machines. Phys Med Biol 49:961–975

Chakraborty DP, Berbaum KS (2004) Observer studies involving detection and localization: modeling, analysis, and validation. Med Phys 31:2313–2330

Chan H-P, Doi K, Vyborny CJ, et al (1990) Improvement in radiologists' detection of clustered microcalcifications on mammograms: the potential of computer-aided diagnosis. Invest Radiol 25:1102–1110

Chan H-P, Niklason LT, Ikeda DM, et al (1994) Digitization requirements in mammography: effects on computer-aided detection of microcalcifications. Med Phys 21:1203–1211

Chan H-P, Sahiner B, Petrick N, et al (1996) Effects of pixel size on classification of microcalcifications on digitized mammograms. Proc SPIE 2710:30–41

Chan HP, Doi K, Galhotra S, et al (1987a) Image feature analysis and computer-aided diagnosis in digital radiography. I. Automated detection of microcalcifications in mammography. Med Phys 14:538–548

Chan HP, Vyborny CJ, MacMahon H, et al (1987b) Digital mammography. ROC studies of the effects of pixel size and unsharp-mask filtering on the detection of subtle microcalcifications. Invest Radiol 22:581–589
Chan HP, Sahiner B, Helvie MA, et al (1999a) Improvement of radiologists' characterization of mammographic masses by using computer-aided diagnosis: an ROC study. Radiology 212:817–827.
Chan HP, Sahiner B, Wagner RF, et al (1999b) Classifier design for computer-aided diagnosis: effects of finite sample size on the mean performance of classical and neural network classifiers. Med Phys 26:2654–2668
Chan HP, Wei J, Sahiner B, et al (2005) Computer-aided detection system for breast masses on digital tomosynthesis mammograms: preliminary experience. Radiology 237: 1075–1080
Chan HP, Wei J, Zhang Y, et al (2008) Computer-aided detection of masses in digital tomosynthesis mammography: comparison of three approaches. Med Phys 35:4087–4095
Cupples TE, Cunningham JE, Reynolds JC (2005) Impact of computer-aided detection in a regional screening mammography program. AJR Am J Roentgenol 185:944–950
Dean JC, Ilvento CC (2006) Improved cancer detection using computer-aided detection with diagnostic and screening mammography: prospective study of 104 cancers. AJR Am J Roentgenol 187:20–28
Dorfman DD, Berbaum KS, Metz CE (1992) Receiver operating characteristic rating analysis. Generalization to the population of readers and patients with the jackknife method. Invest Radiol 27:723–731
Drukker K, Horsch K, Maryellen LG (2005) Multimodality computerized diagnosis of breast lesions using mammography and sonography. Acad Radiol 12:970–979
Elmore JG, Carney PA (2004) Computer-aided detection of breast cancer: has promise outstripped performance? J Natl Cancer Inst 96:162–163
Elmore JG, Nakano CY, Koepsell TD, et al (2003) International variation in screening mammography interpretations in community-based programs. J Natl Cancer Inst 95: 1384–1393
El-Naqa I, Yang Y, Nishikawa RM, et al (2002) A support vector machine approach for detection of microcalcifications. IEEE Trans Med Imag 21:1552–1563
El-Naqa I, Yang Y, Nishikawa RM, Wernick MN (2003) Content-based image retrieval based on learned similarity measures. IEEE Trans Med Imag (in review)
El-Naqa I, Yang YY, Galatsanos NP, et al (2004) A similarity learning approach to content-based image retrieval: application to digital mammography. IEEE Trans Med Imag 23:1233–1244
Fenton JJ, Taplin SH, Carney PA, et al (2007) Influence of computer-aided detection on performance of screening mammography. N Engl J Med 356:1399–1409
Fieg SA, Sickles EA, Evans WP, et al (2004) Re: changes in breast cancer detection and mammography recall rates after the introduction of a computer-Aided detection system. J Natl Cancer Inst 96:1260–1261
Freer TW, Ulissey MJ (2001) Screening mammography with computer-aided detection: prospective study of 12,860 patients in a community breast center. Radiology 220:781–786
Fujita H, Doi K, Fencil LE, et al (1987) Image feature analysis and computer-aided diagnosis in digital radiography. 2. Computerized determination of vessel sizes in digital subtraction angiography. Med Phys 14:549–556
Gavrielides MA, Kallergi M, Clarke LP (1997) Automatic shape analysis and classification of mammographic calcifications. Proc SPIE 3034:869–876
Georgian-Smith D, Moore RH, Halpern E, et al (2007) Blinded comparison of computer-aided detection with human second reading in screening mammography. AJR Am J Roentgenol 189:1135–1141
Getty DJ, Pickett RM, D'Orsi CJ, et al (1988) Enhanced interpretation of diagnostic images. Invest Radiol 23:240–252
Giger ML (1993) Computer-aided diagnosis. In: Haus AG, Yaffe MJ (eds) Syllabus: a categorical course in Physics. Technical aspects of breast imaging. RSNA Publications, Oak Brook, IL, pp 272–298
Giger ML (2004a) Computer-aided diagnosis in diagnostic mammography and multimodality breast imaging. In: Karellas A, Giger ML (eds) Advances in breast imaging: Physics, technology, and clinical applications. Radiological Society of North America, Oak Brook, IL
Giger ML (2004b) Computerized analysis of images in the detection and diagnosis of breast cancer. Semin Ultrasound CT MR 25:411–418
Giger ML, Doi K (1985) Investigation of basic imaging properties in digital radiography. 3. Effect of pixel size on SNR and threshold contrast. Med Phys 12:201–208
Giger ML, Doi K, MacMahon H (1988) Image feature analysis and computer-aided diagnosis in digital radiography. 3. Automated detection of nodules in peripheral lung fields. Med Phys 15:158–166
Giger ML, Huo Z, Kupinski MA, et al (2000) Computer-aided diagnosis in mammography. In: Sonka M, Fitzpatrick JM (eds) Handbook of medical imaging. The Society of Photo-Optical Instrumentation Engineers, Bellingham, WA, pp 915–1004
Giger ML, Huo Z, Vyborny CJ, et al (2003) Results of an observer study with an intelligent mammographic workstation for CAD. In: Peitgen H-O (ed) Digital mammography IWDM 2002. Springer, Berlin, pp 297–303
Giger ML, Chan HP, Boone J (2008) Anniversary paper: History and status of CAD and quantitative image analysis: the role of Medical Physics and AAPM. Med Phys 35:5799–5820Gilbert FJ, Astley SM, McGee MA, et al (2006) Single reading with computer-aided detection and double reading of screening mammograms in the United Kingdom National Breast Screening Program. Radiology 241:47–53
Gilbert FJ, Astley SM, Gillan MGC, et al (2008) Single reading with computer-aided detection for screening mammography. N Engl J Med 359:1675–1684
Granader EJ, Dwamena B, Carlos RC (2008) MRI and mammography surveillance of women at increased risk for breast cancer: recommendations using an evidence-based approach. Acad Radiol 15:1590–1595
Gromet M (2008) Comparison of computer-aided detection to double reading of screening mammograms: review of 231,221 mammograms. Am J Radiol 190:854–859
Gur D, Sumkin JH (2006) CAD in screening mammography. AJR Am J Roentgenol 187:1474
Gur D, Sumkin JH, Hardesty LA, et al (2004a) Re: computer-aided detection of breast cancer: has promise outstripped performance? J Natl Cancer Inst 96:717–718; author reply 718

Gur D, Sumkin JH, Rockette HE, et al (2004b) Changes in breast cancer detection and mammography recall rates after the introduction of a computer-Aided detection system. J Natl Cancer Inst 96:185–190
Gur D, Wagner RF, Chan HP (2004c) On the repeated use of databases for testing incremental improvement of computer-aided detection schemes. Acad Radiol 11:103–105
Hadjiiski L, Chan HP, Sahiner B, et al (2004) Improvement in radiologists' characterization of malignant and benign breast masses on serial mammograms with computer-aided diagnosis: an ROC study. Radiology 233:255–265
Hadjiiski L, Sahiner B, Chan HP (2006a) Advances in computer-aided diagnosis for breast cancer. Curr Opin Obstet Gynecol 18:64–70
Hadjiiski L, Sahiner B, Helvie MA, et al (2006b) Breast masses: computer-aided diagnosis with serial mammograms. Radiology 240:343–356
Hall FM (2007) Breast imaging and computer-aided detection. N Engl J Med 356:1464–1466
Hand W, Semmlow JL, Ackerman LV, et al (1979) Computer screening of xeromammograms: a technique for defining suspicious areas of the breast. Comput Biomed Res 12:445–460
Helvie MA, Hadjiiski L, Makariou E, et al (2004) Sensitivity of noncommercial computer-aided detection system for mammographic breast cancer detection. Radiology 231:208–214
Horsch K, Giger ML, Vyborny CJ, et al (2006) Classification of breast lesions with multimodality computer-aided diagnosis: observer study results on an independent clinical data set. Radiology 240:357–368
Horsch K, Giger M, Metz CE (2008) Potential effect of different radiologist reporting methods on studies showing benefit of CAD. Acad Radiol 15:139–152
Huo Z, Giger ML, Wolverton DE, et al (2000) Computerized analysis of mammographic parenchymal patterns for breast cancer risk assessment: feature selection. Med Phys 27:4–12
Huo Z, Giger ML, Olopade OI, et al (2002a) Computerized Analysis of Digitized Mammograms of BRCA1 and BRCA2 Gene Mutation Carriers. Radiology 225:519–526
Huo Z, Giger ML, Vyborny CJ, et al (2002b) Effectiveness of computer-aided diagnosis—Observer study with independent database of mammograms. Radiology 224:560–568
Jesneck JL, Lo JY, Baker JA (2007) Breast mass lesions: computer-aided diagnosis models with mammographic and sonographic descriptors. Radiology 244:390–398
Jiang Y (2002) Computer-aided diagnosis of breast cancer in mammography: evidence and potential. Technol Cancer Res Treat 1:211–216
Jiang Y, Metz CE, Nishikawa RM (1996) An ROC partial area index for highly sensitive diagnostic tests. Radiology 201:745–750
Jiang Y, Nishikawa RM, Schmidt RA, et al (1999) Improving breast cancer diagnosis with computer-aided diagnosis. Acad Radiol 6:22–33
Kallergi M (1999) Evaluating the performance of detection algorithms in digital mammography. Med Phys 26: 267–275
Karssemeijer N (2002) Detection of masses in mammograms. In: Strickland RN (ed) Image-processing techniques in tumor detection. Marcel Dekker, New York, pp 187–212
Karssemeijer N, Hendriks J (1997) Computer-assisted reading of mammograms. Eur Radiol 7:743–748
Karssemeijer N, Otten JD, Rijken H, et al (2006) Computer aided detection of masses in mammograms as decision support. Brit J Radiol 79 Spec No 2:S123–126
Karssemeijer N, Hupse A, Samulski M, et al (2008) An interactive computer aided detection support system for detection of masses in mammograms. In: Krupinski EA (ed) International workshop on digitial mammography 2008. Springer, New York, pp 273–278
Kegelmeyer WP, Jr., Pruneda JM, Bourland PD, et al (1994) Computer-aided mammographic screening for spiculated lesions. Radiology 191:331–337
Khoo LA, Taylor P, Given-Wilson RM (2005) Computer-aided detection in the United Kingdom National Breast Screening Programme: prospective study. Radiology 237:444–449
Kimme C, O'Loughlin BJ, Sklansky J (1975) Automatic detection of suspicious abnormalities in breast radiographs. In: Fu KS, Kunii TL, Klinger A (eds) Data structures, computer graphics, and pattern recognition. Academic, New York, pp 427–447
Ko JM, Nicholas MJ, Mendel JB, et al (2006) Prospective assessment of computer-aided detection in interpretation of screening mammography. AJR Am J Roentgenol 187: 1483–1491
Kupinski M, Giger ML (1998) Automated seeded lesion segmentation on digital mammograms. IEEE Trans Med Imag 17:510–517
Kupinski M, Giger ML (1999) Feature selection with limited datasets. Med Phys 26:2176–2182
Leichter I, Fields S, Nirel R, et al (2000) Improved mammographic interpretation of masses using computer-aided diagnosis. Eur Radiol 10:377–383
Loo L-N, Doi K, Metz CE (1984) A comparison of physical image quality indices and observer performance in the radiographic detection of nylon beads. Phys Med Biol 29:837–856
Loo L-N, Doi K, Metz CE (1985) Investigation of basic imaging properties in digital radiography. 4. Effect of unsharp masking on the detectability of simple patterns. Med Phys 12:209–214
Malich A, Fischer DR, Bottcher J (2006) CAD for mammography: the technique, results, current role and further developments. Eur Radiol 16:1449–1460
Martin JE, Moskowitz M, Milbrath JR (1979) Breast cancers missed by mammography. AJR Am J Roentgenol 132: 737–739
McClish DK (1989) Analyzing a portion of the ROC curve. Med Decis Making 9:190–195
Metz CE (1978) Basic principles of ROC analysis. Semin Nucl Med 8:283–298
Metz CE (1996) Evaluation of digital mammography by ROC analysis. In: Doi K, Giger ML, Nishikawa RM, et al (eds) Digital mammography '96. Elsevier Science, Amsterdam, pp 61–68
Metz CE (2000) Fundamental ROC analysis. In: Beutel J, Kundel H, Van Metter R (eds) Handbook of medical imaging. SPIE, Bellingham, WA, pp 751–770
Miglioretti DL, Smith-Bindman R, Abraham L, et al (2007) Radiologist characteristics associated with interpretive performance of diagnostic mammography. J Natl Cancer Inst 99:1854–1863

Morton MJ, Whaley DH, Brandt KR, Amrami KK (2006) Screening mammograms: interpretation with computer-aided detection-prospective evaluation. Radiology 239: 375–383

Muramatsu C, Li Q, Suzuki K, et al (2005) Investigation of psychophysical measure for evaluation of similar images for mammographic masses: preliminary results. Med Phys 32:2295–2304

Nishikawa R, Yang Y, Huo D, et al (2004) Observers' ability to judge the similarity of clustered calcifications on mammograms. Proc SPIE 5372:192–198

Nishikawa RM (2002) Detection of microcalcifications. In: Strickland RN (ed) Image-processing techniques in tumor detection. Marcel Dekker, New York pp 131–153

Nishikawa RM (2003) Computer-aided detection in digital mammography. In: Pisano ED, Yaffe MJ, Kuzmiak CM (eds) Digital mammography. Lippincott Williams & Wilkins, Philadelphia, PA, p 231

Nishikawa RM (2007) Current status and future directions of computer-aided diagnosis in mammography. Comput Med Imag Graph 31:224–235

Nishikawa RM, Pesce L (2008) Computer-aided detection evaluation methodologies are not created equal. Radiology (accepted for publication)Nishikawa RM, Yarusso LM (1998) Variations in measured performance of CAD schemes due to database composition and scoring protocol. Proc SPIE 3338:840–844

Nishikawa RM, Giger ML, Doi K, et al (1994) Effect of case selection on the performance of computer-aided detection schemes. Med Phys 21:265–269

Nishikawa RM, Schmidt RA, Metz CE (2007) Computer-aided screening mammography. N Engl J Med 357:83–85

Paquerault S, Petrick N, Chan HP, et al (2002) Improvement of computerized mass detection on mammograms: fusion of two-view information. Med Phys 29:238–247

Park SC, Sukthankar R, Murnmert L, Satyanarayanan M, Zheng B (2007) Optimization of reference library used in content-based medical image retrieval scheme. Med Phys 34:4331–4339

Pisano ED, Gatsonis C, Hendrick E, et al (2005) Diagnostic performance of digital versus film mammography for breast-cancer screening. N Engl J Med 353:1773–1783

Qian W, Song D, Lei M, et al (2007) Computer-aided mass detection based on ipsilateral multiview mammograms. Acad Radiol 14:530–538

Reiser I, Nishikawa RM, Giger ML, et al (2004) Computerized detection of mass lesions in digital breast tomosynthesis images using two- and three dimensional radial gradient index segmentation. Technol Cancer Res Treat 3:437–441

Reiser I, Nishikawa RM, Giger ML, et al (2006) Computerized mass detection for digital breast tomosynthesis directly from the projection images. Med Phys 33:482–491

Roque AC, Andre TC (2002) Mammography and computerized decision systems: a review. Ann NY Acad Sci 980:83–94

Sahiner B, Chan HP, Petrick N, et al (2000) Feature selection and classifier performance in computer-aided diagnosis: the effect of finite sample size. Med Phys 27:1509–1522

Sahiner B, Chan HP, Hadjiisk L (2008) Classifier performance prediction for computer-aided diagnosis using a limited data set. Med Phys 35:1559–1570

Sampat MP, Markey MK, Bovik AC (2005) Computer-aided detection and diagnosis in mammography. In: Bovik AC (ed) The handbook of image and video processing. Elsevier, New York, pp 1195–1217

Semmlow JL, Shadagopappan A, Ackerman LV, et al (1980) A fully automated system for screening xeromammograms. Comput Biomed Res 13:350–362

Singh S, Tourassi GD, Baker JA, et al (2008) Automated breast mass detection in 3D reconstructed tomosynthesis volumes: a featureless approach. Med Phys 35:3626–3636

Sklansky J, Tao EY, Bazargan M, et al (2000) Computer-aided, case-based diagnosis of mammographic regions of interest containing microcalcifications. Acad Radiol 7:395–405

Soo MS, Rosen EL, Xia JQ, et al (2005) Computer-aided detection of amorphous calcifications. AJR Am J Roentgenol 184:887–892

Spiesberger W (1979) Mammogram inspection by computer. IEEE Trans Biomed Eng 26:213–219

Sumkin JH, Gur D (2006) Computer-aided detection with screening mammography: improving performance or simply shifting the operating point? Radiology 239:916–917; author reply 917–918

Swett HA, Fisher PR (1987) ICON: a computer-based approach to differential diagnosis in radiology. Radiology 163: 555–558

Swett HA, Fisher PR, Cohn AI, et al (1989) Expert system controlled image display. Radiology 172:487–493

Swett HA, Mutalik PG, Neklesa VP, et al (1998) Voice-activated retrieval of mammography reference images. J Digit Imag 11:65–73.

Taylor P, Champness J, Given-Wilson R, et al (2005) Impact of computer-aided detection prompts on the sensitivity and specificity of screening mammography. Health Technol Assess 9:1–70

Taylor PM (2007) A review of research into the development of radiologic expertise: implications for computer-based training. Acad Radiol 14:1252–1263

Timp S, Karssemeijer N (2006) Interval change analysis to improve computer aided detection in mammography. Med Image Anal 10:82–95

Timp S, Varela C, Karssemeijer N (2007) Temporal change analysis for characterization of mass lesions in mammography. IEEE Trans Med Imag 26:945–953

Tourassi GD, Harrawood B, Singh S, et al (2007) Evaluation of information-theoretic similarity measures for content-based retrieval and detection of masses in mammograms. Med Phys 34:140–150

van Engeland S, Karssemeijer N (2007) Combining two mammographic projections in a computer aided mass detection method. Med Phys 34:898–905

Velikova M, Samulski M, Lucas PJ, et al (2009) Improved mammographic CAD performance using multi-view information: a Bayesian network framework. Phys Med Biol 54: 1131–1147

Vyborny CJ, Giger ML (1994) Review. Computer vision and artificial intelligence in mammography. AJR Am J Roentgenol 162:699–708

Wagner RF, Metz CE, Campbell G (2007) Assessment of medical imaging systems and computer aids: A tutorial review. Acad Radiol 14:723–748

Warren-Burhenne LJ, Wood SA, D'Orsi CJ, et al (2000) Potential contribution of computer-aided detection to the sensitivity of screening mammography. Radiology 215: 554–562

Wee WG, Moskowitz M, Chang NC, et al (1975) Evaluation of mammographic calcifications using a computer program. Radiology 116:717–720
Wei LY, Yang YY, Nishikawa RM, et al (2005) A study on several machine-learning methods for classification of malignant and benign clustered microcalcifications. IEEE Trans Med Imag 24:371–380
Wei Q, Dansheng S, Minshan L, et al (2007) Computer-aided mass detection based on ipsilateral multiview mammograms. Acad Radiol 14:530–538
Winsberg F, Elkin M, Macy J, et al (1967) Detection of radiographic abnormalities in mammograms by means of optical scanning and computer analysis. Radiology 89: 211–215
Yousef WA, Wagner RF, Loew MH (2004) Comparison of non-parametric methods for assessing classifier performance in terms of ROC parameters. pp 190–195
Yousef WA, Wagner RF, Loew MH (2005) Estimating the uncertainty in the estimated mean area under the ROC curve of a classifier. Patt Recog Let 26:2600–2610
Yousef WA, Wagner RF, Loew MH (2006) Assessing classifiers from two independent data sets using ROC analysis: a non-parametric approach. IEEE Trans Patt Anal Mach Intell 28:1809–1817
Yuan Y, Giger ML, Li H, et al (2008) Identifying corresponding lesions from CC and MLO views via correlative featre analysis. In: Krupinski EA (ed) International workshop on digital mammography 2008. Springer, New York, pp 323–328
Zheng B, Shah R, Wallace L, et al (2002) Computer-aided detection in mammography: an assessment of performance on current and prior images. Acad Radiol 9:1245–1250
Zheng B, Good WF, Armfield DR, et al (2003) Performance change of mammographic CAD schemes optimized with most-recent and prior image databases. Acad Radiol 10: 283–288
Zheng B, Leader JK, Abrams GS, et al (2006a) Multiview-based computer-aided detection scheme for breast masses. Med Phys 33:3135–3143
Zheng B, Lu A, Hardesty LA, et al (2006b) A method to improve visual similarity of breast masses for an interactive computer-aided diagnosis environment. Med Phys 33: 111–117
Zheng B, Mello-Thoms C, Wang XH, et al (2007) Interactive computer-aided diagnosis of breast masses: computerized selection of visually similar image sets from a reference library. Acad Radiol 14:917–927

Softcopy Reading

7

Elizabeth A. Krupinski

CONTENTS

KEY POINTS

Softcopy reading of mammographic images has many aspects in common with traditional screen-film viewing but also many aspects that are very different. Viewing images is the core diagnostic task, and one can consider it from two perspectives. On the one hand, it is the technology used to display the images and how *technical factors* such as luminance and display noise affect the *quality* of the image and, hence, the *perception and interpretation* of features in that image. On the other hand, there are the human observers relying on their *perceptual and cognitive systems* to process the information presented to them to render a *diagnostic decision.* Therefore, to maintain low *error rates,* careful consideration of both sides of the reading equation is required to optimize the interpretation of softcopy mammographic images.

7.1 Introduction

Practical technology dedicated to breast imaging was first developed in 1960 (Egan 1960), but modern mammography can really be traced to the late-1960s when the first commercial X-ray units dedicated to breast imaging were available (Gold et al. 1990). One of the first mammography studies (Strax et al. 1973) in the 1960s using this technology was called the Health Insurance Program of New York Project (HIP). Over 60,000 women were enrolled, having either four consecutive years of screening mammography or the standard of care at that time (e.g., physical exam). Women who underwent yearly screening mammography had a 29% mortality reduction at 9 years and a

Elizabeth A. Krupinski, PhD
Department of Radiology, University of Arizona 1609 N. Warren
Tucson, AZ 85724, USA

23% mortality reduction at 18 years. Screening populations had decreased mortality, and the morbidity associated with a diagnosis of breast cancer was reduced. By 1976, screening mammography became a standard practice, and research continued to explore ways to develop safer and better mammograms.

It has only been in the past decade or so, however, that the most dramatic change in mammography practice has occurred. Since its inception, mammography was film-based, and the images were interpreted using hardcopy film on dedicated mammographic view boxes. In the 1990s, digital mammography was fully developed, and in 2000, the Food and Drug Administration approved for marketing the first Full-Field Digital Mammography (FFDM) system (FDA 2008a). Initially, the digitally acquired images were printed to film for reading, but softcopy was approved soon thereafter and very quickly mammographers were interpreting images using computer displays. This transition from analog to digital reading was facilitated in part by a large prospective multi-institutional study done by the American College of Radiology Investigational Network (ACRIN) called the Digital Mammography Imaging Screening Trial (DMIST). In this study, the overall accuracy of the two modes was equivalent, but mammographers were more sensitive at detecting cancer in certain patient subsets using digital than screen film mammography (Pisano et al. 2005). Other studies have shown similar results (Lewin et al. 2001, 2002; Skanne and Skjennald 2004; Skanne et al. 2003; Cole et al. 2004; Del Turco et al. 2007), and digital mammography use has increased significantly in the twenty-first century. According to the FDAs MQSA (Mammography Quality Standards Act) Facility Scorecard, as of 1 October 2008, there were 3,774 certified FFDM facilities and 5,729 accredited FFDM units in the United States alone (FDA 2008b).

Digital mammography is only one breast imaging modality that is read using softcopy images and computer displays. Magnetic resonance imaging (Lehman et al. 2007; Smith 2007), ultrasound (Yang and Dempsey 2007), nuclear medicine (Van der Ploeg et al. 2008), and molecular imaging (Franc and Hawkins 2007) are all part of the arsenal being used to detect breast cancer today. Although all these modalities use softcopy images, it is digital mammography that requires special consideration in display and workstation design and selection given the unique nature of these images. In the long run, however, one must consider the workstation of the future in which all modalities likely to be available at the same time and data (image and text) from a variety of sources must be perceptually and cognitively processed and integrated to render an accurate and complete diagnosis.

7.2 Softcopy Image Quality

Clearly, the quality of the image acquisition and transmission systems influences the quality of softcopy images, and there are guidelines to help ensure that quality is maintained throughout these two phases (Williams et al. 2006; Avrin et al. 2006). These issues are addressed in Chapters 2 and 3 of this book. Of most concern perhaps to the radiologist interpreting softcopy mammograms is the quality of the display. Quality control measures are also in place for display of softcopy mammograms and should be followed whenever practical and feasible (Siegel et al. 2006).

7.2.1 Display Optimization

Some of the basic minimum quality specifications for softcopy reading are relatively easy to implement and are relatively noncontroversial. The ratio of the maximum to minimum luminance should range between 250 and 650 (including ambient light), with a minimum of 250 cd/m^2 (450 cd/m^2 preferred). A minimum of 8-bit luminance resolution is required. Although 9-bit-depth or 10-bit-depth resolution might be useful with higher luminance ratios, there is considerable debate about this given that most evidence indicates that the human visual system can detect only about 1,000 gray levels at luminance levels currently used in most monitors, so displaying more gray levels may not be useful (Krupinski et al. 2007; Barten 1992, 1999). Display noise should be as low as possible (2–2.5% ideally). Veiling glare and reflections should be kept at a minimum and can be avoided with proper ambient lighting. Ambient light levels should be set at approximately equal to the level of the average luminance of a clinical image being displayed (about 50 lx). Mammographers should also allow their visual systems to dark adapt to the surrounding light levels before they start reading images. Full dark adaptation takes about 30 min, but adequate adaptation can occur in about 10 min. Displays should be calibrated using the DICOM (digital imaging and communications in medicine) grayscale standard display function (GSDF) (NEMA 2001).

One of the more controversial areas regarding softcopy mammography displays is resolution and size. Digital mammography detectors have been designed with very small pixel pitches so that microcalcifications on the order of 100–150 μm in size can be depicted. Therefore, the detectors used in current devices vary in pixel pitch from 50 to 100 μm, yielding an acquisition matrix ranging from about 8 to 30 million pixels per image. Although there was a 9-MegaPixel display on the market briefly and there are currently 6 MegaPixel medical-grade displays available, they are color and it is not clear yet if they are appropriate for use in mammography. At a minimum, it is recommended that a monochrome 3 MegaPixel display be used (SIEGEL et al. 2006) and many would argue that a 5-MegaPixel display should be used. Additionally, the monitor should have a diagonal of about 21 in. to be most effective.

From a perceptual point of view, the issue can be regarded as one of what the human visual system is capable of perceiving. The typical viewing distance for softcopy reading is 30–60 cm. At about 60 cm, contrast sensitivity of the eye drops to zero above 2.5 cycles/mm with a peak sensitivity at 0.5 cycles/mm. Based on this, display pitch or pixel size should be about 200 μm. Most 3 MegaPixel displays have a pixel pitch of about 207 μm and most 5 MegaPixel displays of about 165 μm. Therefore, a 3-MegaPixel display is most likely sufficient for softcopy mammography reading.

There is a complication with this recommendation. Decision-making, at least for experienced observers, depends on an initial gist or global impression (KUNDEL 1975, 2007). For example, in a study by KUNDEL et al. (2007), eye position was recorded as experienced and less experienced radiologists searched mammograms for lesions. The time required to first locate a lesion was determined. The median time for all observers to fixate a lesion, regardless of the decision, was 1.13 s (range = 0.68–3.06 s), with experienced radiologist typically fixating earlier during search than less experienced radiologists. This rapid initial fixation of true abnormalities is evidence of a global perceptual process that analyzes the visual input of the entire retinal image and finds the location of abnormalities. For this global processing to be most effective, it is quite likely that the entire image should be available for processing in this rapid gist impression. On a 3 MegaPixel monitor, it is not possible to show the full resolution image at full size all at once (nor is it possible on a 5 MegaPixel display). Therefore, it is recommended that for routine viewing, to view the entire image at once and avoid possible disorientation, if the whole image is not displayed, images should be viewed at approximately 2:1 or 200% size. To access the full resolution data, zoom and pan should be used – much like magnifying glasses were used with film-screen images. Zooming is always recommended over leaning in closer to the display device.

7.3 Softcopy "Hanging Strategies"

Softcopy viewing of mammograms by necessity involves all of the same tasks that were required with hardcopy film viewing and this involved at a minimum four images – the CC (craniocaudal) and MLO (mediolateral oblique) views of the right and left breasts. If previous and/or diagnostic images are available, the number rises to eight images or more. With hardcopy images on a view box, this did not present any real challenges. A set of four or eight 8″ × 10″ films could easily be viewed at 30–60 cm with minimal head movements – just scanning of the eyes. Displaying even the minimum of four softcopy images at 2:1 resolution on digital displays typically requires four displays, which clearly leads to the necessity to move the head considerably in order to view all images. Most workstations utilize only two monitors for image viewing. To accomplish the basic tasks associated with interpreting mammograms – comparing left and right breasts for symmetry, locating potential lesions in both views, comparing current and prior images for change, and examining each image for subtle lesions and characterizing them as benign or malignant – softcopy viewing requires changes in viewing patterns when compared with hardcopy.

7.3.1 Typical Hanging Options

Viewing images sequentially at full resolution (or even 2:1) is one "hanging strategy" that can be used, and if presets are used that automatically bring up the relevant pairs (e.g., right and left MLO for current symmetry comparison, then right and left CC etc.) it can greatly facilitate workflow. However, this limits the user to viewing only a pair of images at a time, potentially taxing short-term memory capacity. Once the mammographers move on to the next pair, it may be difficult to remember fully what was in the previous pair and where it was. Another very common hanging strategy is called "fit to viewport." With this protocol, a number of images are displayed at the same time for direct comparisons. This of course gets back to the problem of not being able to view the images at full resolution, but it does facilitate the gist or global impression to a large extent. Evidence from perceptual studies suggests that basic patterns and image properties such as symmetry, gross deviations from normal, and overall context can be detected in the global view (Kundel 1975), and since the mammography interpretation task utilizes this type of information, the "fit to viewport" may actually facilitate the global impression. Once perturbations in the images are detected in the global view, detailed scanning of the individual images (by clicking on them to bring to full resolution) can ensue.

A problem often arises, however, even with the "fit to viewport" mode. Patients often have mammograms performed on different units and these units often have different acquisition matrix sizes. The result is that the breast will display at different sizes, making comparisons (especially with regard to symmetry) difficult. This problem was recognized early on with softcopy viewing and today most dedicated mammography workstations account for different acquisition matrices and scale images appropriately to be of the same size. Workstations that utilize the image information contained in the DICOM header regarding type of image, acquisition size, pixel size, etc. and the profiles developed as part of the IHE (Integrated Healthcare Enterprise) regarding ways that heterogeneous information can be integrated properly tend to display the images correctly (Channin 2001; IHE 2008a).

The main IHE profile relevant to the softcopy hanging protocol issue is (MAMMO) Mammography Image. This profile specifies how mammography images and evidence objects are created, exchanged, used, and displayed. The DICOM Part 10 (Media Storage and File Format for Media Interchange) defines a file format with a preamble that contains the information typically exchanged during association prior to DICOM network communications, and this information can be utilized by the MAMMO profile. The MAMMO profile takes FFDM and CR (computed radiography) and creates MG objects. These objects have 40 attributes with specific requirements that are mandatory to comply with the profile. The IHE Mammography Handbook (IHE 2008b) describes how to utilize the profile to deal with a variety of digital mammography issues, including different sized images, different image formats, and hanging strategies. Many of the problems with images from different vendors have been addressed in recent years with the IHE profile, making softcopy reading and workflow much improved.

7.4 Viewing Strategies and Perception

To optimize softcopy mammogram viewing, it is important to consider the mammographers and the perceptual and cognitive capabilities and strategies involved in image interpretation and diagnostic decision- making.

7.4.1 Vision Basics

The eye is a complex organ, but there are certain parts of this highly specialized organ that deserve a brief description. The main function of the eyes is photoreception or the process by which light from the environment produces changes in the photoreceptors or nerve cells in the retina called rods and cones. The retina is located at the back of the eye, so light travels through the pupil, the lens, and the watery vitreous center before it reaches the retina. The retina contains about 115 million rods and 6.5 million cones. Rods sense contrast, brightness, and motion and are located mostly in the periphery of the retina. Cones are responsible for fine spatial resolution, spatial resolution, and color vision and are located in the fovea and parafoveal regions. Pigments in the rods and cones undergo chemical transformations as light hits them, converting light energy into electrical energy that acts on a variety of nerve cells connecting the eye to the optic nerve and subsequent visual pathways that extend to the visual cortices in the brain itself. We have two eyes, accounting for our ability to see depth or for the radiologist to generate the perception of depth from two-dimensional images. The transformation of electrical nerve signals generated in the early stages of vision to the perception of the outside world takes place in a number of brain regions that are equally specialized for visual perception.

Especially important for viewing mammographic images is spatial resolution, or the ability to see fine details (e.g., microcalcifications and fine tendrils or spiculations extending from masses). It is highest at the fovea, but declines quite sharply toward the peripheral regions of the retina. The result is that mammographers must search or move the eyes around the image to detect fine, subtle lesion features with high-resolution vision. As people age, spatial resolution generally degrades and most people require corrective lenses. Important for softcopy viewing are glasses specifically designed for computer viewing and these should be considered when a having trouble viewing softcopy images on digital displays.

Visual acuity also depends on contrast or differences in color and brightness that permit one to distinguish between objects (masses and microcalcifications) and background in an image. Tests have been developed to determine the contrast levels that are perceptible by the human eye using a sinusoidal grating pattern (alternating black and white lines where the average luminance remains the same but the contrast between the light and dark areas differ). Grating discrimination is described in terms of cycles per degree or the grating frequency). Contrast sensitivity peaks in the mid-spatial frequency range around 3–5 cycles/degree. This means that low contrast lesions can often go undetected, especially when viewing conditions are not optimal as can happen with softcopy viewing if the display and environment are not properly addressed.

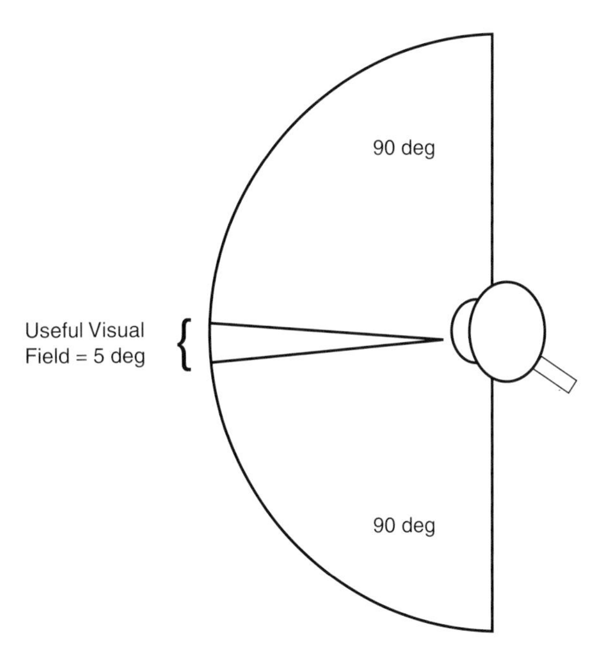

Fig. 7.1. The useful visual field and peripheral vision. The entire visual field covers 180–90° to either side. High-resolution foveal vision extends about 2.5° to either side for a total useful visual field of about 5°. Peripheral vision operates outside this region

7.4.2 Visual Search

Whether hardcopy or softcopy, detection of microcalcifications and masses in mammograms generally requires examination of the image with high-resolution foveal vision because of the small size and subtlety of the lesions. As already noted, to accomplish this, the mammographer must search the image. To characterize visual search in radiology, a number of eye-tracking studies have been done.

Some abnormalities can be detected with peripheral vision (Kundel and Nodine 1975), but normal anatomy (or structured noise) significantly decreases detection the further away the lesions are located from the axis of gaze (Kundel 1975). This is one reason why the number of monitors is important in softcopy viewing. If there are too many monitors, the visual system cannot take in the entire field even with peripheral vision (Fig. 7.1), making it nearly impossible to detect anything in far images with peripheral vision.

One of the main assumptions in recording eye position is that the mammographer is not only directing high-resolution foveal vision to areas within the image, but they are directing their attentional and information-processing resources there as well to extract and process information about to render a diagnostic decision.

Information-processing theory provides the basis for interpreting visual search data (Crowley et al. 2003; Haber 1969; Nodine and Kundel 1987; Nodine et al. 1992; Krupinski et al. 1998). The initial glance at an image produces a global impression of image that includes the processing and recognition of content such as anatomy, symmetry, color, and grayscale (Fig. 7.2). The information gathered in this global impression is compared with information contained in long-term memory that forms the viewer's cognitive schema (or expectations) of what information is in an image. In some cases, the target of search "pops out" in this global impression and the viewer makes a quick decision (Kundel 2007). This early stage perception takes place in less than 250 ms. Information processed in this global percept guides subsequent search with high-resolution foveal vision, where the extraction of feature details from the complex image backgrounds takes place.

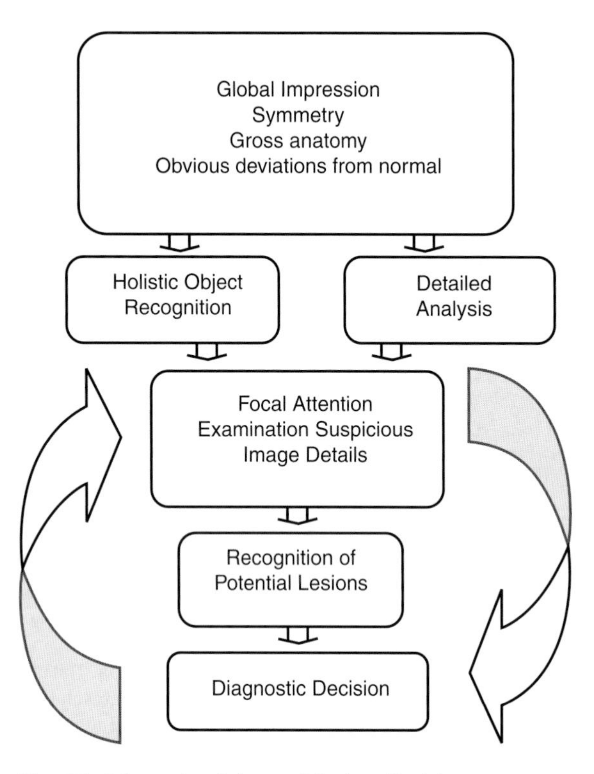

Fig. 7.2. Schematic of the model of medical image perception. Perception starts with an initial global view followed by focal scanning of potential lesion features detected in this global or gist impression. Scanning continues until the radiologist has rendered a final decision

Kundel et al. (2008) found further evidence to support the existence of this holistic model of perception for the detection of cancers on mammograms. They analyzed data (400 records) from three different eye-position recording mammography studies to determine the time required to first fixate a cancer on a mammogram. This time to first fixate the cancer was used as an indicator of the initial perception of cancer. Mixture distribution analysis was used to partition the distribution of times into two normally distributed components. The analysis revealed that 57% of cancers had a 95% chance of being fixated in the very first second of viewing. The other 5% took between 1.0 and 15.2 s. For most of the readers, the true-positive fraction was larger for those lesions fixated within that first second of search. Lesions that took longer to first fixate had a lower chance of eventually being reported. The initial global impression is clearly a significant component in lesion recognition in mammography. Optimizing the display of the softcopy image to be the best possible image for this global impression is clearly one way to help reduce errors of interpretation.

7.4.3 Interpretation Errors

One important question in mammography is: why are errors made – especially when lesions missed initially can often be easily detected when viewed a second time. It is estimated that the miss rate is about 20–30% (false-negatives) with a false-positive rate of about 2–15% (Bird et al. 1992). False-positives are sometimes easier to understand from a perceptual point of view. False-positives often occur because the overlaying anatomic structures can mimic disease entities. For many false-positives, there is clearly something in the image (overlapping tissue or dust) that attracts attention and leads to the false impression that a mass or microcalcification is present.

False-negatives can be more complicated, but are generally more important to understand in terms of why they occur. With digital mammography, technical reasons (e.g., over and underexposure) are practically eliminated, so the cause of many errors most likely resides in the reader. Tuddenham and Calvert (1961)were some of the first researchers to suggest that lesions may be missed due to inadequate search of images and significant inter-observer variability in search strategies. Kundel et al. (1978) used eye-position recording to classify the types of errors made during search. Implicit in studies of visual search is the assumption that focal attention or the so-called "Useful Visual Field" extends to pictorial or image features only as far away as 2.5° from the center of a fixation (Fig. 7.3). Peripheral vision extends beyond the Useful Visual Field to encompass the entire visual field and also processes useful visual information but not at the high level of resolution that foveal vision can process details. In other words, to detect, recognize, and interpret correctly potential lesion features and separate them from overlapping and confusing background tissue structure, the mammographers need to direct foveal gaze to specific locations within the image.

False-negatives have been classified into three categories (Kundel et al. 1978) based on how long lesions are dwelled on or fixated. Each category accounts for about one-third of false-negative errors. The first category is known as a search error because the observer never fixates the lesion with high-resolution foveal vision and therefore cannot process the information at that location in the image. The second error is called a recognition error, because these lesions are fixated with foveal vision, but not for very long. Inadequate fixating time reduces the likelihood

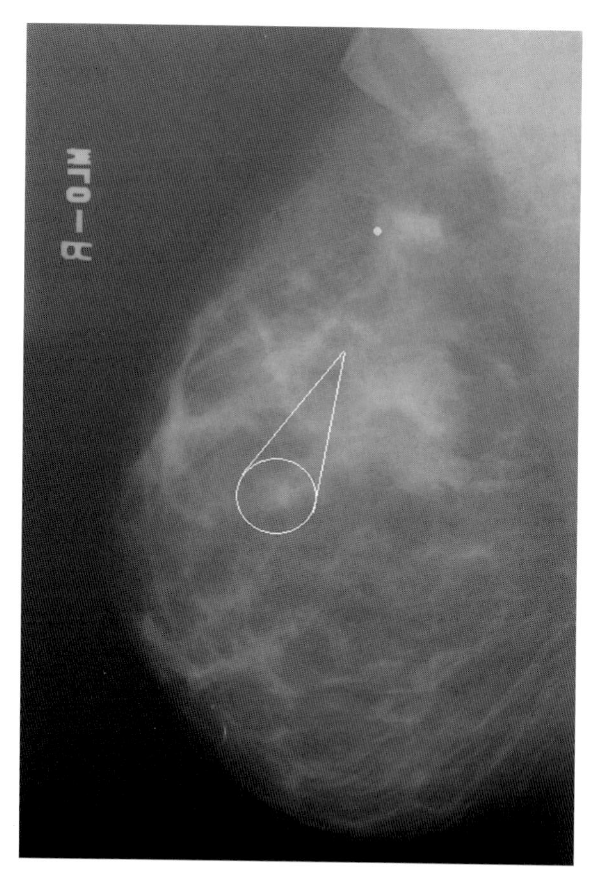

Fig. 7.3. Schematic of the "useful visual field" as a mammographer scans a breast image. The useful visual field (*the white circle*) extends about 2.5° from the center of the axis of gaze

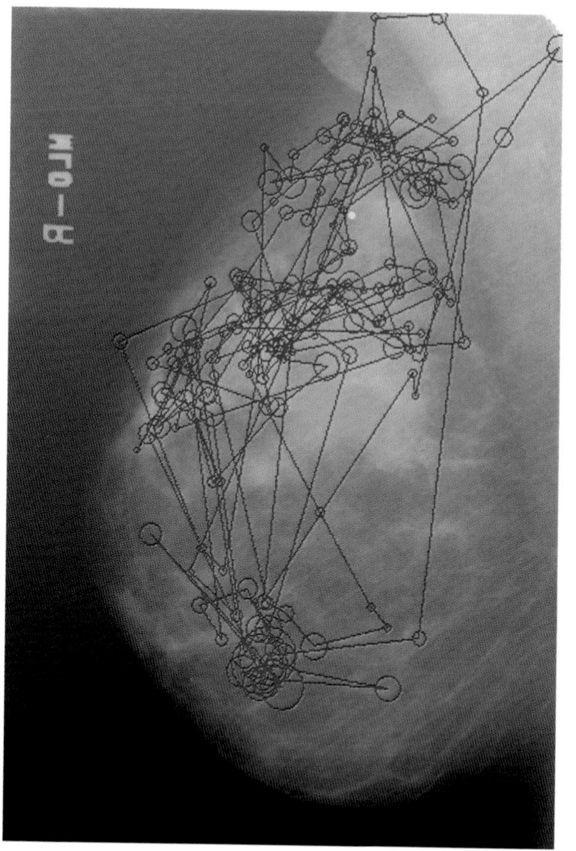

Fig. 7.4. Typical example of a search pattern generated by a mammographer searching a breast image for lesions. Each *small circle/dot* represents a fixation or where high-resolution foveal gaze is directed. The *lines* represent saccades or jumps the eye makes between fixations and indicates the order in which the fixations were generated

that lesion features will be detected or recognized. Decision errors comprise the final group and occur when the observer fixates the lesion for long periods of time, but either does not consciously recognize the features as those of a lesion or actively dismisses them.

One of the first detailed studies of visual search in mammography was done with hardcopy images (Krupinski 1996), but the results have been confirmed in later studies that used either hardcopy or softcopy images (Nodine et al. 1999; Mello-Thomas et al. 2005). These studies confirmed what had been found in chest images. True- and false-positive decisions were associated with long gaze durations and true-negative decisions with relatively short gaze durations. Some false-negatives were not even looked at (search errors), while the rest were associated with gaze durations falling between the true positives and true negatives (recognition and decision errors).

Figure 7.4 shows a typical scanning pattern of a mammographer searching a breast image. Each dot represents a fixation or where the eye directs foveal vision. The lines represent saccades or the jumps that the eyes make between fixations and reflect the order in which the fixations were generated.

7.4.4 What Attracts Attention?

Clear relationships between lesion subtlety and eye position parameters have been found. For example, in one study (Krupinski 2005), six radiologists viewed a set of mammograms each with at least two lesions (masses and/or microcalcifications) while their eye position was recorded. The images were from a database in which lesion subtlety had been rated by experienced mammographers. The eye position data revealed that lesions with lower subtlety ratings were detected later in search than the more obvious ones – whether they ended up being reported or not. When the subtle lesions were reported, the dwell times associated with them were longer than for the obvious lesions. When they were fixated but went unreported, the dwells were shorter (recognition and decision errors). The fact that the readers did fixate the false-negatives indicated that there were some features there attracting visual attention – more than those areas without any lesion at all. Softcopy reading, more than hardcopy was ever able to, is clearly the situation where image processing (Chap. 5) and computer-aided detection and diagnosis tools (Chap. 6) can be of significant help in directing the mammographers' attention to these subtle features and perhaps enhancing them so they are more visible.

Interestingly, similar results have been found using a very different approach to examining lesion conspicuity and what attracts visual attention. Mello-Thomas (2006a, b) used spatial frequency analysis of image locations and correlated that information with search patterns and diagnostic decisions. There were no significant differences in the spatial frequency content of the background image areas sampled before and after the eyes first fixated or landed on the location of a mass that was reported. Reported masses are most likely detected in the global view and thus bias search patterns and where mammographers sample background areas for comparison purposes. However, if the mass is not reported or if the mammographers make a false-positive decision, there are significant differences in the spatial frequency content of background areas sampled before and after the lesion location is fixated. For the false–negatives, it seems that viewers are trying to actively compare the new-found potential lesion location with other background areas (unsuccessfully since it still goes unreported).

What was more interesting perhaps were differences observed as a function of experience. The experienced mammographers tended to detect lesions much earlier in search than did the residents. The residents also spent more time and covered much more of the image than did the radiologists, while still making more errors. This confirms the idea that radiologists, especially those with more experience, take in and process a significant amount of information in the global view and make accurate and efficient decisions based on this initial gist view.

Similar coverage rates have been found with softcopy images (Mello-Thoms et al. 2008). Experts covered on average 17% of the area of a two-view digital mammogram with foveal vision. Coverage was slightly smaller in cases where a true-positive decision was made (15%) than when a false-negative (19%) or false-positive was made (19%). Optimizing the mammographic display, as noted earlier, may be even more important for less experienced readers than for those with experience who are better able to extract information from images in the first place.

The importance of the gist view has been confirmed in another study of breast cancer detection by expert breast radiologists (Mello-Thomas et al. 2005). For cases containing a malignant lesion, if that lesion was detected in the gist view, then in 74% of those cases the mammographers correctly reported it and did not make any false-positives. If, however, the initial view detected a false-positive, then in only 5% of these cases was the mammographer able to "undo" this perception and go on to detect and report the true lesion. For these cases, these actual lesions were fixated much later in search but did not reach the threshold for recognition.

7.4.5 Reader Variability and Expertise

It is clear that expert mammographers and novice residents differ significantly in their ability to correctly and efficiently interpret both hardcopy and softcopy mammograms. However, even expert mammographers render different decisions on the same images, leading to considerable inter-observer variability rates in mammography. Although some studies on variability do not really accurately replicate the clinical reading environment (Berg et al. 2000; Ciccone et al. 1992; Elmore et al. 1994; Kerlikowske et al. 1998), others have been quite large and do a better job at replicating clinical reality (Beam et al. 2003). Having prior images, a reliable clinical history, and information from other modalities as with a diagnostic workup all improve accuracy and reduce variability (Berg et al. 2000; Burnside et al. 2002; Houssami et al. 2003, 2004). With softcopy reading, it has taken quite a while for vendors to develop the software required to make the "hanging strategies" amenable to easy comparison of prior images, compared with how easy it was with hardcopy viewing.

Another factor contributing to reader variability in mammography is sheer volume of reading (Beam et al. 2003; Leung et al. 2007; Miglioretti et al. 2007). For example in one study (Smith-Bindman et al. 2005), physicians who interpreted a high annual volume (2,500–4,000) of screening as opposed to diagnostic mammograms found more cancers with fewer false-positive responses than those who read fewer cases. Another study carried out by Esserman et al. (2002) compared British with United States radiologists. They used a standardized test set (PERFORMS 2 = PERsonal PerFORmance in Mammography Screening) designed to provide feedback to radiologists each year they take the test.

They grouped the 60 US radiologists as low-volume (<100 cases/month), medium-volume (101–300 cases/month), and high-volume radiologists (>301 cases/month). The UK group had 194 high-volume radiologists. The average sensitivity at a specificity of 0.90 was 0.785 and 0.756 for the high volume UK and US readers, respectively; and 0.702 for medium-volume US radiologists; and 0.648 for low-volume US radiologists. At this specificity, the low-volume US readers had significantly lower sensitivity than either group of high-volume or medium-volume radiologists.

Clearly, reader volume is an important determinant of mammogram sensitivity and specificity.

In terms of softcopy reading, Skaane et al. (2008) found very similar results for hardcopy and softcopy reader variability. They had six radiologists read 232 cases acquired with both screen-film and FFDM. There were 46 cancers, 88 benign lesions, and 98 lesion-free cases. The readers had between 4–24 years hardcopy experience and 2–4 years softcopy experience. Overall, they read 2,500–12,000 cases per year. In both scenarios, there was significant inter-observer variability but in the end there were no significant differences between hardcopy and softcopy reader variability. Inter-observer agreement for hardcopy versus softcopy for all possible pairs of readers showed a slightly higher kappa value for hardcopy in ten and a slightly higher kappa for softcopy in five of the pairs of readers. The mean weighted kappa score for hardcopy was 0.74 (range = 0.68–0.81) and for softcopy was 0.71 (range 0.61–0.82).

Three of the readers had relatively low concordance between the two modalities for microcalcifications and there were overall more false-positives on the lesion-free cases on softcopy than hardcopy. The authors suggest that proper training is very likely required when transitioning from hardcopy to softcopy reading. This makes sense when the volume and expertise data discussed early are considered. It takes a considerable volume of cases before someone becomes an expert in mammographic interpretation. With only 2–4 years of softcopy experience, compared with 4–24 years of hardcopy experience, it is likely that the differences in softcopy image appearance compared to hardcopy have yet to be fully learned or appreciated by the readers in the Skaane study. The question that has yet to be addressed directly with softcopy mammography reading is whether the learning curve is the same for softcopy if one already has significant hardcopy experience. It is likely, given the ease with which most mammographers are making the transition, that although there is obviously a softcopy learning curve, it is relatively low and quickly overcome for most experienced mammographers.

7.5 Reading Environment

There are a number of factors that impact diagnostic accuracy, visual search, and interpretation efficiency. Some of these factors relate to the environment in which the softcopy workstation is located. Some issues related to the environment that should be considered when designing a softcopy reading room include: how much heat does the workstation produce, and how much noise does it produce? If the workstation produces too much heat, it may be necessary to improve airflow both for the computer and the mammographer. Fan-cooled systems are common and generate noise levels that might be distracting, so water-cooled systems or systems with high-performance, low-noise fans should be considered.

Workflow is clearly important, as radiologists are faced with more and more cases to read and fewer radiologists going into mammography. The ergonomics of workflow include personnel, equipment, and environmental components (Rau and Trispel 1982) and there are even modeling tools available to simulate and design workstation setup and reading rooms (Ratib et al. 2000). Chapters 5 (Image Processing), 6 (Computer-Aided Diagnosis), and 8 (Digital Workflow) provide clear examples of how softcopy mammography workflow can be improved with the implementation of these various tools and interpretation aids.

7.5.1 Reader Fatigue

One topic that is often overlooked, however, when digital reading and workflow are discussed is reader fatigue. Close work for hours on end can overwork the eyes, resulting in eyestrain or asthenopia (Ebenholtz

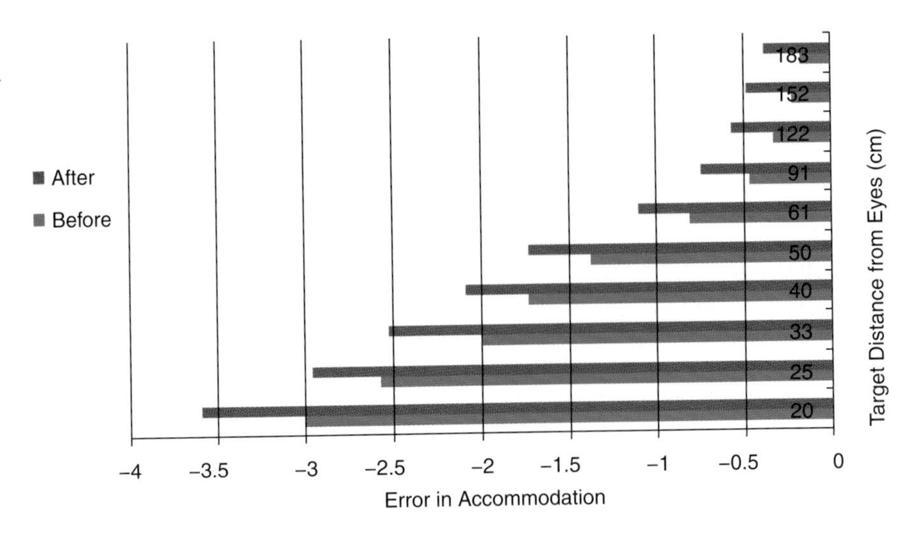

Fig. 7.5. Error in accommodation measures at near and far distances for six radiologists before and after a day of near viewing reading from computer displays

2001; MACKENZIE 1843). With nonmedical tasks, just 4 h of computer display viewing is sufficient to produce asthenopia (SANCHEZ-ROMAN et al. 1996) and there is some evidence that prolonged computer use may even induce myopia in computer users (KOMIUSHINA 2000; MUTTI and ZADNIK 1996). Oculomotor fatigue caused by close work with digital displays may add to the effects of extended workdays and aging eyes (HERON et al. 1999).

One common objective measure of visual fatigue is accommodation. The lens of the eye alters the refractive index of light entering the eye to focus images on the retina. It is covered by an elastic capsule whose function is to mold the shape of the lens – varying its flatness and therefore its optical power. This variation in optical power is called accommodation, and it occurs as the eye focuses on a close object. We are currently measuring accommodation using the WAM-5500 Auto Refkeratometer from Grand Seiko with radiologists before and after long hours of reading softcopy images.

We are using a series of near and far targets (a large asterisk) for the subject to fixate while accommodation measurements are made. The target distances for near-viewing are 20, 25, 33, 40, and 50 cm from the eye (these correspond to 5, 4, 3, 2.5, and 2 diopters, respectively). The target distances for far-viewing are set to 61, 91, 122, 152, and 183 cm from the subject's eye. The targets are kept at the same height and orthogonal position relative to the subject's eye throughout the testing procedure. Subjects to date have been three radiologists (ages 47, 52, 68) and three radiology residents (ages 28, 28, 30). All three radiologists wore glasses (two bifocals, one readers only) and one of the residents wore glasses (near sighted).

The results are shown in Fig. 7.5. The near target distances (20 cm) start on the left and get progressively farther away to the right (ending at 183 cm). The *y*-axis shows the ability of the subject to accommodate where the "0"-line represents good accommodation. For all distances, the accommodation measures for after a day of near distance reading are worse (do not accommodate as well) than for the before reading measurements.

An Analysis of Variance (ANOVA) revealed that there was a significant difference ($F = 1{,}188.36, p < 0.0001$) in the data as a function of target distance, as a function of time of day ($F = 316.10, p < 0.0001$), and radiologists vs. residents ($F = 271.47$, $p < 0.0001$). There was a significant interaction effect for distance by time, with all accommodation measures being better before a day of reading than after at all target distances. There was a significant interaction effect for distance by radiologist vs. resident, with the radiologists having poorer accommodation than the residents at all distances. There was also a significant interaction for time of day by radiologist vs. resident, with the radiologists having poorer accommodation at both points in time than the residents.

It is clear that after a long day of reading softcopy images, radiologists are less able to accommodate or fixate on a specific target location than at the beginning of the day. The next step in this series of studies is to conduct a Receiver Operating Characteristic (ROC) study using hardcopy and softcopy images both before and after a long day of reading to determine if this reduced ability to accommodate impacts observer performance.

7.6 Conclusions

It is clear that there are many things to consider when optimizing the reading of softcopy mammographic images. The display, the room, the images, and especially the mammographer must be taken into account to get a complete picture of the image interpretation process. Even more important is to realize that our understanding of these factors and how the mammographer performs is not a static question. Technology and the nature of the mammogram itself have changed dramatically in recent years and these will continue to change in the near future.

For example, a recent study has demonstrated that stereoscopic FFDM improves significantly (by 23%) the detection of lesions while dropping significantly (46%) the false-positive detections (Getty et al. 2008). Digital breast tomosynthesis (Chap. 13) will also provide the mammographers with depth information never before accessible even with FFDM (Good et al. 2008), but it is very likely that significant thought and experimentation will have to go into determining what will be the optimal way to present these even newer types of softcopy images to the mammographer. The same issues of accurate yet efficient interpretation will continue to arise as these new modalities and ways of collecting and presenting image data are developed. These types of changes will present new challenges to the clinician' perceptual and cognitive systems, so our continued exploration into medical image perception and the way the clinician interacts with medical image will always be important.

The ultimate benefit of softcopy FFDM and softcopy reading is really to the patient. Digital acquisition of images and high-speed digital networks allow mammographers to read high-quality softcopy images acquired nearly anywhere and transmitted to a reading center for interpretation. Telemammography is extending softcopy reading around the world, providing women with timely and accurate breast care (Weinstein et al. 2007).

References

Avrin D, Morin R, Piraino D, et al (2006) Storage, transmission, and retrieval of digital mammography, including recommendations on image compression. J Am Coll Radiol 3:609–614

Barten PGJ (1992) Physical model for contrast sensitivity of the human eye. Proc SPIE 1666:57–72

Barten PGJ (1999) Contrast sensitivity of the human eye and its effects on image quality. SPIE Press, Bellingham, WA

Beam CA, Conant EF, Sickles EA (2003) Association of volume and volume-independent factors with accuracy in screening mammogram interpretation. JNCI 95:282–290

Berg W, Campassi C, Langenberg P, et al (2000) Breast imaging reporting and data system: inter- and intraobserver variability in feature analysis and final assessment. Am J Roentgenol 174:1769–1777

Bird RE, Wallace TW, Yankaskas BC (1992) Analysis of cancers missed at screening mammography. Radiology 184:613–617

Burnside E, Sickles E, Sohlich R, et al (2002) Differential value of comparison with previous examinations in diagnostic versus screening mammography. AJR 179:1173–1177

Ciccone G, Vineis P, Frigerio A, et al (1992) Inter-observer and intra-observer variability of mammogram interpretation: a field study. Eur J Cancer 28A:1054–1058

Channin D (2001) Integrating the healthcare enterprise: a primer, part 2. Seven brides for seven brothers: the IHE integration profiles. RadioGraphics 21:1343–1350

Cole E, Pisano E, Brown M, et al (2004) Diagnostic accuracy of Fischer Senoscan digital mammography versus screen-film mammography in a diagnostic mammography population. Acad Radiol 11:879–886

Crowley RS, Naus GJ, Stewart J, et al (2003) Development of visual diagnostic expertise in pathology: an information-processing study. J Am Med Inform Assoc 10:39–51

Del Turco MR, Mantellini P, Ciatto S, et al (2007) Full-field digital versus screen-film mammography: comparative accuracy in concurrent screening cohorts. AJR Am J Roentgenol 189:860–866

Ebenholtz SM (2001) Oculomotor systems and perception. Cambridge University Press, New York

Egan R (1960) Experience with mammography in a tumor institution: evaluation of 1,000 cases. AJR Am J Rotengenol 75:894–900

Elmore J, Wells C, Lee C, et al (1994) Variability in radiologists' interpretations of mammograms. N Engl J Med 331: 1493–1499

Esserman L, Cowley H, Carey E, et al (2002) Improving accuracy of mammography: volume and outcome relationships. JNCI 94:369–375

Food and Drug Administration Department of Health and Human Services (2008a) Mammography. http://www.fda.gov/cdrh/mammography/ Last accessed 31 October 2008

Food and Drug Administration Department of Health and Human Services (2008b) Mammography http://www.fda.gov/cdrh/mammography/scorecard-statistics.html Last accessed 31 October 2008

Franc BL, Hawkins RA (2007) Positron emission tomography, positron emission tomography-computed tomography, and molecular imaging of the breast cancer patient. Sem Roentgen 42:265–279

Getty DJ, D'Orsi CJ, Pickett RM (2008) Stereoscopic digital mammography: improved accuracy of lesion detection in breast cancer screening. In: Krupinski EA (ed) Digital mammography. Springer, Berlin, Germany, pp 74–79

Gold RH, Bassett LW, Widoff BE (1990) Highlights from the history of mammography. RadioGraphics 10:1111–1131

Good WF, Abrams GS, Catullo VJ, et al (2008) Digital breast tomosynthesis: a pilot observer study. AJR 190:865–869

Haber RN (1969). Information-processing approaches to visual perception. Holt, Rinehart and Winston, New York

Heron G, Charman WN, Gray LS (1999) Accommodation responses and ageing. Invest Ophthal Vis Sci 40:2872–2883

Houssami N, Irwig L, Simpson JM, et al (2003) The contribution of work-up or additional views to the accuracy of diagnostic mammography. Breast 12:270–275

Houssami N, Irwig L, Simpson JM, et al (2004) The influence of clinical information on the accuracy of diagnostic mammography. Br Cancer Res Treat 85:223–2288

IHE Mammography Handbook (2008a) http://www.ihe.net/Mammo/ Last accessed 4 November 2008

Integrated Healthcare Enterprise (2008b) www.ihe.net Last accessed 4 November 2008

Kerlikowske K, Grady D, Barclay J, et al (1998) Variability and accuracy in mammographic interpretation using the American College of Radiology Breast Imaging Reporting and Data System. J Natl Cancer Inst 90:1801–1809

Komiushina TA (2000) Physiological mechanisms of the etiology of visual fatigue during work involving visual stress. Vestnik Oftalmologii 116:33–36

Krupinski EA (1996) Visual scanning patterns of radiologists searching mammograms. Acad Radiol 3:137–144

Krupinski EA (2005) Visual search of mammographic images: influence of lesion subtlety. Acad Radiol 12:965–969

Krupinski EA, Nodine CF, Kundel HL (1998) Enhancing recognition of lesions in radiographic images using perceptual feedback. Opt Eng 37:813–818

Krupinski EA, Siddiqui K, Siegel E, et al (2007) Influence of 8-bit vs 11-bit digital displays on observer performance and visual search: a multi-center evaluation. J Soc Info Display 15:385–390

Kundel HL (1975) Peripheral vision, structured noise and film reader error. Radiology 114:269–273

Kundel HL, Nodine CF (1975) Interpreting chest radiographs without visual search. Radiology 116:527–532

Kundel HL, Nodine CF, Carmody DP (1978) Visual scanning, pattern recognition and decision-making in pulmonary tumor detection. Invest Radiol 13:175–181

Kundel HL, Nodine CF, Conant EF, et al (2007) Holistic component of image perception in mammogram interpretation: gaze-tracking study. Radiology 242:396–402

Kundel HL, Nodine CF, Krupinski EA, et al (2008). Using gaze-tracking data and mixture distribution analysis to support a holistic model for the detection of cancers on mammograms. Acad Radiol 15:881–886

Lehman CD, Gatsonis C, Kuhl CK, et al (2007) MRI evaluation of the contralateral breast in women with recently diagnosed breast cancer. N Engl J Med 356:1295–1303

Lewin J, Hendrick E, D'Orsi CJ, et al (2001) Comparison of full-field digital mammography with screen-film mammography for cancer detection: results of 4,945 paired examinations. Radiology 218: 873–880

Lewin J, Hendrick E, D'Orsi CJ, et al (2002) Comparison of full-field digital mammography and screen-film mammography for detection of breast cancer. AJR Am J Roentgenol 179:671–677

Leung JWT, Margolin FR, Dee KE, et al (2007) Performance parameters for screening and diagnostic mammography in a community practice: are there differences between specialists and general radiologists? AJR Am J Roentgenol 188:236–241

MacKenzie W (1843) On asthenopia or weak-sightedness. Edinburgh J Med Surg 60:73–103

Mello-Thoms C (2006a) How does the perception of a lesion influence visual search strategy in mammogram reading? Acad Radiol 13:275–288

Mello-Thoms C (2006b) The problem of image interpretation in mammography: effects of lesion conspicuity on the visual search strategy of radiologists. Br J Radiol 79: S111–S116

Mello-Thoms C, Ganott M, Sumkin J, et al (2008) Different search patterns and similar decision outcomes: how can experts agree in the decisions they make when reading digital mammograms? In: Krupinski EA (ed) Digital mammography. Springer, Berlin, Germany, pp 212–219

Mello-Thomas C, Hardesty L, Sumkin J, et al (2005) Effects of lesion conspicuity on visual search in mammogram reading. Acad Radiol 12:830–840

Miglioretti D, Smith-Bindman R, Abraham L, et al (2007) Radiologist characteristics associated with interpretive performance of diagnostic mammography. J Natl Cancer Inst 99:1854–1863

Mutti DO, Zadnik K (1996) Is computer use a risk factor for myopia? J Am Optomet Assn 67:521–553

National Electrical Manufacturers Association (2001) Digital imaging and communications in Medicine (DICOM) part 14: grayscale standard display function. National Electrical Manufacturers Association, Roslyn, VA

Nodine CF, Kundel HL (1987) Using eye movements to study visual search and to improve tumor detection. Radiographics 7:1241–1250

Nodine CF, Kundel HL, Toto LC, et al (1992) Recording and analyzing eye-position data using a microcomputer workstation. Behav Res Meth Instrum Comp 24:475–485

Nodine CF, Kundel HL, Mello-Thoms C, et al (1999) How experience and training influence mammography expertise. Acad Radiol 6:575–585

Pisano E, Gatsonis C, Hendrick E, et al (2005) Diagnostic performance of digital versus film mammography for breast-cancer screening. N Engl J Med 353:1773–1783

Ratib O, Valentino DJ, McCoy MJ, et al (2000) Computer aided design and modeling of workstations and radiology reading rooms for the new millennium. RadioGraphics 20: 1807–1816

Rau G, Trispel S (1982) Ergonomic design aspects in interaction between man and technical systems in medicine. Med Prog Technol 9:153–159

Sanchez-Roman FR, Perez-Lucio C, Juarez-Ruiz C, et al (1996) Risk factors for asthenopia among computer terminal operators. Salud Publica de Mexico 38:189–196

Siegel E, Krupinski E, Samei E, et al (2006) Digital mammography image quality: image display. J Am Coll Radiol 3: 615–627

Skanne P, Skjennald A (2004) Screen-film mammography versus full-field digital mammography with soft-copy reading: randomized trial in a population-based screening program-the Oslo II study. Radiologgy 232:197–204
Skanne P, Young K, Skjennald A (2003) Population–based mammography screening: comparison of screen-film and full-field digital mammography with soft-copy reading-Oslo I study. Radiology 229:877–884
Skaane P, Diekman F, Balleyguier C, et al (2008) Observer variability in screen-film mammography versus full-field digital mammography with soft-copy reading. Eur Radiol 18:1134–1143
Smith RA (2007) The evolving role of MRI in the detection and evaluation of breast cancer. N Engl J Med 356:1362–1364
Smith-Bindman R, Chu P, Migloretti DL, et al (2005). Physician predictors of mammographic accuracy. J Natl Cancer Inst 97:358–367
Strax P, Venet L, Shapiro S (1973) Value of mammography in reduction of mortality from breast cancer in mass screening. Am J Roentgenol Radium Ther Nucl Med 117:686–689
Tuddenham WJ, Calvert WP (1961) Visual search patterns in roentgen diagnosis. Radiology 76:255–256
Van der Ploeg IM, Valdes Olmos RA, Kroon BB, et al (2008) The Hybrid SPECT/CT as an additional lymphatic mapping tool in patients with breast cancer. World J Surg 32:1930–1934
Weinstein RS, Lopez AM, Barker GP, et al (2007). The innovative bundling of teleradiology, telepathology, and teleoncology services. IBM Syst J 46:69–84
Williams MB, Yaffe MJ, Maidment AD, et al (2006) Image quality in digital mammography: image acquisition. J Am Coll Radiol 3:589–608
Yang W, Dempsey P (2007) Diagnostic breast ultrasound: current status and future directions. Radiol Clin North Am 45:845–861

Digital Workflow, PACS, and Telemammography

8

Chantal Van Ongeval, Erwin Bellon, Tom Deprez, André Van Steen, and Guy Marchal

CONTENTS

Chantal Van Ongeval, MD
André Van Steen, MD
Guy Marchal, MD, PhD
Department of Radiology, University Hospitals Leuven, Herestraat 49, 3000 Leuven, Belgium
Erwin Bellon, MSc, PhD
Department of Information Technology, University Hospitals Leuven, Herestraat 49, 3000 Leuven, Belgium
Tom Deprez, MSc
Departments of Radiology and Information Technology, University Hospitals Leuven, Herestraat 49, 3000 Leuven, Belgium

KEY POINTS

The transition from a film-based to a completely digital organization cannot be accomplished by merely replacing individual components. In contrast, the whole flow of operations needs reengineering. This affects not only the way individual physicians or radiographers interact with workstations or imaging modalities, but also the overall process of medical cooperation.

The individual components of PACS, RIS, and other information systems must be integrated to a higher degree than one may expect. This requires new strategic decisions regarding the deployment of information systems and regarding the relationship of the radiology department to the technology infrastructure.

Information technology opens up completely new applications such as telemammography. Merely introducing technology will not provide the expected results, as concepts and technology fundamentally depend on the organization of telecooperation.

8.1 Introduction

In this chapter, focus is on organization and workflow in digital imaging. Formally, "workflow" refers to the description of the sequence of operations and interactions to be performed by different actors (persons or groups of persons, or an organization, or machines) to arrive at some predefined goal. This term is used in the science aimed at decomposing and analyzing the work in detail. Such deep analysis is not our goal, though. The first goal of this chapter is to emphasize that different tasks and steps in the process relate to each other. This term is also often used within the context of "workflow support": how the work can be supported by the use of (among other means) technology. The second goal of this chapter is to illustrate how information systems such as PACS (Picture Archiving and Information Systems) and concepts such as telemammography can support the overall process.

This chapter is divided into three main sections. In the first section, emphasis is on *the impact of digital mammography on the workflow of radiographers, radiologists, and physicians*. We emphasize that the different steps in the process are dependent on each other (the workflow aspect) and that the work of individual users in one stage of the process may be influenced by novel digital technology at other stages in the process. We also point out how these tasks are influenced by the supporting information systems (the workflow support aspect).

In the second section, we focus on these information systems and on ensuring that *PACS, RIS* (*Radiology Information System*), and the *medical record* effectively can support our processes instead of merely adding complexity. Emphasis in this section is not on the tasks of the different users but on overall organization. A general knowledge of these systems is assumed. However, we point out aspects of effective deployment that are too often overlooked in practice and we emphasize elements in the use of technology for which a strategic decision is indicated.

Digital technology enables us to establish workflows that extend beyond the local institution. This imposes additional challenges on the supporting information systems. In the third section of this chapter, we discuss technology and organization for *telemammography*. As this is a relatively new concept, there is a danger that miracles are expected from technology. One of our goals is to illustrate that "digital workflows" that are adapted to one organization may turn out useless when blindly applied to a different organization.

In all three sections, *screening* mammography has a somewhat special status (Bick et al. 2008). Screening mammography consists of only four standard mammographic views, while in diagnostic mammography, other views may be obtained. From the perspective of imaging workflow, the diagnostic mode exhibits more interaction between image acquisition and interpreting, for example the radiologist asking for additional exposures or even adding another type of exam such as ultrasound or magnetic resonance imaging (MRI). This mode also requires multimodality viewing. In screening mammography, instead, the workflow is more streamlined, but high throughput is required. From the perspective of supporting technology such as PACS, the question arises whether a single system should be used for screening mammography and for diagnostic work, thereby bringing advantages from integration at the expense of not providing the optimal environment for the specific tasks in screening. Telematics can modify the workflow of mammography screening dramatically, but the options available to exploit this relatively new technology depend fundamentally on how screening is organized. In particular, an approach that may be suited for a region-wide organization dedicated to breast screening may not work at all in a situation in which general radiology practices or hospitals take part in the screening program.

8.2 Workflow in Digital Mammography

8.2.1 Technology

This subsection is about selected technology aspects of the image generating equipment and the workstations with which the end users come in direct contact; the next subsections focus on organization of the work. We limit ourselves here to the interaction between the user and the individual systems. How these systems interact among them, or how different components could be integrated, is a topic for the section on PACS (Sect. 8.3).

8.2.1.1 Image Generating Equipment

Mammography was the last modality to become digital because of the required high resolution over a large area combined with the need to render subtle differences in radiodensity. These requirements also influence the technology that is deployed in subsequent stages of the process: individual mammograms have a high-spatial resolution and take more bytes than other X-ray images and display systems must be able to render small steps in intensity as well as small spatial details.

Digital mammography acquisition can be performed using computed radiography (CR) or direct radiology (DR). Smaller departments tend to implement the CR technology, larger hospitals more often DR mammography. For CR, the workflow is very similar to screen-film mammography (SFM).

The quality requirements on image generating equipment have led to Quality Control (QC) programs that are generally more rigorously implemented in the frame of screening activities than for general X-ray radiography. Although technical performance of these systems can be compared on paper, hands-on experience is needed to assess their efficiency of use and the impact of individual features on overall workflow.

8.2.1.2 Display Systems

The high quality of mammography images translates into extraordinary requirements for the display technology. Diagnostic displays are covered in QC programmes and in most acceptance tests for mammography screening. Displaying the images is one task of reporting workstations, while a second task is providing a user interface to interact with these images. The latter task is discussed in Sect. 8.2.1.3, as it is about reading efficiency rather than image quality.

We deliberately use the somewhat convoluted term "display system" because image quality is determined by the whole chain of components that translate pixel values into intensities in our visual system. The most obvious component is the computer screen that generates light according to input voltage. The component that generates this driving voltage from the pixel values is the display card. Typical computer software talks to the display card using 8-bit numbers, which implies that at most 256 different intensity levels can be generated by the next components in the chain. While this is plenty for most applications also in radiology, software dedicated to mammography may want to improve on this by using nonstandard display cards and dedicated software.

A first requirement for a mammography display is its *spatial resolution*, i.e., the number of separate pixels it can render. In mammography "5 Megapixel" monitors are usually deployed (Bick and Diekmann 2007). This is higher than typically used for general radiology. One reason to aim for high resolution is efficiency: one may be able to detect a subtle detail using lower resolution too, but it may take just a bit longer or might require an additional magnification action (Bacher et al. 2006). This high resolution by itself is not the reason that mammography displays are particularly expensive. It is the combination of a high resolution with the next requirements.

A second requirement, indeed, is that for radiological diagnosis the user must be able to *distinguish many intensity differences*. Because of the characteristics of the human visual system, this translates into the need for high light output, or high luminance. Informally, the term high brightness is often used even as brightness technically refers to the response elicited in the observer's visual system. Providing high luminance is a technological challenge, especially in combination with the other requirements. The display must also provide many shades in between dark and bright.

A third requirement, typical for radiological displays, is *consistent presentation of image values*. A characteristic grey value pattern in the image must appear as the radiologist expects it from his or her mental frame of reference, regardless which display system is used. It simply is not possible to ensure that the same pixel value in the image results in the exact same light output on any display, however. This is not needed either, except for the displays on the same workstation. Rather, equal pixel differences in the dark parts of the image and in the bright parts should elicit the same response in the human visual system. This requires the display to be calibrated according to the characteristics of the human visual system. A typical curve for visual response in function of luminance has been agreed on in the standard DICOM GSDF ("Gray Scale Display Function"). This standard enables software providers to make abstraction of the exact type of display system and end users to more freely select the displays for their workstations.

8.2.1.3
Diagnostic Workstations and User Interfaces

For general diagnosis, it has become clear that soft-copy reading is more efficient overall than reading film on a lightbox. Some elements that increase efficiency are dynamic navigation over a huge number of slices and elimination of the time needed to sort and hang the films. Exactly for screening mammography, however, these elements do not have so much value. As there are only a few images in the study, manipulations of the images on the screen often reflect limitations in human–computer interaction rather than providing new functionality. In screening, films were often prepared on an alternator by supporting personnel. On a preprepared mammography alternator, reading productivity can be between 50 and 80 screening examinations per hour. Different studies point out that with current workstations time needed for interpretations is longer, by a few tens of seconds per case or roughly a factor 1.5 overall (Ishiyama et al. 2009; Haygood et al. 2009; Berns et al. 2006) or nearly a factor 2 if strictly the viewing task is considered (Ciatto et al. 2006). These measurements are about the specific screening situation in which reading speed traditionally was optimized to the extreme – for diagnostic work with much more positive cases, reading times are probably not significantly different (Pisano et al. 2002). Anyhow, up to now it seems that, for screening, the transition to digital imaging resulted in time savings in other parts of the process but at the cost of decreased efficiency of the radiologists.

This difference in reading efficiency is not related to the fact that the images are presented on a computer display instead of on a frosted glass plate, but to the way the radiologist interacts with the images. For example, navigating to another image and zooming into a region using the mouse simply takes longer than moving one's eyes over the lightbox. As is discussed into more detail in Chap. 7, current user interfaces for screening mammography try to increase efficiency by exploiting the structure present in the viewing task by using dedicated "hanging protocols" and enabling switching to the next phase in viewing by just pressing a button on a dedicated keyboard. Further efficiency gains can be expected from tools with additional automation such as for synchronized zooming and panning.

A factor that does not exactly help in developing efficient viewing software is the size of the images in mammography. Fetching up to eight images from storage, adapting their size and blasting them to the display within a few seconds, is pushing current technological limits. It clearly can be done, but at a cost in designing the system. First, application-specific tricks must be exploited to provide fastest response, resulting in a design that is less general and more difficult to extend to other applications. Second, in decisions about overall architecture of the system flexibility may be sacrificed for speed. For example, many systems dedicated to mammography screening assume that the images are preloaded onto the individual workstation they will be viewed at instead of retrieving the images from a central location.

These considerations give rise to reading stations and associated image management systems dedicated to screening mammography. The department may need to decide whether screening should be supported by such a dedicated system or, instead, by the same environment that is already in place for other radiological disciplines. We come back to this in Sect. 8.3.4.

8.2.1.4
Setups for Clinical Review

It is generally accepted that computer displays for users outside the image generating department need not adhere to the same high quality requirements than diagnostic displays. In that setting, convenience of digital access to images usually is medically more relevant than theoretical presentation quality. Clinical users still must be able to discern subtle lesions in the images, but they typically can afford spending a few seconds more in magnifying suspicious areas and panning through the image. In contrast, these users appreciate the flexibility of being able to look at images on any workstation throughout the hospital. It simply is not realistic to provide expensive diagnostic screens on all these computers, not only for reasons of cost but also because special monitors would severely limit the freedom of the hospital in selecting hardware and software infrastructure. Besides, clinical users generally cannot do their other activities in the dark, so even the best diagnostic monitor will not perform as intended. Applications in the operating theater, where viewing conditions are abysmal anyhow, do not require studying of subtle details, while instead the ability to look at images from a considerable distance is an advantage.

Specific for mammography it remains important that a few mammograms can be displayed simultaneously in sufficient detail, possibly even together with ultrasound and MRI. Therefore, the availability of dual monitor systems at selected locations is helpful while typically not placing too much strain on costs, installation freedom, and space.

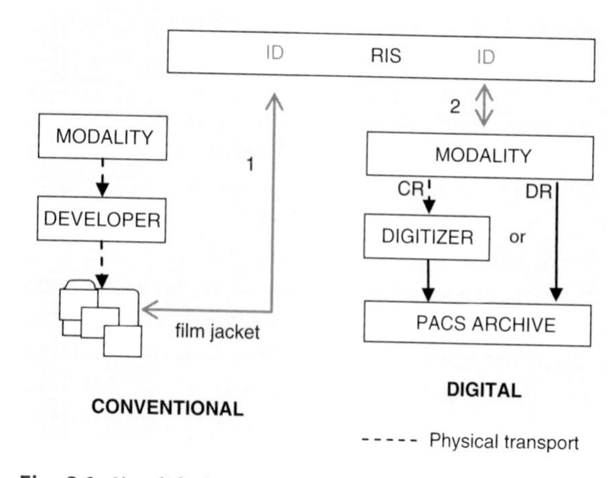

Fig. 8.1. Simplified view of the workflow for radiographers in FSM (left) and FFDM (right). In the FSM workflow quite some steps require physical transport of cassettes or film jackets (dotted lines), which becomes unneeded with FFDM, especially with DR

8.2.2 Impact on the Work of the Radiographers

Radiographers have a key role in any image acquisition process. Radiographers who are performing mammography are typically well trained in all aspects related to patient positioning for breast-related imaging and biopsy procedures. However, the actual positioning is more or less the only part of the job that is not influenced by the transition to digital acquisition. All other parts are influenced more strongly than one may at first suspect. In this section, we assume that not only the imaging modality is digital but that no films are used in further parts of the process either.

8.2.2.1 Working Efficiency

An evident difference with the film-screen organization is that a lot of film handling has disappeared, especially in DR mammography. There is no need any more to fetch film out of the developing machine, put it in jackets or even load a lightbox. In some institutions, films even had to be scanned to apply CAD. With digital imaging, the workload for the radiographers is likely to decrease (Ciatto et al. 2006). This is illustrated also in Fig. 8.1, showing the film handling steps in the process that disappears with the digital workflow.

8.2.2.2 Increased Complexity

Although some manual actions have disappeared, complexity of the job increases, as various computer systems around the modality must be manipulated. The transition from a paper-based workflow already required radiographers to obtain more information from the RIS and to enter information such as that the exam was performed (gray arrow 1 in Fig. 8.1). Now that imaging itself has also become digital, the integration with the RIS becomes even tighter. Radiographers must understand why information from the RIS must be loaded into the modality (gray arrow 2 in Fig. 8.1), as detailed in Sect. 8.3.2.2. In addition, radiographers may need to use computers of the PACS and learn the related tools.

In any workflow, errors introduced at one stage will have an impact on some later stage. With digital workflows, however, the odds are higher that such errors are not immediately noticed and only become apparent later in the workflow, where they will do more harm. To understand and solve resulting errors, radiographers not only need to master new technology, they also need to have a better understanding of the overall process.

8.2.2.3
Stricter Procedures

Although some manual actions have disappeared, radiographers have to spend more attention than before. Procedures must be followed more strictly than in the analog age. Previously, the patient name was entered using a level of detail sufficient for interpretation by humans, as images were retrieved using the film jacket. Now, it has become essential that patient and exam IDs are entered correctly, as images can only be retrieved using these IDs. This need not require more time, as these IDs can be automatically provided using a DICOM modality work list out of which the radiographer selects the current exam (see Sect. 8.3.2.2). To reduce errors, the radiographer must more strictly follow this digital procedure.

If an unplanned or emergency exam must be performed, the exam should first be entered into the RIS before starting the work on the imaging modality, because only then the correct exam ID is available that is needed for later stages of the process. By skipping this task at the start of image acquisition, a more costly correction will be required at a later stage in the process.

A certain amount of later corrections is inevitable, however, even with mechanisms such as modality work lists. For example, the modality operator could have selected the wrong entry out of the work list. This can occur particularly easily if a number of related exams for the same patient are different for the computer system but are in practice performed on the same modality. The tools available to perform this correction combine information from PACS and RIS (see Sect. 8.3.2.1). It is not always easy to relate the images to the right exam in the computer system, and by the nature of this task the computer system cannot check for new errors anymore. Therefore, operators who only occasionally operate these tools run the risk of introducing new errors. On the other hand, if this task is assigned to only a limited number of experienced operators it may be difficult to ensure coverage during night and weekend shifts.

8.2.2.4
Pitfalls

With DR, it becomes quite easy to generate a new exposure if positioning was suboptimal. The ambition of the radiographer to provide the best-quality images might therefore lead to increase in radiation dose. On the other hand, digital modalities provide new ways to follow up number of exposures and radiation dose (see also Sect. 8.3.2.5). One procedural way to detect a tendency of making too many retakes is to only allow radiographers to delete an image after the radiologist has approved this. It can be discussed if it is good practice to delete any image at all or to prevent any image from being sent to the PACS.

On some modalities the order in which the images are obtained is important to have the images displayed in the right layout on the radiologist's workstation, as a workaround for the modality not including orientation-related information in the images. When scanning film, this information must be entered manually and order of images is definitely important.

8.2.2.5
Improved Functionalities

Digital imaging provides tools that permit radiographers to further increase quality and service. One new possibility with DR is to control technical quality of the images immediately, within the examination room itself (Pisano et al. 2007). The patient can immediately leave the waiting room and the risk is reduced that a patient has to be called back for retakes. A second new possibility is obtaining useful information regarding optimal settings from the prior images, as they are probably available in the PACS.

These functionalities come at the expense of additional work, however, and there is definitely a learning curve so time must be spent first in mastering these tools. It is not technically straightforward to provide the optimal setup either, such as computer screens that permit sufficient detail to be discerned in unfavorable lightning conditions and viewing angles, or PACS work lists that automatically suggest prior images without the radiographer having to manually enter search criteria for each patient.

8.2.3
Impact on the Work of the Radiologist

8.2.3.1
Image Viewing

Again, we assume in this section a filmless department in which radiologists read images on a workstation. A few general aspects on workstations are already mentioned in Chap. 7. When thinking about a workstation most people intuitively focus on image presentation and tools for interactive manipulation of and navigation over these images. That image-centric view is too limited, however. The efficiency of a workstation is determined by more subtle aspects of organization, workflow, and integration. We first go into the "micro workflow" of reading a single mammography exam, expanding to other tasks in later subsections.

First, the workstation should know that the radiologist is reading a mammography exam and by default it should present the study in the appropriate structured layout for example, with four views. This requires the definition of "hanging protocols" (or "default display protocols"). Hanging protocols should not be implemented too rigidly and the workstation should be able to cope with exams that not strictly adhere to a predefined category, for example, if a mammography exam has more than four views.

Automated hanging protocols rely on information included with the images, by the modality. The optimal level of standardization in this information has not yet been reached. Therefore, in practice, the workstation must be "taught" the different kinds of exams used in the department, and be able to select the right set of rules according to, e.g., free form exam descriptions. This is not difficult if the department deploys one or just a few modalities but quickly becomes problematic if exams are regularly imported from other centers (see also Sect. 8.4.4.2). It is characteristic for digital workflows that information entered at one stage of the process is deployed at some later stage, and that the different components in the process must be more tightly connected to each other.

Second, the ideal workstation knows about the preferred sequence of actions the radiologist performs with the images and anticipates by enabling easy transition to the next stage. Especially in mammography screening, such sequences can be defined relatively easily.

Third, the workstation should automatically find relevant prior images instead of requiring the user to explicitly search for them, e.g., according to body part. In the general situation, the user can be expected to select the prior exams to be displayed out of the short list suggested by the workstation. For mammography screening, the workstation can be expected to automatically include the previous X-ray screening exam in the hanging protocol. If older exams are only available in a slow part of the archive, the PACS may need to know in advance which new exams are scheduled to make these prior exams available for fast retrieval. As discussed in Sect. 8.3.2.1 this requires PACS and HIS/RIS to be integrated.

It may be clear from these requirements that already for the very specific task of image viewing, a diagnostic workstation cannot function as an isolated component. This need for integration is even more pronounced if other tasks of the radiologist are included (see Sect. 8.3).

8.2.3.2 Organization of Image Reading

Efforts in making image viewing and manipulation efficient must not be counterbalanced by time lost in searching for the next exam to read. In the film-based organization, the order of film jackets on a pile indicated which exam was to be read next. The film jacket also served as the instrument to distribute the work between radiologists: the first one to grab the jacket took ownership of the exam. In screening mammography, the images were even already laid out on the alternator. In the digital era, this work organization has disappeared. What remains is a work list generated by the PACS and presented on the diagnostic workstation.

It is not obvious to make this reading list contain the relevant entries in order of priority, and to include mechanisms to schedule the work. A naïve immediate translation from the film-based organization to the digital one will therefore introduce inefficiencies. Moreover, a PACS generally cannot build reading lists based solely on information in the images but also requires information from other information systems such as the RIS. It may even be preferable to have the work list presented by the RIS instead of the PACS workstation, as that system has more management and workflow information available. In the latter organization, the viewing station refrains from offering management functions and only provides an image presentation service on request by the RIS (see Sect. 8.3.2.3).

Selection of exams remains inefficient if film-based and digital organizations are mixed. If the previous exam is only available on film, the electronic work list becomes useless: either you pick the next exam from the computer but then you have to search through the pile of paper jackets to find the corresponding previous exam or you pick the next film jacket from the pile and then have to search in the computer for the same patient. This surely is a problem in screening mammography. If the timing for the transition period is not in the hands of the own department because of the need to read screening exams acquired elsewhere, it may be considered to have these exams digitized upfront and thus entered into the digital workflow (Trambert et al. 2006a).

8.2.3.3 The Global Reporting Workflow

If image reading is supported by digital technology, it was a missed opportunity to not also integrate reporting into the digital workflow. Options range from dictating into the computer so a transcriptionist can later call up this sound file, over speech recognition, to directly entering information into structured reports. The completely digital system even provides new, although still experimental, opportunities to create multimedia reports. Examples are including selected images within the text of the report, presenting marks the radiologist put next to lesions in the images on a scheme of the breast, or generating reports in which sections are explicitly hyperlinked to locations in the images (Bellon et al. 1994).

We want to point out two opportunities to improve on the overall process. Firstly, substantial efficiency gains can be achieved by *directly linking the report to the entry in the work list*. For example, a transcriptionist who received a tape must first listen to the patient information, search for that patient in the information system, open a blank report, and only then can start typing. If a report must be typed with priority even more time is lost in searching the relevant dictation on the tape. It is a relatively small step to replace the tape recorder by a system in which the radiologist directly dictates into the computer (digital dictation) such that the voice file is electronically linked to the study. The transcriptionist only needs to select an entry from a list and the computer can automatically open the corresponding case. Urgent cases can easily be listed first.

The second opportunity is to shorten reporting turnaround dramatically by enabling the radiologist *to directly finalize the report*. For screening, this can be accomplished by filling in an electronic form. For diagnostic work, one option is deploying speech recognition with proofreading immediately by the radiologist. This shifts some work to the radiologist that previously was performed by the transcriptionist. In our experience that is more than compensated for, however, by elimination of other tasks such as proofreading the transcript some time later and bringing the case back into memory for that.

Debate on the best approach to support reporting will continue for quite some time, but the digital workflow at least provides new options, together with new decisions to be taken.

8.2.4 Impact on the Referring Physicians

8.2.4.1 Image Viewing

Although digitally acquired images could be printed on hard copy for distribution to referring clinicians, such organization was to forfeit the benefits of the digital organization. Hospital-wide direct electronic access makes images available more timely and simultaneously from different locations, even from home when on call. Logistic and medical processes become more efficient. It even turns out that radiologists are now less interrupted by calls from clinical colleagues inquiring for results.

One result of soft-copy reading that may be particularly relevant for mammography is that not everyone shares the same physical image with the same quality. The most important factor here is that the monitors used by clinicians typically provide lower image quality in terms of resolution and primarily in terms of the number of discernable grey values. In our experience, consumer type monitors are sufficient for most clinical review and for radiological–clinical discussion provided the radiologist takes part in the multidisciplinary team.

Most clinical users appreciate intuitiveness of the user interface more than utmost efficiency for any particular task. In contrast, they need the ability to view many different types of exams using the same user interface. They need easy navigation between images on the one hand, and items such as the radiological report on the other hand. Clinical viewing software can offer features adapted to these specific needs. One example is easy navigation to an image that was tagged before, a feature that is particularly useful in clinical conferences. Another example is the possibility to compose an overview of images for use in the operating theatre, as in that environment possibilities for subsequent interaction with the images on screen are limited.

8.2.4.2 Integrating Image Viewing into the Electronic Medical Record

In our experience, referring physicians look at images in a fundamentally different way than radiologists. Images are only part of the information they need to integrate. It has little value to be able to look at "just" the images but not at other clinical information in the computer system. Already for that reason images should conceptually be considered an integral part of the patient file, and they should be accessible from the electronic medical record (EMR). A typical setup nowadays is that the clinical viewer of the PACS is controlled by the EMR, thereby giving the impression of a single information system (see Sect. 8.3.2.4).

This integration also serves workflow, for example, by providing task-oriented work lists. Indeed, reading lists for clinical users are not based on the sheer availability of images and therefore simply cannot be provided by the PACS. In addition, by shifting selection of the case to the EMR, patient-based access control to images can be implemented, which typically is not a concern within the image-generating department.

8.2.4.3 Integrating Images into Hospital-Wide Processes

When discussing image reading organization within radiology we stressed that in high-volume diagnostic work, mixing hard- and soft-copy viewing considerably decreases efficiency because of the hassle involved in selecting related exams. In the clinical review setting, loosing a few seconds in selection presents less of a problem. Instead, the problem with hard copy is that the film is unlikely to be available when and where it is needed. Requests to digitize hard copy, therefore, are primarily motivated by the wish to have the study available in the PACS. It most often is acceptable if the film is only digitized for later use after the physician viewed the film.

Exams arriving on CDs are a completely different matter. It is detrimental to efficiency during outpatient consultations if the physician has to load each CD manually, wait for the software to be ready (assuming that security restrictions do not prohibit the viewer from starting in the first place), and then struggle with the always changing user interfaces provided by all kinds of exotic CDs. In our experience, CDs must be imported into the PACS and into the EMR upfront, as part of the logistic workflow around the patient. It may be clear, also from previous parts of this chapter, that, in the digital organization, medical and logistic workflows are intertwined.

8.3 PACS and Integrating Images into Overall Informatics

8.3.1 Motivation

In this section, we focus on the *information systems* that support the digital workflow. As emphasis is on images, the most relevant system in this discussion is the PACS. PACS is often regarded as a direct replacement for the traditional film-based technology, with workstations replacing lightboxes and digital archives taking over the storage function of film. A limitation with this perspective is that it does not capture the fact that changes in the organization are required in order to deploy PACS successfully. Even worse, we now have to take aspects into consideration that did not matter before.

In the first subsection, we illustrate the need to integrate the different components of the information system. This need was less pronounced in the film-based workflow because humans are extremely good at picking up the information needed to drive the workflow. They do so informally, often without even being aware of the source they used. Computers, in contrast, must be fed all pieces of information explicitly, often by other information systems. The individual systems simply cannot function in isolation. We illustrate this and introduce options available for integration.

In the second subsection, we descend to the level of information technology buried within the individual systems. There is enormous potential in integrating the technological infrastructure over different systems, for example, in providing a common storage system. This requires strategic decisions which we argue a radiological decision maker cannot stay out of.

In the third subsection, we discuss pros and cons of deploying a single PACS for different applications, for example, screening and diagnostic mammography, instead of using dedicated systems. This question relates to the need for integration discussed in the previous sections. In the digital organization, the decision to deploy a dedicated component for one function can surprisingly easily result in information islands in which large parts of the workflow become separated. Therefore, this becomes a strategic decision.

It is not a coincidence that "integration," at different levels, is the common theme throughout these sections. Integration simply is a key element in successfully supporting the digital organization. *In the*

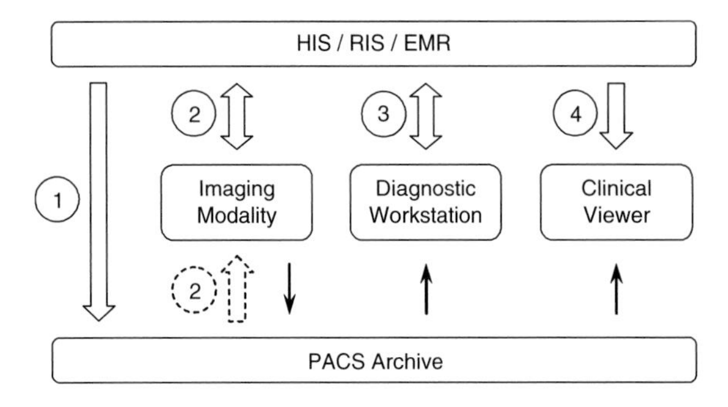

Fig. 8.2. Interaction between the components involved in the digital imaging workflow is usually depicted in terms of image flow (*thin arrows*). In this section, in contrast, we focus on additional interactions and data exchange between these components (*block arrows*)

last section, we briefly position technical integration in the context of hospital management.

8.3.2 Integrating the Components that Support the Image Workflow

The need for integration between the different components in the system is frequently underestimated. Simply putting together even the most expensive high-tech components will result in a workflow that is less efficient than the film-based one. In the next subsections, we first discuss system interactions using the simplified diagram in Fig. 8.2, and then situate the DICOM standard and the IHE initiative.

For our discussion, distinction between HIS (Hospital Information System), RIS (Radiology Information System) and EMR (Electronic Medical Record) is usually immaterial, and so we either use a general term or choose the one that happens to be most convenient.

8.3.2.1 Integrating the PACS with the HIS

The images in the PACS obviously must be linked to a patient and are usually also linked to a request and a report in the RIS. In the analog era, patient and exam IDs appeared on the film jacket, usually after having been printed by the RIS. The modality operator attributed these IDs to the images by inserting the films into that jacket, often not caring too much whether the identifying information appeared absolutely correct on the films themselves. In the digital workflow, instead, these IDs must be explicitly entered at the modality and transmitted as part of the image data.

The odds are high that a typing error gets introduced in such an ID. Moreover, in the digital organization, such identification errors are particularly disrupting. For example, as reading work lists are not based anymore on a pile of film jackets waiting to be picked up by a radiologist, the wrongly identified exam may simply not get read.

A key principle is that such errors should be detected as soon as possible, ideally right when the exam arrives at the PACS. Such a check can be implemented quite effectively by providing the PACS with a list of expected exams (arrow 1 in Fig. 8.2). This list can also be of help when manual corrections of wrongly identified images are needed. There may be additional reasons to provide the PACS with information about scheduled exams, for example, to trigger automatic "prefetching" of prior images from a slow bulk archive.

Already for this reason, there must be a digital connection between the overall information system and the PACS. A PACS cannot function as an isolated system.

8.3.2.2 Integrating the Imaging Modality with the HIS

Even if identification errors can be detected automatically, the problem remains that correcting them takes quite some time and may even become a source of new errors – of a kind that cannot be detected automatically. In addition, typing the patient and exam IDs at the modality takes time. Thus, blindly replacing the film-based organization by the modern digital one actually lowers productivity.

The solution is to have the overall information system provide the modality with a list of exams to be performed and their IDs (block arrow 2 in Fig. 8.2) from which the operator only needs to select the exam. The importance of this function for the digital workflow cannot be overstressed. This work list may even contain additional information such as the type of exam to be performed or the body part, thereby enabling the modality to suggest predefined settings or a specific sequence of images.

As discussed in Sect. 8.2.2.3, there must be a request for the exam in the RIS before it can appear on a work list. Each individual modality may get its own work lists (provided resource allocation is performed into sufficient detail in the computer system) or all modalities of a group may share the same list.

Sharing a list makes it easier to move an exam to another modality, but very long lists may make it harder to select the right exam.

The functionality for a modality to receive a work list has been standardized in the DICOM framework. As most HISs do not speak DICOM natively and given that the PACS may require lists of scheduled exams for other purposes as well (see Sect. 8.3.2.1) a common approach is for the HIS to communicate this information to the PACS and to have the PACS internally distribute work lists to the individual modalities (dotted block arrow 2 in Fig. 8.2). Communication from HIS to PACS is most often using the HL7 standard.

8.3.2.3 Integrating the Diagnostic Viewing Workstation with the RIS

Using a diagnostic workstation isolated from other information systems is a questionable practice. First, the RIS is likely to contain relevant medical information such as clinical history or allergies to contrast. Second, whereas the imaging workstation can only present work lists based on which images arrived in the PACS, the RIS may provide more powerful reading lists. As a minimum, the list provided by the RIS will contain an exam even if the images were not sent to the PACS due to an error. Third, if the report can be directly hooked to the right exam, substantial efficiency gains are possible at later stages of the reporting process (see Sect. 8.2.3.3).

Even if RIS and PACS are separate systems, they can appear as a single system to the radiologist by synchronizing selection (block arrow 3 in Fig. 8.2). Most viewing workstations accept commands from an external system about which exam to present, promoting the RIS to the master system. An alternative is to make the viewing station the master and have that push the current exam to ancillary systems. That organization is not often used to have a RIS synchronize its context, but is a popular way to integrate a speech recognition system.

Increasingly, PACS and RIS are not separate systems anymore but are provided as a bundle. One advantage of such a tightly integrated system is that multimedia reporting can be brought a step nearer, e.g., by linking a section in the report to key images or even to locations within an image. If integration between RIS and PACS comes at the price of less integration between RIS and EMR, one should think twice, however.

Although not technologically challenging, there is little standardization for this type of integration. In practice the options depend on the flexibility of the vendors of the individual information systems. We increasingly select image viewing software based on its ability to integrate rather than on individual functions. Indeed, the advantages of individual viewing features are too often more than offset by inefficiencies in actions such as selecting the next case or searching for additional information in a separate RIS.

8.3.2.4 Integrating Clinical Image Viewing Within the EMR

Outside the image-generating department, the image viewer is most often integrated into the EMR. Clinical users must be able to open a patient file in the EMR and from there call up an image exam with a few mouse clicks, just as other types of data (see also Sect. 8.2.4.2). There is no high-level standard for this integration, but many clinical viewers can listen to commands to present a specific exam (block arrow 4 in Fig. 8.2).

Delegating exam selection to the EMR may even be a requirement to comply with access policies, especially if access control is at the level of individual patients. In addition, many organizational principles that were already available in the EMR can now be reused, for example, to retrieve image exams from task-oriented work lists (Fig. 8.3).

8.3.2.5 DICOM

The DICOM (Digital Imaging and Communication in Medicine) standard originated from the need to standardize the content of radiological images and to make agreements about how computer systems should exchange them. There is no single "DICOM standard" but rather a huge number of separate standards for different purposes. DICOM is primarily about images but includes such aspects as adding notes or presentation directives to images, and standardizes principles involved in presenting an image onto a diagnostic display (see Sect. 8.2.1.2). A thorough discussion is outside the scope of this text (Oosterwijk 2005); we just provide a few key elements focused on the workflow within radiology.

A first purpose of DICOM is to define the data elements that make up an image. This includes the actual pixels but also information such as patient

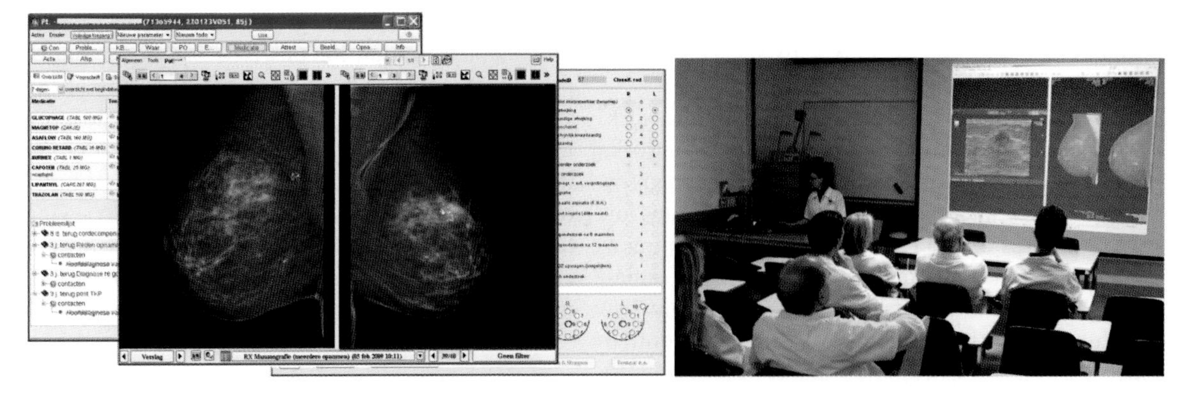

Fig. 8.3. Integrating the clinical image viewer as a module in the EMR provides efficient access to *all* available types of information (*left*). Moreover, already available organizational principles can now also be exploited for image viewing, including retrieving cases from a list with all patients in the ward, or with patients expected on an outpatient consultation, or with cases scheduled for a clinical conference (*right*)

demographics, pixel size, slice distance, slice orientation within the scanner, and acquisition parameters. In practice, one can assume that the imaging modality includes the most important information in the images, but there is still room for improvement. For example, DICOM provides standard definitions for patient orientation or laterality but not all image generating modalities provide these.

A second purpose of DICOM is to specify the messages that computers must exchange to transmit or request information. This communication takes the form of a request for one of the specified "services," for example, to store an image or to print images using a supplied layout. A system component need not provide all services and can even limit a service to a particular type of images.

When purchasing a radiological modality, the device must as a minimum be able to act as a "service user" for the following services:

- *Storage service*: enables sending of images to a component such as a PACS archive or a processing workstation.
- *Modality worklist*: enables the modality to query an information system for information on exams to perform, typically used to present work lists to the modality operator.

Two services one can do without but that are gaining importance are Storage Commitment and Modality Performed Procedure Step (MPPS). Storage Commitment enables the sender to be sure that all transmitted images have arrived (and thus could be deleted locally) and the receiver to be sure that no more data will arrive for this exam. The MPPS service enables the modality to notify the HIS/RIS that a certain exam was just performed, enabling further steps in the workflow to be initiated automatically, and enables the modality to supply additional information about the exam ranging from number of images or administered medication (useful for billing or to automate supply management) to information in radiation dose (useful for QC or to maintain received dose in the patient record).

DICOM focuses on images that can be readily presented on a computer screen, in direct analogy with film on a lightbox. Advanced processing and visualization techniques in ultrasound, instead, are based on raw data from the imager. DICOM does not standardize the raw data but a DICOM image may include it as a private element, in a proprietary format. In X-ray mammography, increasing use of all kinds of processing to visually enhance the images became a serious nuisance to CAD. To cope with this, mammography equipment may generate two type of images: "For Presentation" images intended for viewing and "For Processing" images that have been subject to minimal processing only intended typically for CAD.

As a concluding note: DICOM is primarily suited for information exchange between clearly detached components such as a modality and a PACS or a PACS and a workstation for surgery planning. A viewing station and an image archive (two components often considered to be "internally" to the PACS) can be connected using DICOM but are most often connected using proprietary protocols. A first reason is speed: a workstation and archive of the same vendor can internally use faster methods to retrieve images. A second reason to use proprietary protocols is that the PACS may implement dedicated work lists or exploit other vendor-specific management functionality. Integration of viewing components within a RIS or an EMR (see Sects. 8.3.2.3 and 8.3.2.4) also is outside the scope of this standard. DICOM is essential in supporting the radiological workflow but is not the only element needed to obtain a highly integrated system.

8.3.2.6 IHE

DICOM does not impose a specific workflow. In contrast, it aims at supporting a wide variety of policies. There also is substantial freedom in the options, for example, which information elements a modality can provide. Thus, in practice, one can only infer the details by studying the DICOM conformance statement of the component into detail.

The IHE initiative (Integrating the Healthcare Enterprise) has quite a different focus. This association between healthcare industry and user groups aims at specifying how medical information systems should interact to support specific predefined parts of the workflow. For example, the IHE "Scheduled Workflow Profile" defines all interactions among HIS, RIS, PACS and imaging modality in the process from ordering a radiological exam, over performing it in the imaging department, to providing the final report, possibly with tracking of the status of the request.

Although the workflows are extended and involve many information systems, the aim is to solve a very concrete problem instead of to define general mechanisms. For each of the steps in the chosen workflow, IHE selects among existing standards (often DICOM or HL7), limiting remaining freedom in those standards so that technical implementations are less costly and every vendor knows more precisely what to expect. There still are a lot of options available–simply stating that a component such as an imaging modality must be able to participate in some IHE profile is not sufficient - but the different options get a name with a very precise meaning (Moore 2003).

The goal is to make it easier for healthcare professionals to select interoperable HIS/RIS, PACS or imaging components that suit their needs, while at the same time making it easier for industry to decide on which exact functions to offer. IHE even provides a guide detailing how to pick options and state them into a request for proposals (IHE 2005). As the IHE initiative is more recent than DICOM, it may not yet be obvious to use this framework when buying components, but momentum is building up. One of the noteworthy aspects of the IHE organization is that efforts at defining a profile are only started when there is demonstrated interest from users and industry, and when it is clear that an agreement can be reached quickly and commercial products will follow in the short term. Every year a few "connectathons" are organized in which companies demonstrate their ability to interconnect.

Useful profiles tailored to mammography can be found in the IHE Mammo Workbook (IHE 2007a) or in more detail in the IHE Technical Framework itself (IHE 2007b) and the mammo-specific supplement (IHE 2008). We mention a few of these just to give an idea of the scope. The "Mammography Image Profile" specifies, among others, which exact information modalities must provide in order to feed hanging protocols at the viewing station and which functions viewing software should have to enable efficient presentation of new and prior mammograms even if these were generated by equipment from different vendors. The "Consistent Presentation of Images Profile" combines, and further specifies, DICOM standards for consistent presentation of images on the display, including options to instruct the viewing software how to initially zoom or window an image or show annotations to convey the radiologist's message.

8.3.3 Strategic Decisions on Integrating Technology Infrastructure

Roughly speaking, an informatics component can be divided into the application layer (the software end users interact with) and everything "beneath" it (standard software and hardware that enables the application software to be executed). The integration discussed in the previous subsections is at the application layer, and therefore something that immediately matters to the radiologist.

The lower layer, in contrast, is traditionally not considered of concern to radiology decision makers. However, in complex digital environments with pervasive deployment of information technology, such as a hospital, that layer determines to a distressing degree the total costs and supportability of the application. Not caring about the technological infrastructure, therefore, becomes a strategic decision in its own right: to leave decisions to some other party in the hope that it shares your interests, or to ignore opportunities to improve the quality/cost ratio. We illustrate this using the example of storage.

8.3.3.1
Storage Consolidation

Storage is one of the most expensive elements in a hospital PACS, even as prices keep dropping. An often-used approach at keeping costs under control is to only maintain recent images on fast storage while older images are in less expensive but slower "near-line" storage such as a tape robot. This introduces a more subtle cost, however, in provisions to automatically migrate these images between the storage levels and in ensuring that the near-line storage media remain readable and do not become subject to technological obsolescence.

However, PACS is not the only storage hungry application nowadays. Many hospitals are storing all kinds of biosignals such as EEGs or ECGs, as well as an increasing amount of video loops obtained during, for example, surgical procedures. The traditional textual data in the medical information system may take up considerably less volume than images, but these data are so critical that multiple backups may be provided online. And then, there are all kinds of office data. In this context, a way to lower storage costs considerably is by pooling all needs and satisfying these using a single ("consolidated") storage system instead of using different systems.

One large system is less expensive than two systems of half the size. An obvious factor in this economy of scale is better utilization of the hardware. An even more relevant factor is the cost for daily maintenance of the system. If two different storage systems are deployed, the investments in manpower and know-how are doubled. In contrast, if the capacity of a single system is doubled, the manpower required to operate it increases only slightly.

In our hospital, the strategic option was made to go to the extreme in consolidating all storage and to simply no longer acquire any computer system that cannot satisfy its needs for permanent storage using the central storage system. The resulting decrease in cost per terabyte made it a relatively easy decision to keep all images online, so exams from a few years ago can be retrieved with the same speed as recent ones. In addition, with the centralized system more efforts and expertise can be spent on options for redundancy than we could previously afford for each individual system.

8.3.3.2
Thinking of Technology as an Infrastructure

Storage consolidation is one example of viewing technical resources in the information system as a commodity, a common and standardized infrastructure much like the power system. While everyone easily accepts that there is only one power supply system, it was already more difficult to impose the idea that there should be a single data network for PACS and all other applications, and such acceptance is not yet obvious for storage (or for sheer computing power).

In this view, PACS becomes largely a software application to be run on the existing infrastructure. In evaluating an offer for a PACS, the abilities and willingness of the vendor to comply with policies in the infrastructure now become an important factor. As the central storage is typically maintained by the hospital's central informatics department, that department must from the start be involved in commercial negotiations and gets the responsibility to organize support.

This may result in a decrease of control over PACS technology by the radiology department. It is up to the radiologists to consider this a threat or instead an opportunity to concentrate on providing better medical services on top of this infrastructure.

8.3.4
Central PACS vs. Departmental or Dedicated Systems

In a hospital setting, the question arises whether all (diagnostic) imaging departments should share a central PACS or, instead, operate dedicated departmental solutions. The same question comes up within the radiology department if that department performs screening as well as diagnostic mammography (Trambert 2006b). In this section, we focus on the latter when possible but many of the arguments also apply at the larger scale. Note that this discussion is in terms of reusing application-level components, the software, and is independent of the discussion in the previous section.

A system tailored to a specific application invariably has advantages for that application which a more general system does not provide. A system designed primarily to support mammography screening may show faster response in image viewing, at the expense of providing less support for other types of images or of imposing a less general architecture, thus making it less suited for general radiology in a large department (see Sect. 8.2.1.3). The advantages of a dedicated system are usually easy to see, and candidate vendors will not fail to stress them. In this section, instead, we provide a few motivations to opt for a more centralized system.

8.3.4.1 Reporting

One obvious advantage of a single system is that the same workstations can be used for the different purposes such as screening and diagnostic mammography, whereas a separate screening workstation may be idle during a considerable period of the day. Even if the different viewing applications could use the same hardware, it remains an advantage if radiologists only have to master one single user interface. Limiting the number of applications increases flexibility. If all workstations are the same, one can perform a task using any of them or ask any colleague for quick advice on a function.

If viewing stations from different vendors are mixed, integration efforts with the RIS (or with components such as speech recognition) must be repeated. Given the high cost of integration, it can be expected that some of these integrations will have less functionality than desired. However, reporting efficiency generally is better served by integration between system components than by the specific functions in any of these components.

8.3.4.2 Image-Management Islands

The reason to consider a dedicated system is probably its attractive workstation. A problem, however, is that dedicated workstations typically require a dedicated image archive as well, thereby creating a complete separate "PACS island." Indeed, workstations of one vendor can usually connect to the image archive of a different vendor (using DICOM services) but only at a price of speed and without dedicated workflow support (see Sect. 8.3.2.5). But these functions were probably the exact reasons to consider dedicated workstations in the first place.

Image-management islands clearly decrease the ease with which images can be distributed and make integration more difficult. For the application of mammography screening that may not seem major considerations. These islands also immediately affect working organization within radiology, however. Imaging modalities deployed for both screening and diagnostic work are unlikely to be able to automatically transmit images to a different destination based on the case. Thus, alternating between screening and diagnostic work requires manual interventions that decrease efficiency and increase the number of errors. Procedures and software for error reconciliation must be duplicated for the different PACS islands, which in itself increases the risk for errors and makes it less likely that knowledgeable operators are available around the clock. Information systems are complex already – all efforts should be directed toward making them more conceptually simple.

8.3.4.3 Embedding into Hospital-Wide Operations

From the hospital-wide perspective, operating a minimum number of (large) systems is not only attractive from economies of scale and integration efforts, but also to ease interdepartmental medical cooperation, as that is best served by common policies and common supporting systems. This may generate tension with the imaging department if that sees its immediate needs better served by a system with dedicated functionality. One underestimated benefit of a central system to the imaging department, however, is that the level of redundancy and support are generally higher than with a departmental system.

A less obvious benefit is the increase of functionality on the long term. Features added to a large system become automatically available to all applications supported by that system. For example, in our hospital, the system used to import screening mammograms over the Internet for second reading (using stringent security) also serves other image transmission applications. As a further example, image CDs that patients bring with them can be imported into the central PACS and the RIS/EMR by any logistic personnel from any computer throughout the hospital, sparing the physicians the hassle of juggling CDs and enabling them to view the images in their usual environment. Implementing this service for different PACS environments was difficult to justify, because of development costs as well as because of the difficulties in training, at a hospital-wide scale, the different procedures for the different environments.

8.3.5 Organizational Limits to Informatics Integration

We have argued that the different aspects of integration required for the digital radiological environment must be subject of an explicit strategy. A considerable part of that strategy is at the departmental level, but other parts extend beyond the own department and can only be realized in cooperation with others. That is easier said than done, however, as the benefits of a common policy to the own department are not always as apparent as are the restrictions. Very probably there even are conflicts of interests, at least in the short term.

Therefore, a fundamental limiting factor in informatics integration is the level of integration at the business management level. A strong central management can give priority to cooperation over medical departments and to functionality that enhances the overall medical process, aspects that are typically better served by integrated informatics, and can give the centralized initiative the essential time needed to prove itself. If, instead, decision power and control over technical personnel is with the individual heads of departments, priority is likely to go to short term local functionality.

If management is very decentralized, the best that is attainable may be a "best of breed" approach, in which individual systems are chosen based on their internal functionality while integration is of secondary concern. The resulting system will have a relatively low level of integration, DICOM and IHE compliance notwithstanding.

Thus, what works and provides benefits in one organization may even simply not be a good idea in another organization. The different options should be considered, though, and the desired level of integration should be decided at strategic management level.

8.4 Telemammography

8.4.1 Situation

Telemammography is an area within the more general domain of teleradiology, which itself is one application of telemedicine. Telemedicine can be defined as the use of telecommunication technology to support a process of medical care delivery, including diagnosis, in which the actors are at a distance. In some applications the patient can be directly involved, but in most applications one healthcare professional asks advice or requests services from a distant one. In radiology, this inevitably requires images to be exchanged. In fact, teleradiology or telemammography are often described as transmitting images over large distances. One of the advantages of digital images is that they can be copied to a distant site without the hassle and the delays involved in physically shipping film.

As radiological image exams, and mammograms in particular, are among the more bulky data types, up to a few years ago long distance transmission used to bump onto technological hurdles. In a growing number of regions, however, sheer network bandwidth is not the bottleneck anymore. In contrast, the relative simplicity of radiological workflow and the highly standardized nature of the data involved (compared to more general application of telemedicine) make teleradiology among the most routinely used applications of telemedicine.

8.4.1.1 Driving Forces

A traditional motivation for teleradiology is to serve sparsely populated areas in which facilities for image acquisition are available but no expert radiologists. The notion of sparsely populated depends on the level of quality one aims for, though: even in regions with many inhabitants subspecialists are not available everywhere, while the demand for ever increasing medical quality fosters increasing subspecialisation. In addition, patients not only expect the best possible quality but also would like to get that close to home. Structural cooperation between general hospitals and tertiary care hospitals can help fulfilling these expectations if these medical teams can work together from a distant much like they can inside one building.

Quality improvements cannot be sustained without keeping costs under control, however. Cutting costs is a second major driving force for teleradiology. Hospital conglomerates deploy PACS and telecommunication technology to concentrate radiologists in a central facility while keeping imaging facilities distributed and close to the patient population. New companies employ economies of scale to offer central

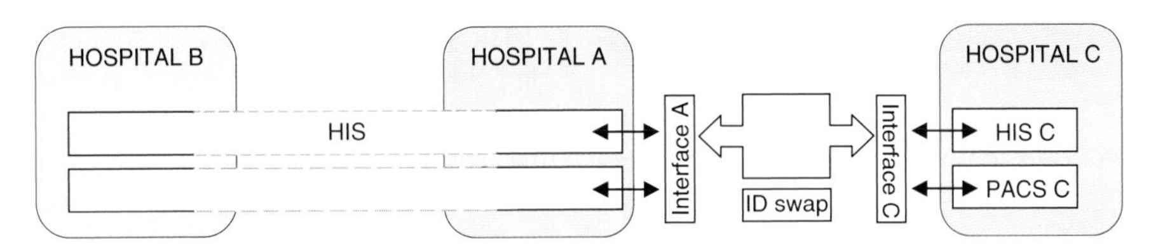

Fig. 8.4. Overcoming geographical distance in an image workflow (*connecting left and middle sites*) is much easier than overcoming organizational distance in which images are exchanged between independent organizations (*middle and right sites*)

radiological reading services to a large number of hospitals, for example, during night shifts. The independence of distance that becomes possible by digital image acquisition can be exploited to the extreme here, as an emergency case can be read by a radiologist located at the other side of the globe in a different time zone.

For mammography, distant reading for pure cost reduction has not taken off yet. The relatively recent introduction of digital mammography acquisition systems combined with large data sizes and the need for specialized workstations surely are factors in this delay.

In contrast, the possibility to bring the highest achievable quality to the whole population is of particular appeal to screening mammography. In breast cancer screening, the purpose is exactly to reach as large as possible a part of the female population (Shaw de Paredes et al. 2007). Often, mammography equipment is brought to the patients using mobile screening units from which images are subsequently transferred to a central reading facility. If screening is organized in a double reading system, there is a structural need to transmit newly acquired and prior images, and thus a clear opportunity to support the image reading workflow using telematics. In Leader et al. (2006), an attempt is done at lowering recalls by enabling radiographers to ask a distant radiologist whether additional exposures should be performed before the patient leaves the imaging center, thus bringing a telematics component even into the core image acquisition workflow.

8.4.1.2
Geographical vs. Organizational Distance

Although so far we stressed geographical distance, this is often the lesser problem. Instead, A distinguishing aspect in most telematics projects, is that cooperation extends outside the own organization. This "organizational distance" typically is more difficult to overcome than geographical distance.

In our situation, we have a number of hospitals connected to a central PACS and RIS while using the same workflow. Physicians will not notice whether an image only needed to travel from the basement or instead was transmitted over a distance of 30 km. However, the radiological workflow of these hospitals is exactly the same and supported by the same information system. Technology makes that these sites seem to be under the same roof, but these sites also work as one single organization.

In contrast, we also cooperate with a hospital just a few kilometres away by sharing an MR imager, but with both hospitals deploying their own PACS and RIS. Instead of sharing one workflow, two separate workflows must be linked. Exams are originally requested in the remote RIS but need to be imported into our RIS as that drives the images acquisition workflow and integrates to the MR imager. The acquired images must now be transferred to the remote PACS. The remote RIS must be explicitly notified about completion of the exam, which was performed outside its control, so it can support the reporting workflow. As both hospitals use different exam IDs and even patient IDs, we must replace our own IDs in the images with the remote IDs before image transmission. All this required adaptations to both systems and dedicated support procedures. Thus, although easy from the perspective of telecommunication technology, the challenges in establishing a combined workflow and providing informatics support for that workflow were considerable (Fig. 8.4).

8.4.1.3
Connections at Many Levels

Supporting a teleradiology workflow clearly requires more than communication technology. Organizational principles and even strategy need to be decided on. In the remaining sections of this chapter, we discuss different levels in connectivity:

- The first section is about transmitting the bytes that make up an image. We go into a few selected aspects of technology involved in communication.
- The second section is about possibilities for organizing image transmission. This is about workflow in the radiological process, but narrowly focused on the images themselves.
- The third section is about integrating images into the overall teleworkflow. We illustrate practical problems you may encounter when linking imaging workflows over different organizations. We use telematics operations in a center for second reading of screening mammograms as an example.
- Whereas much of the previous focused on cooperation with a remote partner, the radiological workflow is not isolated from the global workflow within the own institution. In the last section, we point out that efforts in improving the teleworkflow risks decreasing efficiency of the local workflow.

We do not go into business strategy, issues about trust, legal aspects including liability, financing, etc. Those elements are essential, though. If organizations want to engage into new forms of cooperation and have a clear business case, the technological and organizational challenges are likely to be overcome. However, just introducing new technology in the hope that this will lead from the bottom up to new forms of cooperation, will predictably fail.

8.4.2 Selected Technology Aspects for Telemammography

8.4.2.1 The Internet as a Source of Bandwidth

The term bandwidth refers to the amount of bytes that can be transmitted over a communication channel per unit of time. For a given data set, a high bandwidth channel implies fast transmission speed. Distance is an enemy of bandwidth. Within a building, high speed networks can be provided at reasonable costs. For long distance communication, however, technology costs rise sharply and there are more contenders for the available bandwidth.

The Internet was an important enabling factor for long distance communication. The Internet is not a separate network; it is a network of networks. Most importantly, it established a set of widely accepted communication standards and principles that are independent from the underlying physical network. This enables reuse of networks that originally served totally different purposes such as carrying voice or television signals. Competition between different providers of bandwidth has in many regions resulted in a dramatic increase of available bandwidth for declining prices.

The standardization of communication principles also enables each party in the telematics cooperation to take abstraction from the network details at the other side. As long as our hospital is connected to the Internet (using any technology we choose) we can exchange bytes with any other party that has an Internet connection (using whatever technology they choose). This provides extreme flexibility and enables applications to focus on organizational and functional aspects. Incidentally, this also provides higher resilience against network problems as the bytes usually can find another path through the Internet cloud.

Since a few years, our hospital every day receives tens of screening mammography exams over the Internet for second reading, not only from other hospitals but also from private radiologists. Radiologist in private practices deploy typical residential Internet connections, be it with commercial offerings in which the usual restrictions in upload speed and upload volume are relaxed. Because a standard screening mammography exam with four images takes about 40 Megabytes as a minimum, transmitting one exam using such a residential connection may require nearly 20 min, but that is not a bottleneck for this application.

8.4.2.2 Security

A disadvantage of the Internet is that it does not provide the least inherent security, neither from technological nor from legal side. A first concern is privacy. Sending medical information over the Internet (e.g., using e-mail) without further precautions is equivalent to sending it on a postcard. The answer to this problem is encryption. A weakness of traditional encryption methods is the need to exchange the secret key, at which moment it could be intercepted. Advanced mathematical methods have provided a solution called public key cryptography in which a pair of keys is used instead of a single one. A message encrypted using the public key of an addressee can only be decrypted using that by addressee's an addressee's private key. The sender does not need that private key, however. In fact, that private key never needs to be communicated.

A more difficult problem is that of authentication, making sure that the one at the other side is indeed the person he or she claims to be. Individuals can authenticate themselves using something they know (a password or pin code), something they are (using, e.g., voice recognition, fingerprint or iris scan), or something they have (a "token" such as a device that generates one-time passwords or a smart card). Passwords are absolutely insecure and could easily be stolen. The most secure but still practical form of authentication we currently have at our disposal is using a physical "token," usually combined with a pin code to activate it.

Organizations that want to exchange information often establish a "virtual private network" (VPN). Although the bits travel over the public Internet, encryption methods ensure that no one can read the information in-between and provide proof that the message originated from the other site. A drawback is that such a VPN must be set up beforehand. While this is no problem for connections between sites that cooperate structurally, as is most often the situation with teleradiology, this is not a solution for ad-hoc communication.

The latter limitation also applies to individuals. Technology to establish secure communication channels is widely available, but a token and pin code for authentication must be distributed beforehand. Many countries issue electronic ID cards that can serve to authenticate any citizen (the techniques for this are outside the scope of this book, but be assured that the level of security is high). Thus, it gradually is becoming possible to establish trusted communication over the Internet without sacrificing flexibility.

8.4.2.3
Image Compression

As transmission bandwidth is costly it is useful to represent each image using as little bits as possible. Reversible (or lossless) compression methods look for patterns among the bits of the image that can be encoded more succinctly. For example, if a number of adjacent pixels have the same value, this fact can be encoded using fewer bits than it takes to list those values a number of times. However, in radiological practice the gain is only moderate, about a factor of 2 or 3 (maybe up to 4 for mammograms). A fundamental limitation is that any real world measurement contains noise. Therefore, even if a group of pixels all represent air and thus should contain the same value, image noise will make them differ a bit.

Irreversible (or lossy) compression methods do not object to modifying the pixel values slightly to obtain higher compression ratios. After all, nature uses noise to modify our pixel values randomly anyhow. The question then becomes which compression factor is acceptable. This can only be answered by experiments in which one evaluates whether an observer is able to see the difference between original and modified image at all, or whether there is an impact on diagnosis. Even this is a moving target, as the impact of compression depends on image characteristics such as pixel size or slice thickness, characteristics which change with imaging technology. Despite growing evidence of safety (Seeram 2006), most hospitals so far did not deploy irreversible compression at the heart of the PACS or for primary diagnosis, partly because at very conservative compression factors the cost savings are not dramatic.

Interest in generally established guidelines is increasing, however. One motivating factor is that new imaging techniques generate so much data that not applying lossy compression may limit their applicability due to costs. A second motivating factor is exactly that teleradiology could open up completely new applications.

A recent large Canadian study aimed at proposing conservative compression ratios indicates that for X-ray mammograms compression using a factor of 20 should be safe (Koff et al. 2008; The Royal College of Radiologists 2008). This result is roughly in agreement with a consensus meeting in Germany in which for mammograms a slightly more cautious factor of 15 was proposed (realizing that some security margin was taken) (Loose et al. 2009). some studies suggest that even significantly higher factors may be acceptable (Seeram 2006; Penedo et al. 2005).

8.4.3
Organizing Image Exchange and the IHE XDS Initiative

8.4.3.1
Image Batch Transmission vs. Remote On Line Access

It can make quite a difference in system setup if a remote radiologist needs the image for further internal processing or instead just needs to "look at" the image from a distance. The latter can be accomplished by viewing software that establishes an on line connection to the local PACS. A typical application is providing a physician who is on call at home with access to the hospital's PACS. A simple viewing application can also be offered over the Internet to, e.g., the general practitioner who referred the patient but who may have little interest in managing a copy of the images locally.

The alternative is to explicitly copy the images from the local system to an independent remote image management system. The receiver can then use the original image data freely within the new context. That flexibility can be a prerequisite, for example if the images are needed for surgery planning at the receiver's site, or to enable a remote radiologist to study the case using the image viewing tools he or she is familiar with. There could also be technical reasons to have the images transmitted before they will be viewed, for example to limit bandwidth needs: on line viewing requires high peak bandwidths as the user expects immediate response on any interaction, even as in between these bursts of transmission activity the communication channel is idle.

8.4.3.2
Pushing Images to the Partner

Copying images to a remote image management system is most often done on initiative of the sender. The sender's system acts as an imaging modality that transmits images to a (remote) PACS. Indeed, image transmission is typically performed according to the DICOM standard. The DICOM protocol is independent from the transmission channel - and from whether that channel is through a local network or over the Internet. However, the complete setup is more involved than for a modality transmitting to a local PACS.

A first aspect to consider is *security*. The communication channel itself can easily provide encryption. Authenticating the sender is more difficult, especially as DICOM currently does not provide support for this. Therefore, in our hospital, we only accept DICOM connections over the VPNs we established with our individual communication partners (see Sect. 8.4.2.2). Even then, we do not expose our PACS directly but catch the images on a dedicated server that forwards them to the PACS. Thus, our PACS only needs to accept connections from computers we have control over.

A second aspect to consider is that images must be *imported into the local image management*. The exam ID included within the images refers to the sender's environment and therefore is meaningless to the receiving PACS. The receiving system may even deploy different patient IDs. Even worse, any of such IDs in the images may happen to correspond to a different local patient or exam. Therefore, for the application of screening mammography we use the gateway computer mentioned in the previous paragraph to also map the patient in the transmitted images to the local patient and to include the exam into our RIS. Sometimes a common ID such as a national social security number is known to both sites even if not used internally in any of these sites. This common ID could be included in the DICOM data set to facilitate patient mapping at the receiver. DICOM even provides a field "Other patient IDs" for such purposes.

Even then the transmitted exam is in the local image management system but not yet included into the local workflow. With structural image transmission, the sheer fact that an image exam arrives from a specific source could be used to trigger further actions. If images are sent ad-hoc, for example to ask a second opinion, the sender may need to also send a structured e-mail message specifying the intended use.

8.4.3.3
Pulling Images from any Partner: The IHE XDS Initiative

The organization described in the previous section assumes that the one who generated the images takes initiative for transmission. However, in most clinical situations, one would like the reverse: the actor currently in charge of the patient is looking for prior images, often even unaware of their source. An approach that has gained popularity over recent years is for a number of cooperating institutions to offer to trusted partners an index describing which data they have available on a patient. Authorized partners can subsequently, on their own initiative, fetch some of those data, all within the access restrictions the originating center imposes.

This approach is standardized in the recent IHE "XDS" profile (Cross Enterprise Document Sharing) (http://www.e-health-insider.com/comment_and_analysis/266/the_xds_factor). In the XDS framework the cooperating sites (Fig. 8.5) share a single "document registry," an index with for each patient a structured description of the medical reports or images

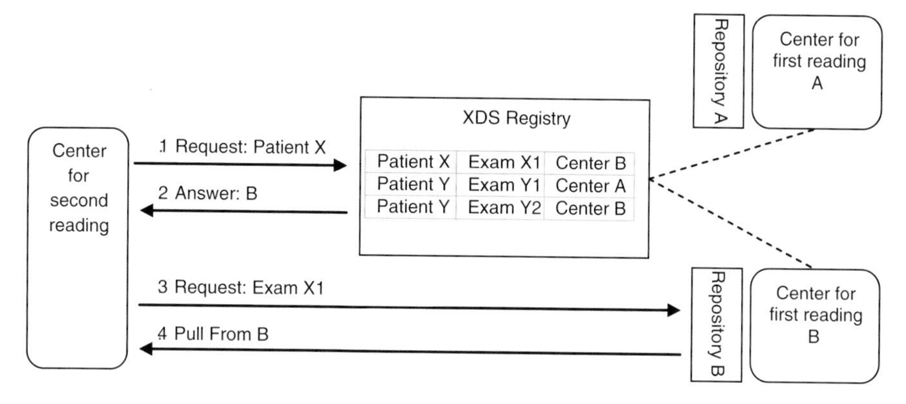

Fig. 8.5. The XDS framework enables a site such as a center for screening (*left*) to query a central registry about availability of exams with other centers and to subsequently request the exam directly from the source (*right*)

each site is willing to share. Distributed among the partners are any number of "document repositories" that hold the actual documents. For diagnostic images XDS incorporates, and improves upon, communication concepts in DICOM.

XDS defines in detail the messages and commands that flow between the various parts of the systems, using established principles and technological standards in the domain of e-business. The framework provides security mechanisms including strong authentication and standardized provisions for logging. XDS is been adapted as the framework for a growing number of large projects in exchanging health care information.

Note that XDS provides an elaborate infrastructure but does not impose a higher-level workflow. For the short term, a viewer that presents an overview of externally available medical results can already bring a lot of benefits; in the long run information systems could more actively request information as part of the workflow.

8.4.4 Integrating Images into a Teleworkflow

8.4.4.1 Centralizing Organization of Screening Using Telecommunication

Structural cooperation in a telematics setting requires more than merely communicating images. The complete workflow is likely to involve steps for reporting and communicating the report, or even for planning the imaging exam. In this and the next subsection we illustrate two radically different approaches to support telemammography for screening.

In a centralized organization, an obvious approach at supporting the workflow is to set up a central information system, including a central PACS. The DigiBOB project being set up in the Netherlands takes this route. Screening images are obtained in a decentralized way, often using mobile mammography units, but they are stored at a central location. Images are not read at that central location but instead in 15 locations distributed over the country. The reading workstations at these centers will be connected to the central image archive, however. Reporting and all other organizational aspects are supported by the central system.

As already mentioned in Sect. 8.4.1.2, communication technology is deployed here to enable a single organization to be distributed geographically. The size of the system does not fundamentally add to the complexity. That about a million images a year must be stored is no argument against centralization but rather an opportunity to benefit from economies of scale. The very high-speed connections between the reading centers and the archive may represent a large investment but enable a conceptually simple design which lowers running costs.

A particular aspect in this setup, to which we come back in the final section of this chapter, is that this organization and the workflow are solely devoted to screening. Diagnostic or therapeutic workflows are in hospitals outside this organization.

8.4.4.2 Organizational Challenges in a Decentralized Screening Organization

In a decentralized organization, supporting the teleworkflow is considerably more complicated. There may be one overall workflow at the conceptual level, but in practice there are independent local workflows that must be connected. Thus, independent information systems must be linked.

We illustrate some of the resulting organizational complexities using our own experiences as a center for second reading of screening mammograms obtained by many independent radiological units (private practices and well as hospitals). Those first readers not only perform screening exams but are active in general radiology as well. For the latter activity, they each have their own local workflow and information systems. Many radiological units already converted to digital acquisition and want to transmit images electronically. We have the images transmitted over the Internet and we try to obtain structured reports by the first readers in electronic form as well.

A first challenge is to cope with the occasional technical problems that result in missing images. Such errors are merely a nuisance if detected at an early stage of the process, but they disrupt the workflow if detected only when a radiologist needs the images. In our local workflow such early checks could be performed automatically by combining information from RIS and PACS. In the teleworkflow—where these checks are more important—there is no overall RIS. The number of images obtained could be included as an information item in the structured report transmitted by the first reader, but many of those send reports on paper. It is indeed more difficult to connect reporting systems than to send images—the difficulty of electronic communication must not be judged by the number of bytes transmitted.

A second challenge in this decentralized organization is the extreme variability in image generating equipment at the first readers. This has a direct effect on the configuration of hanging protocols at the reporting workstations for second reading, as these protocols may depend in subtle ways on the image generating modality and even on settings used by the first reader.

A third challenge is the huge variability in system setup at the sending sites. Images may be directly sent from the modality, or through a local mini PACS, or forwarded by the central PACS of a hospital. Complex setups make it difficult to locate errors. Indeed, at a certain day, it turned out that all images sent by one partner appeared in lower quality on our workstations, although image quality seemed fine at the sending side and even the stringent quality assurance programs did not point to a problem at the source. It turned out that the simple modification of the name of a modality stopped the PACS in applying a lookup table correction to the images before being transmitted. That was not apparent at the source, however, as that PACS was only involved in image routing but not in the local mammography reading workflow.

The previous difficulties can be situated at the technology level. However, even organizational variations between the transmitting sites directly translate into difficulties for the receiver. Efficient reading requires electronic work lists based on structured information entered in previous stages of the process. In our setting the only information from the previous stage is the image data set itself, so we have to rely on free text descriptions of the procedure or the name of the modality to construct reading lists. Reconfigurations at the remote site may make such lists incomplete, resulting in similar disruptions as if there had been a technical problem.

The setup is further complicated by the fact that the different first readers do not use a single patient ID. In our country, a common ID has been introduced for screening, but as the first readers also have other radiological activities besides screening they cannot afford using that ID as the primary index in their local information systems. Thus, the local information systems had to “artificially add” the common screening ID into the images (see also Sect. 8.4.3.2).

8.4.5 Choosing Between Integration into Central or Local Workflow

In the previous section and throughout this chapter, we stressed the difficulties of supporting an overall workflow that is composed out of partial local workflows. Merging these partial workflows into a single common workflow would solve many of the difficulties, but this implies that all who are involved follow the same procedures, adhere to the same policies, and refrain from any attempt at locally optimizing their work independent from the others. In short, this would require the organizations to melt together, which probably was not the intention.

Independent institutions could decide, though, that they should merge just that *part* of their workflows that is directly related to the telematics cooperation. They could thus share a common information system for that part only. However, now each institution deploys separate information systems for the teleworkflow and for the remaining parts of the local workflow. An unanticipated but inevitable result of improved integration with the telematics partner is lower integration into local operations.

This consideration gives a different perspective on our example of supporting a decentralized screening organization (Sect. 8.4.4.2). In contrast to the centralized screening example from Sect. 8.4.4.1, all the participants in this decentralized screening network are also active in general radiology. We could try and implement a common information system for reporting. However, our hospital has explicitly chosen to integrate its local screening and diagnostic workflows and to deploy a common platform for both these applications because of the advantages this integration brings (see Sect. 8.3.4). The private radiologists or other first readers probably would see little advantages either of entering screening exams in a separate reporting system as used for their other radiological work, or using two separate systems for planning, or loosing integration with local systems for billing, management decisions, or quality assurance.

A first consequence is that it makes no sense to compare technical solutions for telecooperation without taking into account the management context in which these solutions are deployed. An ideal solution for an organization that only performs screening may simply not be suitable for an institution that mixes screening and general radiology.

A second consequence is that in any form of telecooperation one should have a clear idea about one's own position. There may not be a clear-cut answer to the question whether one should give precedence to integration into the local setting or instead to integration into the telematics setting but then only for part of the operations. However, one cannot have it all.

References

Bacher K, Smeets P, De Hauwere A et al (2006) Image quality performance of liquid crystal display systems: influence of display resolution, magnification and window settings on contrast-detail detection. Eur J Radiol 58(3):471–479

Bellon E, Van Cleynenbreugel J, Suetens P et al (1994) Multimedia e-mail systems for computer-assisted radiological communication. Med Inform 19(2):139–148

Berns EA, Hendrick RE, Solari M et al (2006) Digital and screen-film mammography: comparison of image acquisition and interpretation times. AJR Am J Roentgenol 187(1):38–41

Bick U, Diekmann F (2007) Digital mammography: what do we and what don't we know? Eur J Radiol 17:1931–1942

Bick U, Diekmann F, Fallenberg EM (2008) [Workflow in digital screening mammography]. Radiologe 48(4):335–344 (article in German)

Ciatto S, Brancato B, Baglioni R et al (2006) A methodology to evaluate differential costs of full field digital as compared to conventional screen film mammography in a clinical setting. Eur J Radiol 57(1):69–75

Haygood TM, Wang J, Atkinson EN et al (2009) Timed efficiency of interpretation of digital and film-screen screening mammograms. AJR Am J Roentgenol 192(1):216–220

IHE (2005) IHE radiology user's handbook. IHE, Chicago

IHE (2007a) IHE radiology: mammography user's handbook. IHE, Chicago

IHE (2007b) IHE radiology technical framework revision 8.0. IHE, Chicago

IHE (2008) IHE radiology technical frame work supplement 2007–08: mammography acquisition workflow. IHE, Chicago

Ishiyama M, Tsunoda-Shimizu H, Kikuchi M et al (2009) Comparison of reading time between screen-film mammography and soft-copied, full-field digital mammography. Breast Cancer 16(1):58–61

Koff D, Bak P, Brownrigg P et al (2008) Pan-Canadian evaluation of irreversible compression ratios ("lossy" compression) for development of national guidelines. J Digit Imaging (Epub ahead of print)

Leader JK, Hakim CM, Ganott MA et al (2006) A multisite telemammography system for remote management of screening mammography: an assessment of technical, operational, and clinical issues. J Digit Imaging 19(3):216–225

Loose R, Braunschweig R, Kotter E et al (2009) [Compression of digital images in radiology - results of a consensus conference] Rofo 181(1):32–37 (article in German)

Moore SM (2003) Using the IHE scheduled work flow integration profile to drive modality efficiency. Radiographics 23(2): 523–529

Oosterwijk H (2005) DICOM: DICOM basics, 3rd edn. OTech. Aubrey, TX

Penedo M, Souto M, Tahoces PG et al (2005) Free-response receiver operating characteristic evaluation of lossy JPEG2000 and object-based set partitioning in hierarchical trees compression of digitized mammograms. Radiology 237:450–457

Pisano ED, Cole EB, Kistner EO et al (2002) Interpretation of digital mammograms: comparison of speed and accuracy of soft-copy versus printed-film display. Radiology 223(2): 483–488

Pisano ED, Zuley M, Baum JK et al (2007) Issues to consider in converting to digital mammography. Radiol Clin North Am 45(5):813–830

Seeram E (2006) Irreversible compression in digital radiology. A literature review. Radiography 12:45–59

Shaw de Paredes E, Lopez FW, Strickland WJ et al (2007) Telemammography: interfacing between primary physicians and experts. Appl Radiol 36(9):26–30

The Royal College of Radiologists (Board of the Faculty of Clinical Radiology) (2008) The adoption of lossy image data compression for the purpose of clinical interpretation. http://www.rcr.ac.uk/docs/radiology/pdf/IT_guidance_LossyApr08.pdf/

Trambert M (2006a) Digital mammography integrated with PACS: real world issues, considerations, workflow solutions, and reading paradigms. Semin Breast Dis 9:75–81

Trambert M (2006b) A perfect match – Integrating digital mammography with RIS/PACS and mammography QA. Image (www.rt-image.com) 19(8)

Digital Mammography Clinical Trials: The North American Experience

9

John M. Lewin

CONTENTS

John M. Lewin, MD
Diversified Radiology of Colorado, PC, 938 Bannock St, Suite 300, Denver, Colorado 80204, USA

KEY POINTS

There have been two major clinical trials of digital mammography in North America. The Colorado-Massachusetts trial was groundbreaking in its design, as the first trial to test the modalities head-to-head, and to consider findings detected by each modality equally. This trial showed a significant decrease in the recall rate for digital and a nonsignificant trend for film in increased cancer detection. As an early trial, it was limited by technical factors that would be improved shortly after the trial. The most apparent of these was the digital workstation used for interpretation. The DMIST trial built on the Colorado-Massachusetts trial and expanded on it with markedly larger numbers of subjects and institutions. It also looked at machines from multiple vendors. This trial found a significant advantage for digital in cancer detection rate and overall performance, as measured by ROC analysis, for young women with dense breasts. For the entire cohort, however, there was no significant difference between the modalities.

9.1 Introduction

The first digital mammography prototype machine, made by Fischer Imaging of Denver, Colorado, USA, was deemed to be ready for testing on human volunteers in trials in 1996. In August of the same year, a group of eight Fischer employees underwent bilateral screening mammography with the new prototype. The images were interpreted by noted breast imager Stephen Feig, MD. Defying all odds, a suspicious abnormality found on one of the studies was shown to be a cancer. More surprisingly, the subject had had her screening mammogram only a few months earlier. These eight employees probably constituted the first clinical trial of digital mammography, and the results were extremely promising. Clearly, the technology was ready for head-to-head comparison with its much older cousin, screen-film mammography.

9.2 The Colorado–Massachusetts Screening Trial

About 10 months after the Fischer test, the first formal clinical trial of digital mammography started, also in Denver. This two-site trial, conducted at the University of Colorado Health Sciences Center and the University of Massachusetts Medical Center, ran from 1997 to 2000. The design was simple, yet not without issues. The goal was to compare digital and screen-film without any bias in favor of either modality. This would allow a two-tailed design so that it could be determined whether digital was better than, equal to, or worse than screen-film. The only way to allow a fair comparison would be to manage the patient clinically based on both the standard screen-film mammogram and the digital mammogram, without favoring either one. Although such treatment seems reasonable today, at the time of the study, the use of this new experimental type of mammography to guide the clinical care was quite controversial. The specific question was whether it was ethical to commit patients to additional imaging and possible biopsy on the basis of an abnormal digital mammogram when the screen-film mammogram was normal. After some consternation, it was determined that the possible benefit of finding additional cancers balanced the possible risk of extra benign ("unnecessary") biopsies. Additionally, any other design would introduce biases for which correction would be impossible. The most problematic bias would be verification bias, in which the findings from one test are more likely to be verified than those from the other. In this case, verification bias would result if findings seen only on digital mammography were not worked up and therefore were not proven to be cancer. Such a bias would lead to screen-film having an artificially high sensitivity relative to digital.

The other issue was that of the increased radiation from exposure of a large number of women to two mammograms. The competing design, adopted at the time by the US Food and Drug Administration (FDA), was to only perform digital on subjects with an abnormal screen-film mammogram. Because most mammographic abnormalities are due to superposition of normal tissue or other benign entities, this design would both artificially raise the relative sensitivity of screen-film and the specificity of digital. For this reason, the FDA's design was rejected for this study. The FDA itself changed their recommended study design to a more balanced one after an early publication by Lewin (1999), based on interim results of the Colorado–Massachusetts trial, which demonstrated the inherent biases in the original FDA design.

The prototype system used in the trial was made by General Electric Medical Systems (Milwaukee, WI). The system used a 18 × 23 cm detector incorporating an amorphous silicon thin-film transistor bonded to a cesium iodide crystal scintillator. Other than the detector and associated electronics, the system was identical to one of the GE's clinical units, the DMR. For this reason, the GE DMR was chosen as the film unit for the trial. The digital and film units had almost identical dimensions, allowing the same patient positioning to be used on both. The only noticeable difference was the thickness of the digital detector on the prototype, compared with the film cassette used on the DMR.

The workstation supplied with the prototype consisted of a Unix-based computer with two 21 in. CRT monitors with 1,000 × 1,500 pixel resolution. While state of the art for the time, these monitors were much less bright and had lower resolution than the 2,000 × 2,500 pixel monitors used clinically today. Additionally, the workstation software was fairly crude, having been written not by software engineers, but by the physicists who had developed the detector and acquisition system.

9.2.1 Results

Interim results were published in 2001 (Lewin et al. 2001) and final results in 2002 (Lewin et al. 2002). Six thousand seven hundred and sixty-eight paired examinations were conducted on 4,521 women (women could re-enroll at the time of their next annual screening) over a 30-month period at the two institutions. Two thousand and forty-eight findings were detected in 1,467 of the examinations on either digital mammography, screen-film mammography, or both examinations. Recall of these findings led to 183 biopsies. Forty-two of the biopsies were positive for cancer.

Nine of these cancers were detected only on the digital mammography interpretation, 15 were detected only on the film interpretation, and 18 were detected on both interpretations. Eight additional cancers were detected clinically within a year of a negative mammogram in the study. These interval cancers were used in the sensitivity calculations. Thus, the sensitivity of film was 33/50 = 66% whereas that of digital was only 24/50 = 48%. Using a McNemar's test, however, the difference in the proportion of cancers detected by the two methods was not statistically significant.

There was a significant difference in recall rate, however, with 15.0% of mammograms being recalled for additional views vs. only 11.9% on digital. Both these rates were unusually high, compared with clinical practice, possibly due to the added scrutiny of a study where every case is double-read, not the standard of care at most US institutions, including the two study institutions. The positive predictive values of the two modalities were identical, reflecting that the trend toward higher sensitivity/true positives for film mammography came at the expense of more recalls/false positives.

Receiver operating characteristic (ROC) curves for the two modalities, based on a continuous likelihood of malignancy scale for each finding from 0 to 100%, were not significantly different ($p = 0.18$). The trend in area under the curve favored film (Fig. 9.1).

A secondary analysis was performed to attempt to determine the reason behind discordant readings between digital and film for each case in which a finding was called on only one modality. For this analysis, two readers, usually the two who originally interpreted the pair of images, reviewed the screen-film and digital mammograms side-by-side and, by consensus, determined a major reason and, optionally, a minor reason, for the difference in interpretation. This analysis was done prior to work-up and, if done, biopsy of the finding. For all findings, the most common reasons were slight difference in positioning, causing overlapping normal tissue to simulate a possible mass on one modality but not the other, and small differences of opinion between the readers (Table 9.1). For the 24 cancers called on only one modality, no single reason was notably more common than another. The only trend was that factors involving the interpretation (as opposed to the appearance of the cancer) played a role in about a third of the cancers called missed on digital but called on film. These factors included differences of opinion, radiologist error, and workstation issues (Table 9.2).

Although no single reason for the relatively poor performance of digital in terms of sensitivity was elicited by the secondary analysis, the authors postulated

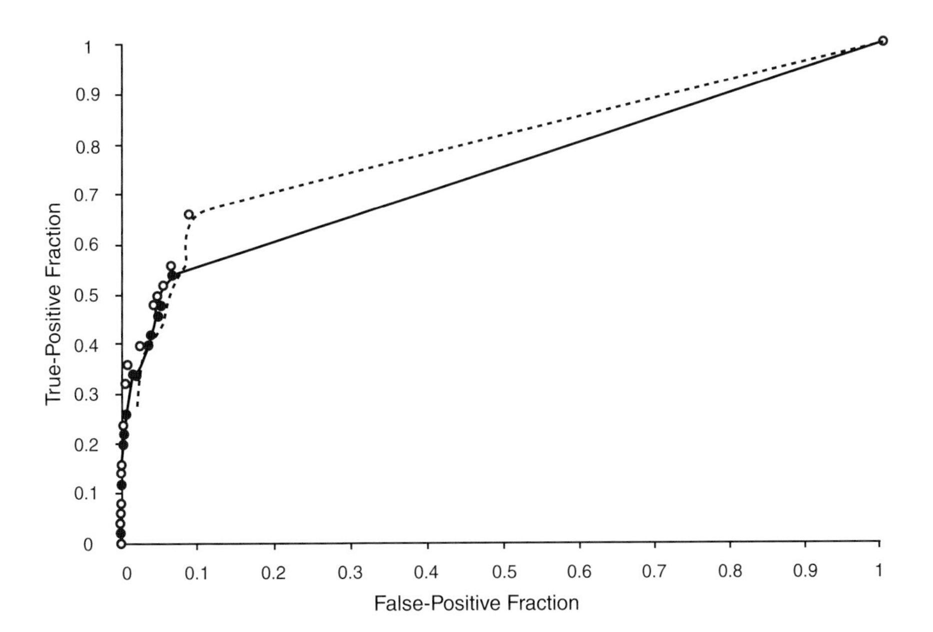

Fig. 9.1. Free-response receiver operating characteristic (ROC) curves for screen-film mammography (*Square boxes, dotted line*) and full-field digital mammography (*closed diamonds, solid line*) based on rating scale of 0–100. Scale for x-axis is probability of false-positive finding occurring on two screen-ing images of given breast and is analogous to false-positive rate in standard ROC experiment. Area under screen-film curve is 0.80; area under digital curve is 0.74. Difference is not statistically significant ($p = 0.18$) (©American Roentgen Ray Society. Reprinted with permission from Lewin et al. 2002)

Table 9.1. Most common major reasons for findings seen only on one modality, either screen-film mammography or digital mammography, in the Colorado–Massachusetts trial (Lewin et al. 2002)

Screen-film-only findings	No.	Digital-only findings	No.
Fortuitous positioning	309	Fortuitous positioning	238
Minor difference of opinion	277	Minor difference of opinion	138
Visibility or contrast difference (unsure of cause)	59	Mass margin or shape more suspicious on digital	62
Technique difference	49	Visibility or contrast difference (unsure of cause)	42
Mass margin or shape more suspicious on SFM	40	Technique difference	35
More calcifications visible on SFM	38	Sharpness greater on digital	34
Error in detection on FFDM	32	Error in detection on film	19
Positioning	31	More calcifications visible on digital	16
Sharpness greater on SFM	28	Error in interpretation on digital	14
Major difference of opinion	20	Positioning	10

A minor reason was also optionally given (not shown)

Table 9.2. Major reasons given for the difference in interpretation for the 24 cancers detected only on one modality, either screen-film mammography or digital mammography, in the Colorado–Massachusetts trial (Lewin et al. 2002)

Detected only on screen-film mammography	No.	Detected only on digital mammography	No.
Minor difference of opinion	4	Fortuitous positioning	2
Visibility or contrast difference (unsure of cause)	2	Sharpness greater on FFDM	1
Fortuitous positioning	2	Ability to magnify on digital workstation	1
Mass margin or shape more suspicious on SFM	2	More calcifications visible on FFDM	1
Sharpness greater on SFM	1	Mass margin or shape more suspicious on FFDM	1
Compression difference	1	Error in interpretation, digital[a]	1
Calcification forms more suspicious on SFM	1	Major difference of opinion	1
Film assessed as more changed from comparison[b]	1	Minor difference of opinion	1
Workstation suboptimal	1		

[a]Evaluation before obtaining diagnosis of cancer
[b]Observer interpreting SFM judged the finding to be more different from the comparison study that did the observer interpreting the FFDM

that the suboptimal workstation, especially the user-unfriendly software and the relatively dim monitors, was a major hindrance in interpreting the digital images. Also cited was the lack of an automated exposure mode, as is uniformly available on screen-film units, on the digital prototype. Such a mode was added to the system prior to its clinical introduction after FDA approval. The relative inexperience of the radiologists with digital, when compared with their years of film mammography experience, was felt to have also contributed to the less than expected performance of digital mammography.

9.3 ACRIN: DMIST

After release of the initial results of the Colorado–Massachusetts trial, it became clear that a larger trial would be needed to detect the relative differences between screen-film and digital mammography in the presence of the inherent reader variability. At this time, the American College of Radiology (ACR) was forming the American College of Radiology Imaging Network (ACRIN) as a cooperative clinical trial group designed to evaluate diagnostic imaging. The model was based on acronym-friendly cooperative oncology groups such as RTOG (Radiation Therapy Oncology Group) and SWOG (Southwest Oncology Group). A previous iteration of a cooperative group in diagnostic radiology, RDOG (Radiology Diagnostic Oncology Group), did one study on image-guided breast biopsy with limited success, but did not survive to do a second trial. It was proposed that ACRIN should sponsor a large trial to compare screen-film and digital mammography. A group of researchers met and considered possible design options, including both a paired design, modeled after the Colorado–Massachusetts trial and a randomized design, in which each subject would undergo either digital or screen-film, but not both. Because of the increased statistical power of a paired design, in which each subject is her own control, it was decided to model the trial after the Colorado–Massachusetts trial, with the exception that it would incorporate digital units from multiple manufacturers.

A calculation of statistical power determined that 50,000 subjects would need to be enrolled to have a chance of detecting the small difference predicted by the Colorado–Massachusetts trial. Even then, a secondary reader study with multiple readers and a cancer-enriched subset of cases would most likely be needed to measure a difference with statistical significance. The plan was to enlist 20 sites to enroll 2,500 subjects each, equally distributed among four makers of digital mammography units: GE, Lorad/Trex, Fischer Imaging, and Fuji. At the time of study initiation, only the GE and Fischer units were approved by the U.S. Food and Drug Administration (FDA) for clinical use in the U.S.

The four machines represented four distinct technologies. The GE machine, the commercial version of the prototype used in the Colorado–Massachusetts trial, used an 18 × 23 cm area ("flatpanel") detector consisting of a cesium-iodide (CsI) phosphor bonded to an amorphous silicon substrate containing a rectangular photodiode array. The CsI phosphor converts the incident x-rays to light, which are then detected by the photodiode array. A thin-film transistor array deep to the amorphous silicon layer processes and transmits the electronic signal to the external electronics. The pitch (size) of a detector element in this system was 100 μm.

The Fischer device was a slot-scanning system in which a 14-mm wide fan-beam scanned across the breast over about 5 s. The image was collected by a rectangular detector consisting of a CsI crystal coupled via a fiber-optic bundle to a linear array of charge-coupled device (CCD) chips. The detector moved with the beam, collecting the signal using a technique unique to CCD arrays called time-delay integration (TDI). TDI mode allows a moving CCD chip to increase its signal by continuously collecting light across different portions of the chip as it moves. This system operated at a 50 μm detector pitch. An advantage to the slot-scanning design was inherent scatter-rejection, eliminating the need for an anti-scatter grid, thus decreasing radiation dose to the patient.

The Lorad/Trex device, which was never commercialized, consisted of a matrix of CCD arrays in two layers bonded to fiber-optic bundles, which in turn were bonded to a CsI crystal. Because of the space needed for the electronics, the CCD chips could not be placed in a contiguous fashion. The resulting image was created by electronically stitching together the images from the multiple CCD chips into a single seemless image. Midway through the accrual phase of the trial, Hologic bought Lorad and the multiple CCD technology was abandoned in favor of a flatpanel design based on an amorphous selenium substrate. This amorphous selenium unit, the same design as in the current commercial Hologic system, was used to acquire most of the Lorad images in the trial. The detector was larger than the GE detector, measuring 24 × 29 cm. The detector pitch in this system was 70 μm. Unlike the other detectors in the trial, the amorphous selenium detector does not require a phosphor to convert the x-rays to light prior to detection; it directly detects the incident x-rays.

The Fuji device used computer radiography (CR) technology rather than direct digital capture to collect its images. The technology was the same as used in Fuji's commercial CR product for chest and other general radiography applications, with the modifications of double-sided phosphors in the CR plates and a smaller laser spot size, 50 μm, in the reader. This spot size determines the size/pitch of the detector elements. This device, or one similar to it, had already been in clinical use in Asia for many years, but had never been compared with film mammography in a controlled fashion and was years away from FDA approval. No doubt Fuji felt, as likely did Hologic, that the ACRIN data would be useful in its FDA submission.

While the GE images, being smaller that the others, could be interpreted in softcopy on a dedicated workstation, the much larger images from the other units could not be readily displayed and manipulated using the computer technology available at the time. The decision was made to let each manufacturer determine in what mode, on a workstation or on printed film, their images would be viewed. Based on the manufacturers' preferences (and their ability or inability to provide a user-friendly workstation), it was decided that all GE studies would be interpreted on softcopy whereas all Fuji studies would be viewed on printed film. Fischer studies would be printed out, but a dedicated workstation would be available to use for additional viewing. The Lorad/Hologic images would be interpreted either on printed film or in softcopy, depending on the preference of the site.

When accrual started, there were about 20 sites selected for participation, but it became clear that the rate of accrual at most sites was not going to allow ascertainment of the accrual goal. For this reason, other sites were allowed to enter. Most of these sites were GE sites, which had purchased GE units for clinical use (there was no money for purchasing units in the trial). In the end, 33 institutions were involved, with two institutions running parallel efforts on two different machines, for a total of 35 sites. Of these, 18 were GE sites, 7 were Fischer sites, 6 were Fuji sites, and 4 were Lorad sites. During the course of the study, one site switched from GE to Lorad and another from Lorad to GE (Pisano et al. 2005b).

9.3.1 Protocol and Data Collection

As in the Colorado–Massachusetts trial, each subject was imaged twice, once on screen-film mammography and once on digital mammography. The order of the acquisitions was randomized so that half the time the screen-film exam was obtained first and half the time the digital exam was obtained first. In general, both exams were performed by the same technologist on the same day. Technical factors were recorded.

Each exam was interpreted by a single radiologist, with old films available, but blinded to the images and results from the other modality. The readers were randomized so that each reader interpreted an approximately equal number of screen-film and digital exams. For each exam, the reader gave three results, a BI-RADS® assessment, a descriptive assessment of likelihood of malignancy on a 7-point scale, and a probability of malignancy on a continuous scale from 0 to 100. The BI-RADS® assessment would guide clinical workup and could be used to calculate a sensitivity and specificity. The latter two ratings would be used for ROC analysis. Data were collected on paper forms and entered into the study database using custom web forms on the ACRIN website.

9.3.2 Results

Accrual lasted 25.5 months. 49,528 subjects were enrolled at 35 sites at 33 institutions (see above). The total number of cases by machine was not reported, but adding the numbers reported to have been accrued at each site (Pisano et al. 2005b), reproduced in Table 9.3, shows that there were approximately 23,000 GE cases, 11,000 Fischer cases, 11,000 Fuji cases, and 4,000 Lorad cases.

The results of the primary study were published in the New England Journal of Medicine in October 2005 (Pisano et al. 2005a). After exclusions, 42,760 cases were available for analysis. The primary reason for exclusion was a lack of follow-up data, but 1,528 women were excluded due to protocol violations, including 1,489 women from one (unnamed) site.

Two hundred and fifty-four cancers were detected either by at least one modality at the time of the index examination or clinically within 365 days of follow-up. An additional 81 were diagnosed over the next 90 days. Most of these were diagnosed based on their next screening mammogram.

For the entire cohort, area under the ROC curve (AUC) using the 7-point scale was 0.78 for digital and 0.74 for film. This difference was not statistically significant. Sensitivity, based on the 5-point BI-RADS® scale, was 0.70 for digital mammography and 0.66 for screen-film, also not a statistically significant difference. Figure 9.2a shows the ROC curves for the entire cohort. Specificity was the same for each modality at 0.92; positive predictive value for each was 0.05.

Although no significant difference was observed in any of the primary measures (sensitivity, specificity, and AUC) for the cohort as a whole, digital was significantly superior in sensitivity and AUC in the subgroups of women under 50, pre-menopausal women, and women with dense breasts (Fig. 9.2b–d). Of note, the complementary groups (i.e., women over 50, post-menopausal women, and women with fatty breasts) did not show a significant benefit for film, although there was a trend in that direction. The overall trend was in favor of digital, however.

With an overall negative result, the results of the subgroup analyses became the big news. Digital was declared superior for younger women (and no worse overall) and sales of digital units soared. With no indication as to which machine(s) contributed most to the positive subgroup results (something still not obtainable from published data), all manufacturers benefitted.

Table 9.3. Enrollment by institution in the DMIST trial. The far right column list the manufacturer of the digital equipment used at that site. Note that two institutions, Mount Sinai Hospital and Elizabeth Wende Breast Clinic, ran the trial with two different machines, essentially constituting two separate sites each. Two sites switched machines during the course of the accrual phase (©Radiological Society of North America. Reprinted with permission from Pisano et al. 2005b)

Institution	Date opened	Enrollment		Digital manufacture
		No. enrolled	Percentage of total	
Allegheny General Hospital (Pittsburgh, Pa)	1 April 2003	526	1.1	GE Medical systems
Beth Israel Deaconess Medical Center	12 December 2001	1,712	3.5	Fischer Imaging
Elizabeth Wende Breast Clinic	March 17 2003	2,315	4.7	Fischer Imaging
Elizabeth Wende Breast Clinic	July 24 2003	667	1.3	Lorad/Hologic
Emory University Hospital	April 2 2002	1,719	3.5	GE Medical Systems
The Johns Hopkins Hospital (Baltimore, Md)	December 27 2001	288	0.6	Lorad/GE Medical Systems
LaGrange Memorial Hospital (III)	January 16 2002	1,677	3.4	Fischer Imaging
Lahey Clinic (Burlington, Mass)	March 3 2003	408	0.8	Fuij Medical Systems
Mallinckrodt Institute of Radiology at Washington University (St Louis, Mo)	October 1 2002	3,347	6.8	GE Medical systems
Massachusetts General Hospital (Boston, Mass)	January 14 2003	1,385	2.8	GE Medical Systems
Memorial Sloan-Kettering Cancer Center (New york, NY)	October 30 2001	2,548	5.1	Fischer Imaging
Moffitt Cancer Center and Research Institute of the University of South Florida (Tampa, Fla)	March 11 2003	197	0.4	GE Medical systems
Monmouth Medical Center (Long Branch, NJ)	December 19 2002	742	1.5	GE Medical systems-Lorad
The Mount Sinai Hospital	March 12 2002	1,146	2.3	Fuij Medical Systems
The Mount Sinai Hospital	May 20 2002	364	0.7	GE Medical systems
New York Presbyterian Hospital-Columbia Presbyterian Campus (New York, NY)	April 23 2002	810	1.6	Lorad/Hologic
Northwestern Memorial Hospital (Chicago, III)	October 29 2001	3,168	6.4	GE Medical Systems
Rhode Island Hospital of Brown University (Providence, RI)	February 4 2003	642	1.3	GE Medical Systems
Shore Memorial Hospital (Somers Point, NJ)	October 15 2002	828	1.7	GE Medical Systems
Thomas Jefferson University Hospital (Philadelphia, Pa)	April 16 2002	816	1.6	Lorad/Hologic
University of California-Davis Medical Center (Sacramento, Calif)	November 6 2001	2,720	5.5	Fuij Medical Systems
University of California Los Angeles Medical Center	January 10 2002	2,174	4.4	Fuij Medical Systems
University of Cincinnati (Ohio)	November 13 2002	993	2.0	GE Medical Systems
University of Colorado Health Sciences Center (Denver)	December 5 2001	1,371	2.8	GE Medical Systems
University of lowa college of Medicine (lowa City)	September 13 2002	1,548	3.1	Lorad/Hologic
University of North Carolina Ambulatory Care Center (Durham)	October 29 2001	2,180	4.4	Fuij Medical Systems
University of North Carolinal Clinical Cancer Center (Chapel Hill)	February 1 2002	488	1.0	Fischer Imaging
University of Pennsylvania (Philadelphia)	January 16 2002	1,461	2.9	GE Medical Systems
University of Texas Southwestern Medical Center at Dallas	November 4 2002	1,757	3.5	GE Medical Systems
University of Toronto, Sunnybrook and Women's College Health Sciences Center-Women's College	January 7 2003	674	1.4	Fischer Imaging
University of Toronto, Sunnybrook and Women's College Health Sciences Center-Sunnybrook Campus	October 29 2001	2,386	4.8	GE Medical Systems
University of Virginia Health System (Charlottesville)	July 23 2002	1,385	2.8	GE Medical Systems
University of Washington Medical Center-Roosevelt Clinic (Seattle)	November 9 2001	2,424	4.9	Fuij Medical Systems
Washington Radiology Associated (Washington, DC)	November 12 2001	1,815	3.7	Fischer Imaging
William Beaumont Hospital (Royal Oak, Mich)	December 17 2002	847	1.7	GE Medical Systems

○ Empirical operating point for film mammography
△ Empirical operating point for digital mammography

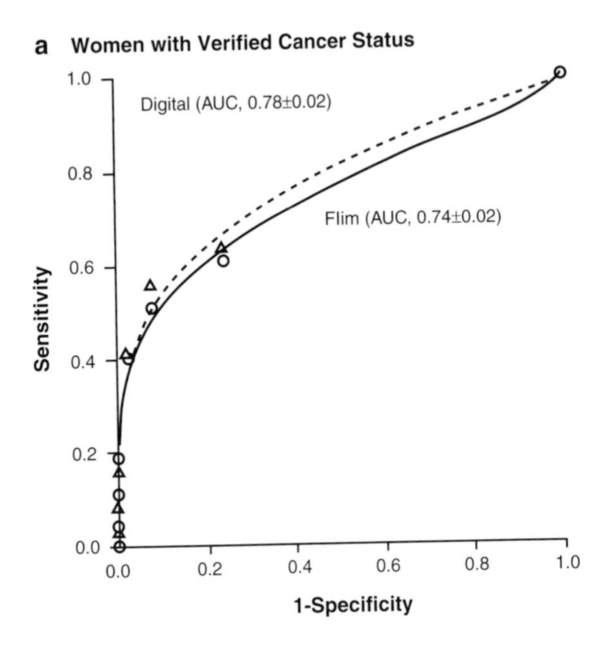

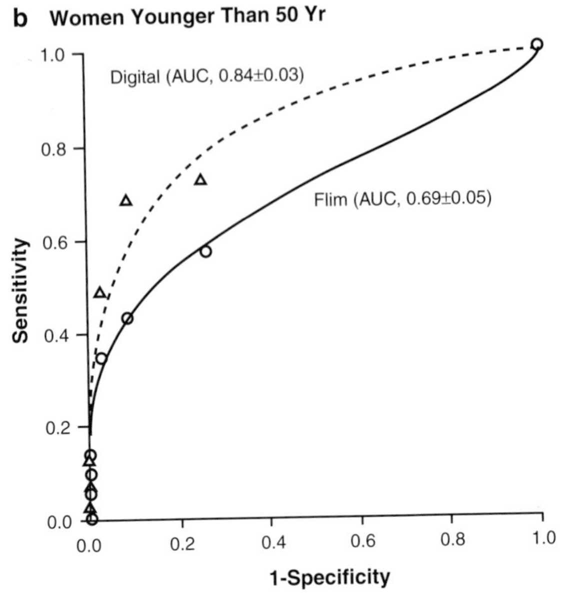

c Women with Heterogeneously Dense or Extremely Dense Breasts

Digital (AUC, 0.78±0.02)
Flim (AUC, 0.68±0.03)
Sensitivity
1-Specificity
1.0
0.8
0.6
0.4
0.2
0.0
0.0 0.2 0.4 0.6 0.8 1.0

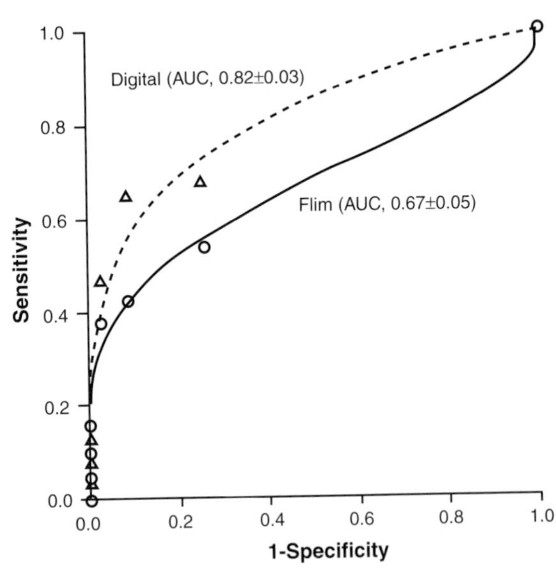

Fig. 9.2. Receiver operating characteristic (ROC) curves for digital mammography (*dashed line*) and screen-film mammography (*solid line*). (**a**) Entire cohort, (**b**) women <50 year, (**c**) women with heterogeneously dense or extremely dense breasts, (**d**) pre- or perimenopausal women. For the entire cohort, the curves are not significantly different, but for each subgroup, the area under the curve (AUC) for digital is larger than that for screen-film, indicating increased accuracy of the test (©Massachusetts Medical Society. Reprinted with permission from Pisano et al. 2005a)

Table 9.4. Relative performance of screen-film mammography and digital mammography in a reader study with an enriched case set taken from the DMIST trial

Digital manufacturer	AUC			Sensitivity		
	Screen-film	Digital	p- value	Screen-film	Digital	p-value
GE	0.82	0.78	0.16	0.53	0.50	0.56
Fischer	0.76	0.73	0.59	0.59	0.56	0.62
Fuji	0.78	0.73	0.09	0.53	0.51	0.61

Separate case sets and groups of readers were used for each machine type. A study was also performed using the Lorad/Hologic machine, but the results were not reported due to an insufficient number of cases. There was a nonsignificant trend toward superiority of screen-film in area under the ROC curve (AUC), sensitivity, and specificity (not shown) for all three manufacturers reported (Hendrick et al. 2008)

In February 2008, a more detailed analysis of the various subgroups was published (Pisano et al. 2008). The same data were used; the difference seemed to be only in the rigor of the statistical analysis. Using a more rigorous level of significance, it was determined that only the group of women under 50 who were also premenopausal and had dense breasts showed superior performance for digital over film. In this group, digital had statistically significant increases in AUC (0.79 vs. 0.54), sensitivity (0.59 vs. 0.27), and PPV (0.033 vs. 0.015).

In women over 65 with nondense breasts, film had higher AUC than digital (0.88 vs. 0.71). Because of the statistical test used, however, the p value of 0.0025 for the difference did not give statistical significance.

The NEJM paper did not disclose the various digital units' performances. A follow-up paper used an enriched test set made up of cancers detected during the accrual period (i.e., excluding those detected at follow-up) mixed with other random cases to perform a reader study (Hendrick et al. 2008). For this study, the readers and cases were separated by machine. Because of the relatively late start for Hologic, there were not enough cancers detected in those subjects to be included in reported results. Thus, the study compared GE, Fischer, and Fuji. Each test set was composed of cancer and benign cases in approximately a 2:3 ratio. The target was to have 50 cancers and 75 controls in each set. Because of the limited number of cases available at the time, the study needed to be performed (owing to funding limitations), the actual numbers were lower than the target. The GE set consisted of 48 cancers and 72 controls, the Fischer set 42 cancers and 73 controls, and the Fuji set 27 cancers and 71 controls. Each reader interpreted half the cases in digital and half in screen-film while blinded to the other. That reader then read each case in the other modality after a 6-week period designed to decrease the reader's recall. The order of reading was randomized for each reader so that on average half the readers interpreted a given case in digital first and the other half in screen-film first. Twelve readers were used for each GE and Fuji case, but only 6 were used for the Fischer cases. Reader experience in standard mammography ranged from 1.5 to 33 years and in digital mammography from 0 to 8 years.

No difference was detected between the film and any of the three digital manufacturers in any of the calculated parameters of AUC, sensitivity, and specificity. For each manufacturer and each parameter, however, the trend was toward the superiority of film (Table 9.4). The authors do not report an analysis of all the cases pooled together. They also state that the results are not intended to compare the manufacturers to each other (raising the question of why they bothered to separate the results by manufacturer). From the data in Table 9.4, the relative performance compared to film is similar across manufacturers. The authors did look at the subset of cases in the groups that showed a benefit for digital in the main trial (<50, premenopausal, dense breasts) and found no difference. The results are not reported, so it is unknown whether the trend in these subgroups favored screen-film or digital. Of note was the large variability in performance among the readers on both screen-film and digital. For example, calculated screen-film sensitivity in the Fischer group spanned from about 0.40 to 0.85 and in the GE group from 0.2

to 0.7. Variability was slightly less for digital sensitivity in each group, but not significantly.

9.4 Conclusion

As with trials performed in Europe, the North American digital mammography trials gave mixed results. The Colorado/Massachusetts trial, like the Oslo I trial conducted a few years later, seemed to slightly favor film over digital. The ACRIN trial showed a definite advantage for digital in younger women with dense breasts, but this was counteracted by relatively poor performance in older women with fatty breasts. A reader study conducted under controlled conditions did not replicate, even qualitatively, any of the results of the primary study, leaving some (Kopans 2008) to wonder whether the benefit of digital, widely circulated in the popular press after being reported in the New England Journal, was just a fluke. A few years later, it seems like a moot point. Digital mammography has become the de facto standard in the U.S., at least in large cities, and is the sole focus of mammography research both by companies and universities. Meanwhile, radiologists, referring physicians, and patients are already looking forward to the next big technological advance.

References

Hendrick RE, Cole EB, Pisano ED, et al (2008) Accuracy of soft-copy digital mammography versus that of screen-film mammography according to digital manufacturer: ACRIN DMIST retrospective multireader study. Radiology 247:38–48

Kopans DB (2008) DMIST results: technologic or observer variability? Radiology 248:703

Lewin JM (1999) Full-field digital mammography. A candid assessment. Diagn Imaging (San Franc) 21:40–45

Lewin JM, Hendrick RE, D'Orsi CJ, et al (2001) Comparison of full-field digital mammography to screen-film mammography for cancer detection: results of 4945 paired examinations. Radiology 218:873–880

Lewin JM, D'Orsi CJ, Hendrick RE, et al (2002) Clinical comparison of full-field digital mammography to screen-film mammography for breast cancer detection. AJR Am J Roentgenol 179:671–677

Pisano ED, Gatsonis C, Hendrick E, et al; Digital Mammographic Imaging Screening Trial (DMIST) Investigators Group (2005a) Diagnostic performance of digital versus film mammography for breast-cancer screening. N Engl J Med 353:1773–1783

Pisano ED, Gatsonis CA, Yaffe MJ, et al (2005b) American College of Radiology Imaging Network digital mammographic imaging screening trial: objectives and methodology. Radiology 236:404–412

Pisano ED, Hendrick RE, Yaffe MJ, et al; DMIST Investigators Group (2008) Diagnostic accuracy of digital versus film mammography: exploratory analysis of selected population subgroups in DMIST. Radiology 246:376–383

Digital Mammography in European Population-Based Screening Programs

10

PER SKAANE

CONTENTS

KEY POINTS

Full-field digital mammography (FFDM) offers several benefits when compared with screen-film mammography (SFM) in breast cancer screening, such as: Elimination of technical failure recalls; simplified archival, retrieval, and transmission of images; reduction of average glandular dose; higher patient work-flow; improved diagnostic accuracy, especially in women with dense breast parenchyma due to higher contrast resolution; implementation of advanced technologies including computer-aided detection (CAD) and tomosynthesis; and the potential for telemammography and teleconsultation. Several European studies comparing FFDM and SFM in population-based breast cancer screening programs have demonstrated a higher cancer detection rate at FFDM approaching borderline significance in some studies, and showing statistically significant higher detection rate in women presenting with Ductal carcinoma in-situ (DCIS) or clustered microcalcifications. The higher cancer detection rate is, however, achieved at the cost of a higher recall rate. Overall, there has been no significant difference in the positive predictive value between FFDM and SFM. The huge challenge of interobserver variability for interpretation in mammography screening may have prevented the advantages from being observed in several studies, and will also, in the future, be a challenge for trials comparing the two imaging techniques.

PER SKAANE, Dr. MED
Department of Radiology, Ullevaal University Hospital, Kirkeveien 166, NO – 0407 Oslo, Norway

10.1 Introduction

Promising results from pioneer studies in the US on periodic mammographic examinations for early detection of breast cancer (Gershon-Cohen et al. 1961; Gold et al. 1990; Shapiro et al. 1971) initiated the interest in Europe for mammography screening. The earliest case-control study started in the Florence district, Italy, in 1970 (Palli et al. 1986), and was followed by two Dutch case-control studies, the DOM project, in 1974 in Utrecht (Collette et al. 1984) and the Nijmegen project in 1975 (Verbeek et al. 1984), and the Edinburgh trial in 1978 (IARC Handbooks 2002; Roberts et al. 1990). The program "Europe Against Cancer" was launched in 1986 and aimed to introduce systematic screening for breast cancer for women 50–69 years of age (Moral Aldaz et al. 1994; De Waard 1994).

The Swedish randomized controlled trials in the late 1970s were the most important for the European concept of organized mammography screening. A pilot project (the Sandviken study) conducted by B. Lundgren in 1974 showed that single view (MLO) mammography has the potential to be a simple and cost-efficient approach for breast cancer screening (Lundgren and Jakobsson 1976). The four Swedish randomized control trials in the late 1970s were designed to evaluate the value of mammography alone (Andersson et al. 1988; Bjurstam et al. 2003; Frisell et al. 1991; Nystrom et al. 2002; Tabar et al. 1985), whereas clinical breast examination was included in the three Anglo-American trials (the HIP trial, the Canadian National Breast Screening Study, and the Edinburgh trial). It is important to keep in mind that the results from the Swedish trials may underestimate the benefit to women undergoing service screening with modern technology for several reasons, such as: Quality below digital technology, use of one view in some and longer intervals in other trials, noncompliance in the intervention group, and contamination in the control group. The benefit in mortality reduction for women actually attending mammography screening has been estimated to be 35–40% or even higher (Duffy et al. 2002; Gabe et al. 2007; IARC Handbooks 2002; Olsen et al. 2005).

Today, the "European guidelines for quality assurance in breast cancer screening and diagnosis" (Perry et al. 2006) serves as guidelines for the member states of the European Union when introducing population-based service screening programs. A new chapter on digital mammography is included in the fourth edition, describing the system demands on digital mammography, regarding image quality evaluation, image receptor, image acquisition, image presentation, and quality control of digital systems. "European guidelines" recommend two-view mammography every 2 years for women aged 50–69 years, but the target population varies from 40 to 74 years of age in the European countries. Women are invited by a personal letter scheduling place and time for the mammographic examination. Multidisciplinary team-work including pre- and postoperative conferences is considered to be an important aspect of screening programs. Ongoing quality assurance is essential, and the performance indicators guide the programs to reach the goal of mortality reduction of about 30%. To keep the adverse effect of false positive findings low, recall rates should be below 5% in the prevalent and below 3% in the subsequent screening rounds (Perry et al. 2006).

Double reading is an important aspect when comparing the results from European screening programs with those in the US using single reading. Most European organized screening programs practice batch reading. Double reading can be independent or nonindependent (second reader being aware of the first reader's opinion, "second reader bias"), there could be unilateral recall (all cases with a positive score by either reader are called-back for assessment), or agreement on decision for recall, either consensus (the two readers reach a consensus) or arbitration (a third reader makes the decision on recall). Consensus or arbitration meetings may include discordant as well as concordant interpretations (Fig. 10.1).

10.2 European Studies Comparing Screen-Film and Digital Mammography in Breast Cancer Screening

A total of 11 European studies comparing screen-film mammography (SFM) and full-field digital mammography (FFDM) in mammography screening have been reported. Five studies have been carried out in Norway (Oslo I study, Oslo II study, Tromsø study, Vestfold County study, and Sogn and Fjordane study), one study in Sweden (Helsingborg study), one study in Italy (Florence study), one in the Netherlands (DSPP study), one in the UK (CELBSS study), one in Ireland (INBSP study), and one in Spain (Barcelona study). The two Oslo studies were prospective trials: Oslo I study had a paired study design and the Oslo II

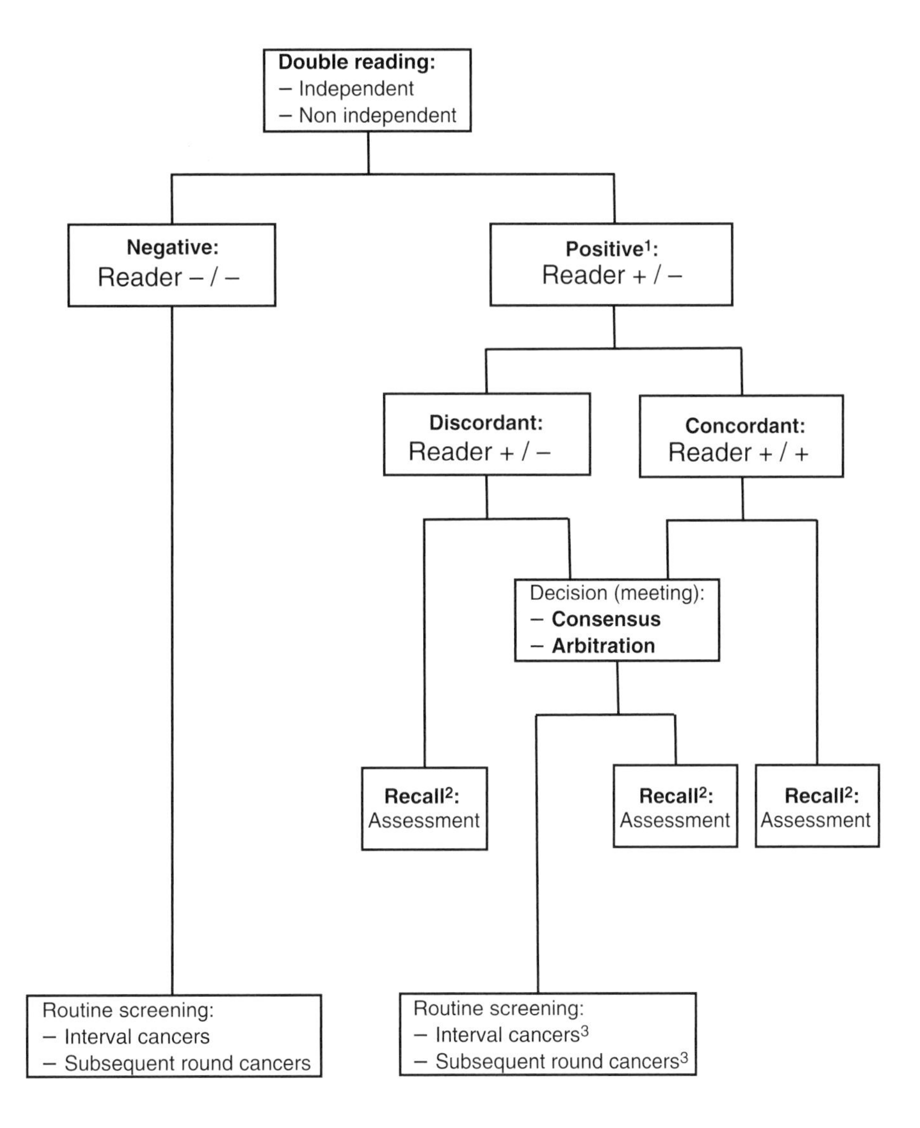

Fig. 10.1. Flow chart shows decision tree for double reading in mammography screening. True positive interpretations may include all positive scores for cancers at baseline reading (*1*), cancers diagnosed at assessment only (*2*), or also including cancers with true positive scores dismissed at consensus (arbitration) meeting and presenting as interval or subsequent round cancers (*3*)

study was a prospective randomized trial. The other nine European studies were retrospective investigations with concurrent cohorts design or allocation by time, area, or a random allocation. The 11 studies are presented chronologically according to publication year and then year of abstract (Table 10.1).

10.2.1 The Oslo I Study

The Oslo I study and the US Colorado–Massachusetts (Co–Ma) study (Lewin et al. 2001, 2002) were the pioneer studies comparing SFM and FFDM in mammography screening. The prospective Oslo I trial was carried out from January to June, 2000 (Skaane et al. 2003). A follow-up analysis of cancers missed at FFDM and subsequent cancers with a true positive score at baseline reading was published 2 years later (Skaane et al. 2005).

Women aged 50–69 years, invited to the Norwegian Breast Cancer Screening Program (NBCSP), were asked when attending the screening unit if they wanted to participate in the study. There were no exclusion criteria. Similar to the two prospective US studies, the Co–Ma study and the DMIST trial (Pisano et al. 2005), the prospective Oslo I trial had a "paired study design", i.e., the women underwent SFM as well as FFDM. Double examination in a population-based service screening program means a heavy burden for the radiographers, and the daily capacity of the radiographers was the limiting factor for including women in the study.

FFDM examinations were carried out with a production-type Senographe 2000D (GE Healthcare, Buc, France). The workstation included two high-resolution

Table 10.1. Studies comparing screen-film mammography (SFM) and full-field digital mammography (FFDM) in breast cancer screening: study period, year presented, study design, age group of study population, and method of reading (interpretation)

Study	Study period (publ./present)	Study design	Age (years)	Reading
European studies:				
(A) Published				
Oslo I	Jan 2000–Jun 2000 (2003)	Prospective; paired	50–69	Double
Oslo II	Nov 2000–Dec 2001 (2004)	Prospective; randomized	45–69	Double
Helsingborg[a]	Jan 2000–Feb 2005 (2007)	Retrospective; allocation by time	46–74	Single and double
Florence	Jan 2004–Oct 2005 (2007)	Retrospective; concurrent cohorts	50–69	Double
Vestfold County	Feb 2004–Dec 2005 (2008)	Retrospective; historic control	50–69	Double
CELBSS[b]	Jan 2005–Jun 2007 (2007)	Retrospective; allocation by area	50–70	Double
Barcelona	Feb 2002–Jan 2007 (2007)	Retrospective; allocation by time	50–69	Double
(B) *Abstracts*				
Tromsø	Sep 2004–Mar 2006 (2006)	Retrospective; allocation by area	50–69	Double
DSPP[c]	Oct 2003–Dec 2006 (2008)	Retrospective; random allocation	50–75	Double
INBSP	Jan 2005–Sep 2007 (2008)	Retrospective; random and area allocation	50–64	Double
Sogn and Fjordane	Jan 2005–Dec 2007 (2008)	Retrospective; allocation by time	50–69	Double
Non-European studies: Published				
Co–Ma	1997–2000 (2001)	Prospective; paired	>40	Single
DMIST	Oct 2001–Dec 2003 (2005)	Prospective; paired	47–62	Single

INBSP Irish National Breast Screening Program; *Co–Ma* The US Colorado–Massachusetts trial; *DMIST* The US Digital Mammographic Imaging Screening Trial

[a]Double reading: 40% of SFM and 65% of FFDM examinations

[b]Central East London Breast Screening Service (CELBSS): FFDM hard-copy reading

[c]Digital Screening Project Preventicon (DSPP): CAD used for FFDM soft-copy reading

2K × 2.5K monitors. The production-type workstation used in this study was improved over the prototype workstation used in the Co–Ma study. The soft-copy interpretation was performed as batch reading, and was carried out in a nondedicated room in which also other activities were going on. Prior analog mammograms had not been scanned for comparison for the digital soft-copy reading. In order to avoid bias, previous mammograms were not offered for the interpretation sessions, neither for SFM nor for FFDM. Prior mammograms were, however, offered at consensus meeting, if available.

All the eight participating radiologists had more than 4 years of experience with screening mammography, but their experience in FFDM soft-copy reading before the project started was limited to a test set of only 100 examinations. There was independent double reading for SFM as well as for FFDM. Thus, the mammographic examinations of each woman were interpreted independently by four different radiologists. There were separate consensus meetings for the two techniques. A 5-point rating scale for the probability of cancer was used in the NBCSP, and all examinations giving a score of 2 or higher (defined as positive) were discussed in the consensus meeting. In the consensus meetings, it was decided whether the woman should be called back for assessment or scheduled for screening in 2 years. In the NBCSP, short-term follow-up was not recommended.

A total of 3,683 women were included in the study. The recall rate for diagnostic work-up after consensus meeting was 3.5% for SFM and 4.6% for FFDM ($p < 0.05$). The cancer detection rate in the study group ($n = 3{,}683$ women) was significantly higher when compared with women not included in the study (Fisher's exact test, $p < 0.05$). This was, however, of no importance for the study itself, as the aim was to compare SFM vs. FFDM in a paired study design. A total of 31 cancers were found in the study group (detection rate 0.84%), of which 28 were detected using SFM (detection rate 0.76%) and 23 using FFDM (0.62%). The difference in cancer detection rate between the two techniques was not statistically significant. A retrospective side-by-side feature analysis of cancer conspicuity performed by an external expert group concluded that there was no difference in cancer conspicuity between the two imaging techniques (Skaane et al. 2003). A follow-up study analyzing the missed FFDM cancers concluded that inexperience in soft-copy reading, suboptimal reading environments, and very quick interpretation might have contributed to the lower detection rate at FFDM (Skaane et al. 2005).

The positive predictive value (PPV_1) was 20.2% for SFM and 11.8% for FFDM. The performance indicators are listed in Table 10.2.

Table 10.2. Studies comparing screen-film mammography (SFM) and full-field digital mammography (FFDM) in breast cancer screening: number of examinations, recall rate, cancer detection rate (including invasive cancers and DCIS), and positive predictive value (PPV_1) (percentage of cancer among women recalled for diagnostic work-up)

Study	Examinations (n)		Recall rate (%)		Cancer detection (%)		PPV_1 (%)	
European studies: (A) Published	SFM	FFDM	SFM	FFDM	SFM	FFDM	SFM	FFDM
Oslo I[a]	3,683	3,683	3.5 *	4.6	0.71	0.54	20.2	11.8
Oslo II	16,985	6,944	2.5 *	4.2	0.38 *	0.59	15.1	13.9
Helsingborg	25,901	9,841	1.4 *	1.0	0.31 *	0.49	21.8 *	47.1
Florence[b]	14,385	14,385	3.5 *	4.3	0.58	0.72	14.7	15.9
Vestfold county[c]	324,763	18,239	4.2	4.1	0.65	0.77	15.1 *	18.5
CELBSS	31,720	8,478	4.4	4.8	0.65	0.68	14.6	14.3
Barcelona	12,958	6,074	5.5 *	4.2	0.42	0.41	7.5	9.7
(B) Abstracts								
Tromsø	12,450	4,890	1.9 *	2.6	0.36 *	0.63	17.9	24.8
DSPP	210,231	26,987	1.3 *	2.4	0.50	0.58	38.4 *	24.2
INBSP	136,438	26,593	3.2 *	4.0	0.55	0.62	16.9	15.8
Sogn and Fjordane	7,442	6,933	2.4	2.5	0.42	0.48	17.7	19.4
Non-European studies: Published								
Co–Ma	6,736	6,736	14.9 *	11.8	0.49	0.40	3.3	3.4
DMIST[d]	42,555	42,555	8.6	8.6	0.41	0.44	4.7	5.1

*Difference statistically significant ($p < 0.05$).
[a]Difference for PPV1 of borderline significance ($p = 0.049$).
[b]Detection rate higher ($p = 0.007$) for cancers manifesting as clustered microcalcifications.
[c]SFM historic data from 18 counties. Cancer detection rate borderline significant ($p = 0.058$).
[d]Performance indicators based on the 42,555 paired examinations.

10.2.2
The Oslo II Study

This prospective randomized trial was carried out from November 2000 to December 2001 and included women aged 50–69 years invited to the NBCSP and women aged 45–49 years invited to the Oslo screening program (Skaane et al. 2004). Stratified randomization on age and residence was performed by the Norwegian National Health Screening Service. It was decided prior to the commencement of the study that about 70% of the invited women should be assigned to undergo SFM and about 30% should have FFDM because of equipment availability (two analog mammography units and one FFDM Senographe 2000D, GE Healthcare, Buc, France). The equipment was the same as for the Oslo I study. The mammographic screening examinations performed in the screening unit in downtown Oslo were sent through a closed network to the Breast Imaging Center for batch reading the following day, similar to the Oslo I study.

All the radiologists had taken part in the prior Oslo I study, and were consequently experienced in FFDM soft-copy reading. There was independent double reading in batch mode for both SFM and FFDM, and consensus meeting for all positive interpretations, similar to the Oslo I study. At the time when the Oslo II study was started, only the results from the Co–Ma study and the Oslo I study were available, and these studies had shown a lower cancer detection rate for FFDM. It was therefore decided to have "double-double" reading, i.e., four independent readers, for all FFDM examinations during the first 4 months, and then an interim analysis with respect to cancer detection rate was performed. In order to avoid bias (two readers for SFM and four readers for FFDM examinations), it was decided that the two readers who daily interpreted the batch first (i.e., first logged into the database for the online reporting) were defined as the "official readers". Women with a positive score by one or both of the "unofficial readers" underwent work-up if decided so at the consensus meeting, but the scores were excluded from comparison to avoid bias. Interim analysis after 4 months showed, however, a higher cancer detection rate at FFDM, and the "double-double reading" was therefore stopped. During this period, one cancer had been detected by the "unofficial" readers but missed by both "official" readers, and this cancer was excluded from analysis. An important difference from the Oslo I study was that reading sessions were carried out in a dedicated room.

A follow-up study including the interval cancers and the subsequent screening round cancers with respect to true positive score at the baseline examination was published in 2007 (Skaane et al. 2007). Rechecking the screening database revealed that several women attending the program during the study period had received a reminder letter following invitation for a scheduled examination date before the randomization started. Thus, these women had not been properly randomized for the trial, and it was decided to exclude these women from the analysis. The follow-up study also included the interval cancers to estimate sensitivity and specificity, and also comprised the baseline mammograms with a true positive score that had been dismissed at consensus meeting (Skaane et al. 2007).

The recall rate for the 16,985 who underwent SFM was 2.5% when compared with a recall rate of 4.2% for the 6,944 women having FFDM, and this difference was statistically significant ($p < 0.05$). Thus, the higher recall rate for FFDM from the Oslo I study was confirmed. It is important to be aware that the recall rate is based on call-back decisions at the consensus meeting. About 70% of the examinations with a positive score at the baseline interpretation were dismissed at the consensus meeting, and this relatively high number can be partly explained by the fact that prior mammograms were not available during the baseline interpretation sessions.

The cancer detection rate was 3.8% for SFM and 5.9% for FFDM, and the difference in favor of FFDM was statistically significant ($p = 0.03$). The detection rate for Ductal carcinoma in-situ (DCIS) was also higher for FFDM, but this difference was not significant. The sensitivity (including the interval cancers with a true positive score at baseline interpretation) was 61.5% (95% CI, 51.5–70.8) for SFM, when compared with 77.4% (95% CI, 63.4–87.3) for FFDM, and this difference was in favor of FFDM, which approached borderline significance ($p = 0.07$). The specificity was 97.9% (95% CI, 97.8–98.1) for SFM, when compared with 96.5% (95% CI, 96.0–96.9) for FFDM, and this difference was significant ($p < 0.05$). The PPV_1 was 15.1% for SFM and 13.9% for FFDM, and this difference was not significant (Skaane et al. 2007). Performance indicators of the Oslo II follow-up study are presented in Tables 10.2 and 10.3.

10.2.3
The Helsingborg Study

This Swedish study, a retrospective analysis of the data from a population-based mammography screening

Table 10.3. Ductal carcinoma in-situ (DCIS) in studies comparing screen-film mammography (SFM) and full-field digital mammography (FFDM): number of examinations, age group of study population, DCIS detection rate, proportion of DCIS among total number of cancers at FFDM, and significance (*p*-value) of DCIS detection between the two imaging techniques

Study	Examinations (*n*)		Age group (years)	DCIS rate (%)		Proportion of DCIS (%)	*p*-value SFM vs. FFDM
European studies (incl. abstracts)	SFM	FFDM		SFM	FFDM		
Oslo II	16,985	6,944	45–69	0.12	0.16	26.8	0.551
Florence[a]	14,385	14,385	50–69	0.12	0.26	27.9	0.007
Vestfold[b]	324,763	18,239	50–69	0.11	0.21	27.1	<0.001
Barcelona	12,958	6,074	50–69	0.10	0.07	16.0	0.56*
DSPP[c]	210,231	26,987	50–75	0.09	0.16	26.8	<0.001
INBSP	136,438	26,593	50–64	0.10	0.14	22.4	0.059
Sogn and Fjordane[d]	7,442	6,933	50–69	0.03	0.09	18.2	*
Non-European studies							
DMIST[e]	42,760	42,760	47–62	0.12	0.14	33.2	0.393

*Small number of DCIS in the Barcelona study (13 and 4, respectively). No *p*-value given for the Sogn and Fjordane due to very small number of DCIS (2 and 6, respectively).
[a]Numbers are given for cancers presenting as clustered microcalcifications.
[b]Prevalent screening round; SFM is mean value of merged data from 18 counties.
[c]CAD used for FFDM only.
[d]Rural area in Norway with low background cancer incidence.
[e]Cancers diagnosed within 455 days after imaging.

program, compared cohorts designed before and after introduction of FFDM (photon-counting system, Sectra Microdose, Linkoping, Sweden). The study period was from January 2000 to February 2005 at the Helsingborg Hospital (Heddson et al. 2007). The study also included evaluation of a computed radiography system, but the results of the three different mammographic systems are separated. The average glandular dose was reported to be 1.1 mGy for SFM and 0.28 mGy for FFDM.

The study material included 25,901 women having SFM, 9,841 women with FFDM, and 16,430 women examined with computed radiography (CR). Only the results comparing SFM and FFDM are presented. The upper age limit of the screening population was 74 years throughout the study period, but the minimum age was gradually reduced from 50 to 46 years during the study period due to political decisions. No prevalent screening examinations were included. The screen-film mammograms represented a historical control group for the comparison with FFDM. There was no random assignment, but the authors had no reason to believe that women were preferentially directed to one technology over another, and it was suggested that the cohorts (SFM, FFDM, and CR) were similar (Heddson et al. 2007).

The images were read by two radiologists highly experienced in SFM screening. Data from the digital system were collected after a learning period of 6 months from the installation of the digital system had been completed, as this period was suggested to be sufficient. Double reading was performed for 40% of the SFM cases and 65% of the digital cases. For cases having double reading, a woman was recalled by consensus among the two radiologists.

The recall rate was 1.4% (372 of 25,901 women) for SFM when compared with 1.0% (102 of 9,841) for FFDM, and this difference was significant ($p = 0.003$). The cancer detection rate for SFM was 0.31% (81 cancers among 25,901 women) and 0.49% (48 cancers among 9,841 women) for FFDM. The higher detection rate for FFDM was statistically significant ($p = 0.01$). As a consequence of the lower recall rate and higher cancer detection rate for FFDM, the PPV_1 was remarkable high for FFDM (47.1%) when compared with SFM (21.8%). Performance indicators from this study are listed in Table 10.2.

10.2.4 The Florence Study

This retrospective Italian study compared the diagnostic accuracy of digital mammography with that of SFM in concurrent cohorts participating in the same breast cancer screening program. Mammographic examinations were carried out in two mobile units of the Florence screening program from January 2004 to October 2005 (Del Turco et al. 2007). One mobile unit was equipped with SFM and the other with FFDM (Senographe 2000D, GE Healthcare, Buc, France). The digital mobile unit was connected to the main center by a wireless digital subscriber-line connection, and the digital images were automatically sent to a dedicated workstation.

The two concurrent screening cohorts of women aged 50–69 years were matched by age and interpreting radiologists, and included 14,385 participants in each cohort. Four radiologists highly experienced in SFM mammography screening interpreted the images. Each reader had more than 5 years of experience in mammographic screening and a workload of more than 10,000 mammograms per year. Their experience with FFDM soft-copy reading before the study started is not given in the publication. Double reading was performed, but was not independent, as the second reader was aware of the first reader's report. Call-back for further assessment was based on suspicion by either reader.

The recall rate for FFDM was 4.29% (618 women) when compared with 3.46% (498 women) for SFM, and this difference was statistically significant ($p = 0.0002$). The higher recall rate at FFDM was mainly evident at incidence screening, but the authors had no explanation for this finding. The recall rate due to poor technical quality was significantly lower for digital mammography (39 vs. 72 women, $p = 0.002$), possibly because the real-time feedback was available. A statistically significant higher recall rate was found for FFDM for the age group of 50–59 years (5.12% vs. 4.17% for SFM, $p = 0.009$), but not for the age group of 60–69 years. An important finding was the remarkable higher recall rate for FFDM for clustered microcalcifications (1.05% for FFDM vs. 0.41% for SFM).

The overall cancer detection rate for FFDM (0.72% or 104 cancers) was higher than that for SFM (0.58% or 84 cancers), but this difference was not significant (Del Turco et al. 2007). However, the detection rate for cancers presenting as clustered microcalcifications was significantly higher for FFDM (0.26% vs. 0.12%, $p = 0.007$).

The PPV_1 was comparable for the two imaging techniques (SFM 14.7% vs. FFDM 15.9%, $p = 0.65$). Performance indicators from the Italian Florence study are listed in Tables 10.2 and 10.3.

10.2.5 The Vestfold County Study

The Vestfold County was the last Norwegian county to join the national breast cancer screening program in February 2004. This retrospective Norwegian study compared the performance indicators of the prevalent FFDM screening in the Vestfold County with the results of the prevalent SFM screening in the other Norwegian counties (Vigeland et al. 2008). The FFDM study group ($n = 18{,}239$) represented women attending the FFDM screening in the Vestfold County, and the "historical" control group ($n = 324{,}763$) was all women attending the SFM prevalent screening in the 18 remaining Norwegian counties. The county-based screening centers in Norway are linked to the Cancer Registry by a closed IT network, and consequently, the same database was used as for the Oslo I and II studies.

Women aged 50–69 years underwent bilateral two-view mammography, and there was independent double reading using a 5-point rating scale for each breast and selection for recall through consensus or arbitration meetings, similar to the Oslo studies. For both the study group and the historic control group, prior screen-film mammograms were offered for comparison in the interpretation sessions, if available. The FFDM population was examined using one of the two stationary Lorad Selenia units (Hologic, Danbury, CT, USA) between February 2004 and December 2005, and the historical control group included SFM mammographic examinations from the NBCSP between November 1995 and August 2003. The screening experience of about 100 radiologists with regard to reading mammograms for the historic control group varied considerably. Five readers in the Vestfold County (FFDM study group) had experience in clinical mammography, but only one reader had experience in organized mammography screening. The FFDM experience for all the five radiologists was limited to 5 months of clinical use before the study started (Vigeland et al. 2008).

The recall rate for FFDM for abnormal mammographic findings (4.09%) did not differ significantly from the SFM control group (4.16%) in this study ($p = 0.645$). However, for recalls owing to technically inadequate mammograms, the recall rate was statistically significant lower for FFDM ($p < 0.001$). The cancer detection rate (including both invasive cancers and DCIS) was higher for FFDM (0.77%) when compared with that for SFM (0.65%), and this difference in favor of digital mammography was of borderline significance ($p = 0.058$). Comparison of the detection rate for DCIS, however, showed a statistically significant ($p < 0.001$) higher detection rate for FFDM vs. SFM (0.21% vs. 0.11%, respectively).

In this study, the PPV_1 based on abnormal mammographic recalls was higher for FFDM (18.5%) when compared with that for SFM (15.1%), and this difference was statistically significant ($p = 0.015$). The performance indicators for FFDM and SFM found in this study are presented in Tables 10.2 and 10.3.

10.2.6 The Tromsø Study

The retrospective Tromsø study included women aged 50–69 years invited to the NBCSP. The study was carried out at the University Hospital of North Norway in Tromsø, between September 2005 and March 2006 (Bjurstam et al. 2006). Women included were allocated by area: Most SFM examinations were carried out in a mobile unit serving rural areas in the two northern Norwegian counties (Troms and Finnmark), whereas the FFDM examinations (Siemens Novation, Siemens, Erlangen, Germany) were carried out on a stationary unit in the "urban" area of Tromsø city.

Similar to the other Norwegian studies, film interpretation included independent double reading using a 5-point rating scale for probability of cancer, and consensus meetings for all examinations with a positive score by at least one of the two readers. Mammograms from previous screening were available for both SFM as well as FFDM reading sessions, as prior analog screening mammograms had been scanned. Five radiologists participated in the study and read mammograms from both the imaging techniques. One reader was highly experienced in SFM screening, whereas the other four radiologists had moderate experience in SFM screening mammography. All the readers had some experience in FFDM soft-copy reading before the study started.

A total of 12,450 women underwent SFM and 4,890 women were screened with FFDM. The attendance rate was remarkably high as usually seen in the rural areas of Norway. The attendance rate was 85% among women invited to the stationary digital unit in Tromsø city, and 81% for women invited to the mobile analog unit. The recall rate due to abnormal mammographic findings was 1.9% for SFM and 2.6% for FFDM, and this difference was statistically significant ($p = 0.006$).

The cancer detection rate was 0.36% for SFM when compared with 0.63% for FFDM, and this higher detection rate for FFDM was also significant ($p = 0.02$). However, PPV_1 was comparable for the two techniques, 17.9% for SFM and 24.8% for FFDM ($p = 0.162$). The authors noted that these performance indicators did not differ significantly between the stationary and mobile unit in the period before digital mammography (Bjurstam et al. 2006). The performance indicators from this study are shown in Table 10.2.

10.2.7 The Central East London Breast Screening Service (CELBSS) Study

The Central and East London Breast Screening Service (CELBSS) at St. Bartholomew Hospital, London, was the first screening unit in the UK National Health Service Breast Screening Programme (NHSBSP) to incorporate FFDM into routine screening. The study was carried out from January 2005 to June 2007 (Vinnicombe et al. 2009). In the population-based NHSBSP, women aged 50–70 years were invited for a two-view mammographic examination once every 3 years.

The digital images were acquired on three Senographe DS mammography units (GE Healthcare, Buc, France) and one Lorad Selenia mammographic unit (Hologic, Danbury, CT, USA). All the mammograms were double read. However, the second reader was aware of the first reader's opinion. Any disagreement was arbitrated independently by a third reader (Vinnicombe et al. 2009). In this study, FFDM was interpreted using hard-copy films on alternator and not as soft-copy reading. The reading conditions including a dedicated roller viewer and the interpretation room was the same for SFM and FFDM examinations. All, but one of the radiologists, taking part in the study were experienced screen readers. Allocation to screening technique (SFM or FFDM) was mainly influenced by residential area. The SFM study population included 31,271 women and the FFDM study population comprised 8,380 women. However, some women were examined twice, and the total number of screening examinations being compared was 31,720 for SFM vs. 8,478 for FFDM.

The recall rate was 4.4% (1,404 of 31,720 mammographic examinations) for SFM vs. 4.8% (406 among 8,478 examinations) for FFDM, and this difference was not statistically significant. A total of 263 cancers were detected. The cancer detection rate for SFM was 0.65% (205 among 31,720 examinations) vs. 0.68% (58 cancers among 8,478 mammographic examinations) for FFDM, and again, this difference was not significant. Furthermore, the PPV_1 (based on cancers diagnosed among recalls of abnormal mammographic findings) was comparable between the two imaging techniques; 14.6% for SFM vs. 14.3% for FFDM. The performance indicators from this study are listed in Table 10.2.

10.2.8 The Barcelona Study

The retrospective Spanish study was carried out at a screening unit in the city of Barcelona. Women aged 50–69 years were invited to have a standard two-view mammography within a population-based screening program with a 2-year interval. Data for SFM were collected between February 2002 and January 2004, and data for FFDM were collected between February 2005 and January 2007. Thus, the study groups were allocated by time, but women represented more than once were excluded from the analysis (Sala et al. 2009).

FFDM was carried out with DM 1000 Agfa (Lorad, Danbury, Conn). Mammograms were read by two radiologists with arbitration by a third radiologist for discordant interpretations. Analysis included only mammograms read by the same team of three radiologists, to reduce reader-related variability (Sala et al. 2009). The readers had more than 6 years of experience in mammography screening, but no previous experience in soft-copy reading of digital mammographic images.

The study groups included 12,958 women having SFM and 6,074 women examined with digital mammography. The overall recall rate was 5.5% (718 among 12,958 women) for SFM and 4.2% (257 of 6,074 women) for digital mammography, and this difference was statistically significant. The false-positive rate was significantly lower in the digital (3.8%) mammography, when compared with the analog (5.1%) group ($p < 0.001$). The cancer detection rate was nearly identical for SFM and digital mammography (0.42 and 0.41%, respectively). However, the cancer detection rate for digital mammography (1.1%) was significantly higher than that for SFM (0.4%) in the first screening round, whereas the cancer detection rate was lower for digital mammography (0.2% vs. 0.4%, respectively) in the successive screening rounds. The overall PPV_1 was 7.5% (54 of 718) for SFM and 9.7% (25 of 257) for digital mammography, but this difference was not significant. However, in the first screening round, the PPV_1 was significantly higher for digital mammography (9.7%) when compared with that for SFM (3.3%).

The performance indicators from the Barcelona study are listed in Tables 10.2 and 10.3.

10.2.9 The Digital Screening Project Preventicon (DSPP) Study

The Dutch mammography screening program was started in 1989, inviting women in the age group of 50–69 years for screening mammography biennially. The target group was changed to 50–75 years of age by 1998. The Digital Screening Project Preventicon (DSPP) was a retrospective study carried out at a regional screening center in the city of Utrecht during the period from October 2003 to December 2006. The preliminary results of this study, carried out in a center that is a part of the nationwide screening program, are presented in 2008 (Karssemeijer et al. 2008).

One of the six conventional mammography units in the center was replaced by one Lorad Selenia FFDM (Hologic, Danbury, CT, USA). Six radiologists were involved in the image interpretation. The readers used a dedicated mammography workstation, and all the mammograms were double read. Computer-aided detection (CAD) was used for all FFDM examinations, but not for the SFM images.

The women were randomly allocated to FFDM or SFM when attending the screening center. A total of 237,218 screening examinations were included in the study, comprising 210,231 SFM examinations and 26,987 FFDM screening examinations. The recall rate for SFM was 2.3% in the initial screening round and 1.1% in subsequent screens, when compared with 4.4% for initial and 1.8% for subsequent screening examinations for FFDM. The recall rate was significantly higher ($p < 0.001$) for FFDM in this study. The cancer detection rate for the initial screening was higher for FFDM (0.76%) when compared with SFM (0.56%), and this higher detection rate for FFDM was of borderline significance ($p = 0.053$). For the subsequent screening examinations, however, the cancer detection rate was similar for FFDM and SFM, 0.51% vs. 0.48%, respectively (Karssemeijer et al. 2008). The fraction of DCIS detected for FFDM was higher and statistically significant when compared with SFM ($p < 0.001$).

The PPV_1 for SFM was 38.4% when compared with 24.2% for digital mammography ($p < 0.001$). The performance indicators of the DSPP study are summarized in Tables 10.2 and 10.3.

10.2.10 The Irish National Breast Screening Program (INBSP) Study

The Irish National Breast Screening Program (INBSP, BreastCheck) introduced FFDM into their mammography screening in 2005 (Hambly et al. 2008a, b). The Breast Check program is a Government-funded breast screening service that provides free mammography to women aged 50–64 years every 2 years. The program screens systematically on an area-by-area basis, including stationary as well as mobile units.

The Irish study was a retrospectively review of FFDM when compared with SFM in the INBSP between January 2005 and September 2007 (Hambly et al. 2008a). A total of 184,730 women were invited to the Breast Check examinations, and 163,031 women attended the screening program. A total of 136,438 women underwent SFM and 26,593 (16.3%) underwent FFDM. The FFDM examinations were acquired with one of the three types of machines: Sectra MDM (Sectra, Linkoping, Sweden), Lorad Selenia (Hologic, Bedford, MA, USA), and GE Essential (GE Healthcare, Buc, France). The women were randomly selected to undergo SFM or FFDM when they attended the screening. FFDM soft-copy reading was carried out using a Sectra PACS IDS5 mammography review workstation. Previous mammograms were reviewed on a viewbox placed adjacent to the workstation. Six dedicated breast radiologists participated in the FFDM interpretation. Two readers independently interpreted each examination using a BIRADS category 1–5. In case both the readers assigned a score of 3–5, the woman was automatically recalled for assessment. Women were also recalled for assessment if any clinical symptoms were noted or the images were technically suboptimal. If there was a discrepancy in the BIRADS category (one reader giving a score BIRADS of 1–2 and the other one giving a score BIRADS of 3–5), the case was discussed at a consensus meeting and it was decided whether the woman should be recalled for diagnostic work-up or listed for routine screening 2 years later (Hambly et al. 2008a).

The recall rate for women who underwent SFM was 3.23% (4,408 of 136,438 exams) when compared with a recall rate of 3.95% (1,050 of 26,593 exams) for FFDM. The higher recall rate for FFDM was statistically significant ($p < 0.05$). If the recall rate was divided into the prevalent and subsequent screening rounds, the recall rate in women undergoing their first screen was 5.65% (2,359 of 41,744) for SFM and 7.17% (527 of 7,351) for FFDM ($p < 0.05$). Furthermore, for women undergoing subsequent screening, the recall rate was significantly higher for FFDM; 2.16% for SFM when compared with 2.72% for FFDM ($p < 0.05$).

The overall cancer detection rate (for both the techniques) during the study period was 0.56%, with a detection rate of 0.71% in women undergoing their first screening round and 0.50% in those undergoing a subsequent screening round. The cancer detection rate for FFDM was 0.62% when compared with 0.55% for SFM, but this difference was not statistically significant. The proportion of DCIS among the detected cancers was 22.4% (37 of 165 malignancies) for FFDM when compared with 19.9% (134 of 747 malignancies) for SFM. The higher detection rate for FFDM for DCIS was of borderline significance ($p = 0.059$). The PPV_1 for SFM and FFDM were comparable for the two techniques. The performance indicators from this study are listed in Tables 10.2 and 10.3.

10.2.11 The Sogn and Fjordane Study

The Sogn and Fjordane County study was carried out in a rural area in the Norwegian west coast from January 2005 to December 2007 (Juel et al. 2008). The women included in the study were all invited to the NBCSP. The County has one stationary unit (located in Førde) and no mobile units. The digital mammographic examinations were carried out with a Sectra MDM unit. Similar to the other rural Norwegian areas, there was a constant high attendance rate in the screening program. The overall attendance rate during the study period was 82.8% (14,375 of 17,352). The women were allocated by time to SFM or FFDM. The attendance rate for SFM during the 18 month period from January 2005 to June 2006 was 83.6% (7,442 of 8,901), whereas the attendance rate for FFDM during the second 18 months study period from August 2006 to December 2007 was 82.0% (6,933 of 8,451 invited women).

Independent double reading using a 5-point rating scale for probability of cancer was carried out, similar to the other Norwegian studies. The two radiologists reading most of the examinations had more than 20 years of experience in diagnostic mammography and were experienced from one screening round before the project started. Both the readers were active during the study period. Analog screening mammograms were scanned, so that prior images were available for the SFM as well as FFDM interpretation sessions. All the examinations with a positive score (2 or higher on the 5-point rating scale) were discussed at the consensus meeting before final decision for recall.

The recall rate due to abnormal mammographic interpretations was 2.4% (175 of 7,442) for SFM and 2.5% (170 of 6,933 women) for FFDM. This difference was not statistically significant (JUEL et al. 2008). The difference in recall rate due to technically insufficient mammograms for SFM was 0.26% (19 of 7,442 examinations) when compared with 0.01% (1 of 6,933), and this difference was significant ($p < 0.01$).

The cancer detection rate for SFM was 0.39% (29 cancers among the 7,442 examinations) and 0.48% for FFDM (33 of 6,933), and this difference was not significant ($p = 0.43$). Conclusions regarding the detection rate of DCIS cannot be drawn from this study, as the numbers were too small, 2 DCIS (detection rate 0.27 per 1,000 screened women) were found on SFM and 6 DCIS (detection rate 0.87 per 1,000 screened women) on FFDM. It should be mentioned that the background cancer incidence in the Sogn and Fjordane County was lower than that in the other Norwegian counties.

The PPV_1 for SFM was 16.6% (29 cancers among 175 recalled women) and for FFDM, it was 19.4% (33 of 170). This difference was not significant ($p = 0.49$). The performance indicators from this study are summarized in Tables 10.2 and 10.3.

10.3 Overall Results and Discussion on the European Studies

All the 11 European studies comparing SFM and FFDM have been carried out within organized mammography screening programs. The five Norwegian studies (Oslo I, Oslo II, Vestfold, Tromsø, and Sogn and Fjordane) were all a part of the population-based NBCSP inviting women of age group 50–69 years every 2 years, and all the five studies used the common database of the centralized nationwide screening program. The Swedish Helsingborg study, the UK CELBSS study, the Dutch DSPP study, and the Irish INBSP study were also a part of the nationwide screening programs. The Swedish screening program was not centralized, and the regions may define their own target population for mammography screening. This explains the significant different age group of 46–74 years in the Helsingborg study. The Italian Florence study and the Spanish Barcelona study were a part of the regional screening programs inviting women aged 50–69 years (Table 10.1). The Vestfold County study included only results from prevalent screening rounds, whereas the other 10 studies compared the results mainly from subsequent (incident) screening rounds.

The 11 European studies had very different design, which partly may explain the conflicting results. The two Oslo studies were prospective trials. Oslo I study had a paired study design, and all the women were examined with SFM and FFDM on the same day. The Oslo II study was a prospective randomized trial. The other nine European studies were all retrospective with different principles for allocating the women to SFM or digital mammography (Table 10.1). Some further important aspects must be kept in mind: CAD was used for FFDM only in the Dutch DSPP study. Hard-copy reading was used for FFDM in the CELBSS study, whereas the other studies used soft-copy reading.

Double reading improved accuracy when compared with single reading, and double reading by consensus or arbitration achieved an increase in cancer detection together with a reduction in the recall rate (DINNES et al. 2001). Double reading was performed in all the 11 European studies, but was carried out for only 40% of the SFM examinations and 65% of the FFDM examinations in the Helsingborg study. Double reading was independent in most of the European studies, but the second reader was aware of the first reader's opinion in the Helsingborg study, the Florence study, and the CELBSS study. Some studies (CELBSS, Barcelona) used arbitration rather than consensus regarding decision for recall, some studies (DSPP, CELBSS, Barcelona, INBSP) used consensus or arbitration only for discordant interpretations, whereas the five Norwegian studies used consensus (or arbitration) for all positive scores by either readers, i.e., also for concordant positive interpretations. In contrast to double reading in the European studies, single reading was the standard in the US studies. The prospective Co–Ma trial had a paired study design using independent single reading for each technique, with the readers being blinded to the results from the other modality. Discordant interpretations were discussed in a "discrepancy evaluation" in which the SFM and FFDM images were evaluated side-by-side (Lewin et al. 2001, 2002). The prospective paired Oslo I study used independent double reading and separate consensus meetings for both the techniques, so that findings for one modality would not influence the decision-making for recall for the other modality (Skaane et al. 2003). The second prospective US study, the DMIST trial, also had a paired design but used single reading with "unilateral recall" for each technique, i.e., a callback if either one of the two readers gave a positive score (Pisano et al. 2005, 2008).

Double reading with arbitration may detect about 11% more cancers, when compared with single reading alone, after taking into consideration the second reader bias (Cornford et al. 2005). Arbitration of discordant double reading would substantially reduce the recall rate with only a small reduction in the cancer detection rate (Ciatto et al. 2005; Shaw et al. 2009). Consensus meeting for all the recalls, also including concordant double reading, has occasionally been recommended (Matcham et al. 2004) and is the standard in the NBCSP. It is important to be aware that cancers will be dismissed at consensus (arbitration) meetings. In the two Oslo studies, about 8% of the cancers discussed at the consensus (arbitration) meetings were dismissed and presented as subsequent cancers (Skaane et al. 2005, 2007).

Interobserver variability is a great challenge in breast imaging, and radiologists differ substantially in their interpretation of screening mammograms (Beam et al. 1996; Berg et al. 2000; Elmore et al. 1994; Skaane et al. 2008). Discordant interpretation was found in 23% of screening-detected cancers in a large nationwide mammography screening program (Hofvind S, Geller BM, Skaane P et al. 2008). The three prospective trials with a paired study design showed a high rate of discordant interpretations for SFM vs. FFDM, although the differences were not statistically significant (Skaane 2009). Problems regarding the positioning and interobserver variability may outweigh any difference between the two imaging techniques (Bick and Diekmann 2007). Consequently, there will be an open question on whether divergent results in studies comparing SFM and FFDM are due to technological differences or caused by interobserver variation (Kopans 2008).

The problem of defining a true positive test (SFM and FFDM) in mammography screening using double reading has not been sufficiently addressed. Basically, there might be different definitions of a true-positive score in a screening program using double reading with consensus or arbitration. The crucial question is whether cancers dismissed at consensus (arbitration) meetings and presenting as interval cancers or screening-detected cancers in the subsequent screening round should be included as "true positives" (Fig. 10.1). In principle, cancers presenting between two screening rounds should be classified as interval cancers. However, for studies comparing the two diagnostic tests (SFM and FFDM), it is reasonable to define the test result as true-positive, if either of the (or both) reader at the double reading gives a true positive score at baseline interpretation, regardless of whether the finding is dismissed at consensus meeting or not. Radiologist-related variability should be kept at minimum in the comparison of the two diagnostic tests, and the test result should not depend on interobserver variability at consensus (or arbitration) meeting. In the DMIST trial, a cancer manifesting within 365 days of a positive screening mammogram was considered to be true-positive (Pisano et al. 2005). Dismissed findings at discrepancy evaluation were still counted as positive in calculations of recall rate and sensitivity in the Co–Ma study (Lewin et al. 2001). It is obvious that follow-up including interval cancers – and subsequent screening round cancers, if possible – is mandatory to include true positives dismissed at consensus (arbitration) meetings. Adjustment for true positives dismissed at consensus/arbitration meeting (or following false negative assessment) and presenting as subsequent cancers has only been performed for the two prospective trials, the Oslo I and the Oslo II study (Skaane et al. 2005, 2007). For the other European studies, only the cancer detection rate at baseline interpretation has been used for comparison.

Recall rates reported in the European studies using double reading are in principle based on decision at consensus or arbitration meeting, unless there is "unilateral" recall for concordant positive scores. In the two Oslo studies, about 70% of the cases having a positive score by one or both readers were dismissed at consensus meetings (Skaane et al. 2005, 2007). The results on recall rates in the 11 European studies are most conflicting (Table 10.2). The recall rate was higher for SFM in 3 of the 11 studies, of which the difference was statistically significant in two studies (Helsingborg study and Barcelona study). The recall rate was nearly identical (4.2% vs. 4.1%) in the Vestfold County study. However, no explanation for the remarkable low call-back rates in the Helsingborg study has been given. In six of the eight studies showing higher recall rate for FFDM, the difference was statistically significant. Thus, the results on recall rate from the European studies differ noticeably from those of the two US trials (Table 10.2).

The cancer detection rate was higher for FFDM for 9 of the 11 European studies, and in three of these nine studies, the difference in favor of FFDM was statistically significant: The Oslo II study, the Helsingborg study, and the Tromsø study (Table 10.2). Preliminary report of the Oslo II study had shown the higher cancer detection rate for FFDM to be of borderline significance (Skaane et al. 2004). However, follow-up evaluation excluding reminders that had not been properly randomized and including cancers with a true positive score dismissed at consensus meeting revealed that the cancer detection rate was significantly higher for FFDM (Skaane et al. 2007). The number of cancers in the Tromsø study was small, and a bias might have been present in this study, as the FFDM examinations were carried out in the "urban" area of Tromsø city, whereas the SFM examinations were performed in a mobile unit in the rural areas in North Norway. The cancer detection rate was only slightly lower for FFDM in the Barcelona study. Except for the Barcelona study, it is of interest that a lower cancer detection rate for FFDM has only been reported in the two "pioneer" studies, the US Co–Ma trial and the Oslo I study (Table 10.2).

Side-by-side feature analysis for cancer conspicuity on the Oslo I material, carried out by external experts experienced in both the techniques, showed no difference in cancer conspicuity between SFM and FFDM (Skaane et al. 2003). Analysis of cancers missed at FFDM soft-copy reading in the paired Oslo I study concluded that learning curve effect, suboptimal reading environments, and sometimes, too hasty interpretations probably caused several cancers to be overlooked (Skaane et al. 2005). Reasons for discrepant interpretations of cancer in the paired Co–Ma trial were equally distributed among those relating to lesion conspicuity, lesion appearance, and interpretation (Lewin et al. 2002). These reasons, together with slightly positioning variability, account for cancers missed using one technique but detected using the other (Fig. 10.2).

The PPV_1, based on cancers among women recalled for diagnostic work-up, showed diverging results in the 11 European studies, with the PPV_1 being higher for SFM in 5 studies and higher for FFDM in 6 studies (Table 10.2). The higher PPV_1 for FFDM was statistically significant in two of the six studies, and the higher PPV_1 for SFM was significant in one of the five studies. The remarkably high PPV_1 for FFDM in the Swedish Helsingborg study and for SFM in the Dutch

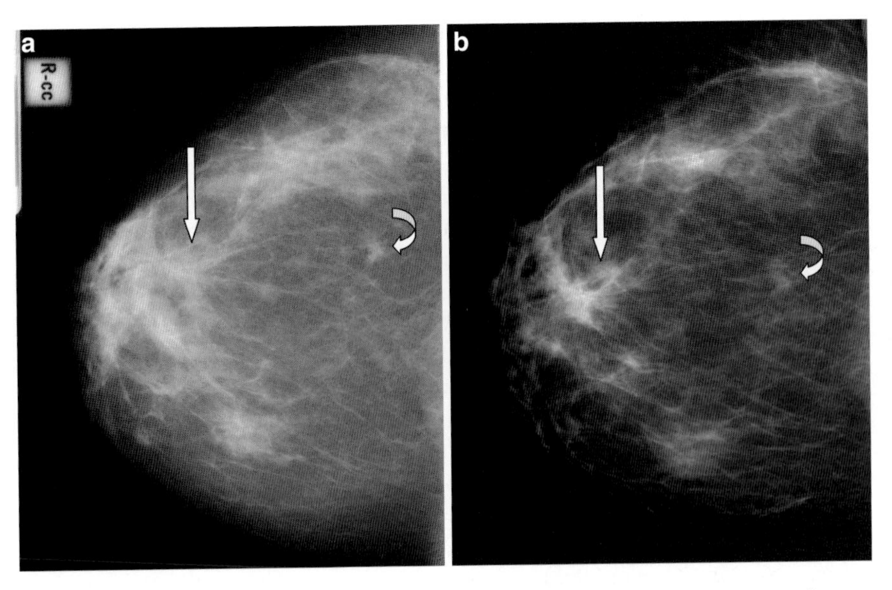

Fig. 10.2. Positioning, contrast, and interobserver variability contributing to one true positive interpretation (both SFM readers having a positive score) and one false negative interpretation (both FFDM readers having a false negative score) in the paired Oslo I study. The CC view of the right breast is shown for SFM (**a**) and FFDM (**b**). The anterior malignant tumor with long and thin spiculations (*arrow*) is well comparable, shown on both imaging techniques, but the small irregular posterior tumor with small spiculae (*curved arrow*) is slightly better demonstrated on SFM. Histology revealed two invasive lobular carcinomas of 11 mm

DSPP study were found to be associated with the very low recall rates. The PPV_1 for both the imaging techniques was considerably lower in the Barcelona study when compared with the other European studies, but was still higher than those found in the US studies (Table 10.2).

The receiver operating characteristic (ROC) curve, a graph of the pairs of true-positive rates and false-positive rates, is a preferred method for comparing the two diagnostic tests. The two US trials presented their results with ROC analysis. In the Co–Ma study, the ROC analysis did not show any significant difference in the diagnostic performance between SFM and FFDM (LEWIN et al. 2002). An important finding of the DMIST trial was a statistically significant higher diagnostic accuracy (ROC analysis) for digital mammography in women under the age of 50 years, women with mammographically dense breasts, and premenopausal or perimenopausal women. For the entire population, however, there was no difference in the diagnostic performance (ROC analysis) between the two techniques (PISANO et al. 2005). It has been pointed out that if the two imaging techniques result in different false-positive recall rates and the full diagnostic performance (ROC curves) are not measured, then a direct comparison of sensitivity (cancer detection rates) is problematic and frequently not valid (GUR 2005). However, ROC analysis for the evaluation of mammography screening is not unproblematic (KEEN 2006; SKAANE and NIKLASON 2006), and is not commonly used for the evaluation of mammography screening in Europe. The 11 European studies used the performance indicators, recall rate, cancer detection rate, and PPV_1 according to the European guidelines for evaluation and comparison of FFDM and SFM as listed earlier.

Sensitivity and specificity including interval cancers have been given for some trials. The sensitivity for SFM and FFDM in the paired Co–Ma study with 1 year follow-up was 66 and 48%, respectively (LEWIN 2006). The DMIST trial revealed a mean sensitivity of 70% for digital and 66% for analog mammography, and the specificity was 92% for both the modalities (PISANO et al. 2005). From the European studies, sensitivity and specificity has been reported for the two Oslo studies. The sensitivity including interval cancers from the 2-year follow-up in the Oslo I study was 76% for SFM and 59% for FFDM, and the specificity was 89% and 84%, respectively (SKAANE et al. 2005, 2009). In the prospective randomized Oslo II study, the sensitivity was 61.5% for SFM and 77.4% for FFDM (p = 0.07), and the specificity was 97.9 and 96.3%, respectively ($p < 0.05$). Cross-sectional studies can be used to compare the relative sensitivities by comparing the cancer detection rates in paired studies, because missed cancers will be common to both tests, but such studies do not provide interval cancer rates for each test, as cancers detected by either test will be treated (IRWIG et al. 2006).

Two reasons for the reluctant attitude regarding digital mammography with soft-copy reading in organized mammography screening programs some years ago was concern about detection and characterization of microcalcifications due to the lower spatial resolution of digital technology and the challenge of soft-copy reading in batch mode. Line-pair resolution, commonly applied for describing the spatial resolution in SFM, is meaningless for digital mammography (BICK and DIEKMANN 2007). It is of interest that six of the seven European studies that reported detection rates of DCIS for both imaging techniques found a higher detection rate for FFDM (Table 10.3). Furthermore, the higher detection rate for DCIS for FFDM was statistically significant in three of these six studies and of borderline significance in one study. Higher contrast in FFDM may account for better detection (Fig. 10.3). Some aspects regarding the DCIS detection rates in these studies should be noted: In the Florence study, the numbers are given for cancers presenting as clustered microcalcifications and not specifically for DCIS. The mean value for SFM listed for the Vestfold County study represents merged data from 18 Norwegian counties. In addition, CAD was used only for FFDM in the DSPP study. The higher DCIS detection rate of borderline significance (p = 0.059) was found in the Irish INBSP study (Table 10.3).

Batch interpretation is the standard in most European mammography screening programs, and has the potential to significantly reduce the recall rates without affecting the cancer detection rate (BURNSIDE et al. 2005; GHATE et al. 2005). However, implementation of FFDM soft-copy reading also signifies a challenge for the batch mode interpretation, due to the large number of single images that are presented. Soft-copy reading usually includes more than ten single layouts for each case. which has to be interpreted in 30–60 s (BICK et al. 2008). Using a hanging protocol with "quadrant zooming" instead of magnification glass means that about 20–32 single images (layouts) are presented per case. It is understandable that the reported interpretation times for FFDM soft-copy reading occasionally has been significantly longer than for SFM hard-copy reading (BERNS et al. 2006;

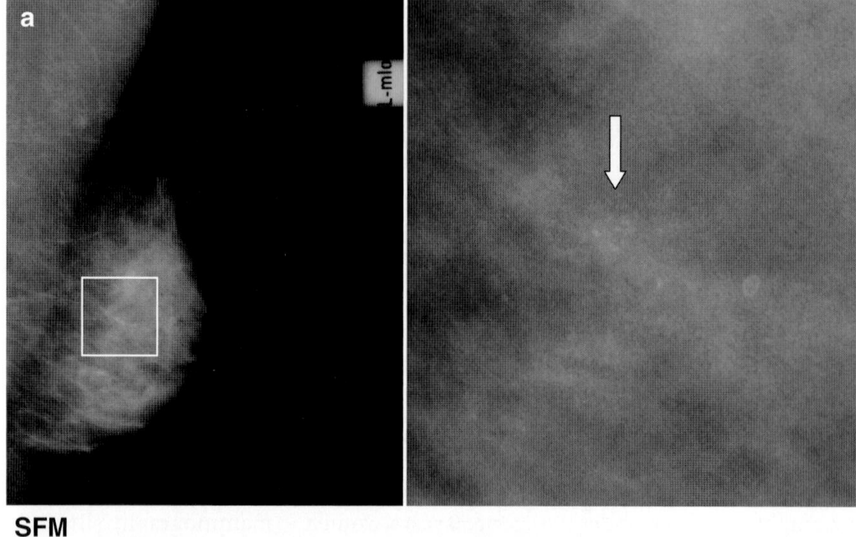

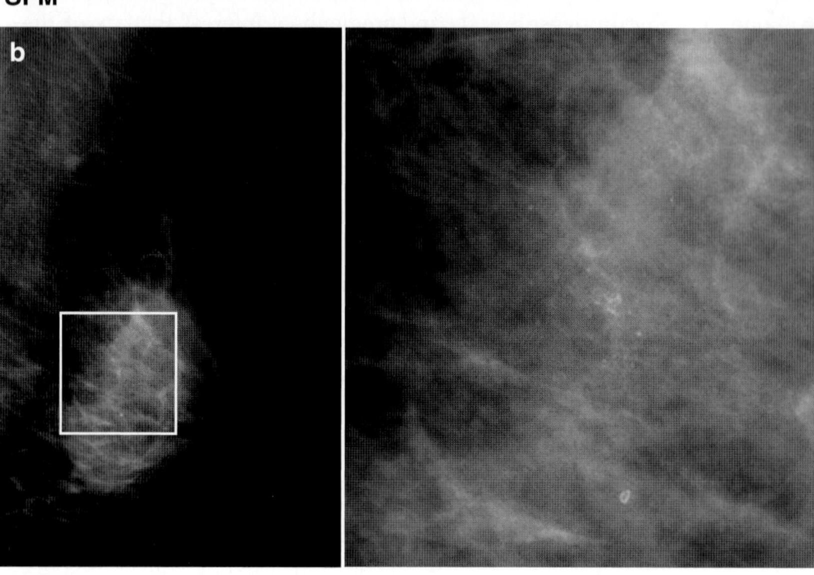

Fig. 10.3. Ductal carcinoma in-situ (DCIS) from the paired Oslo I study. (**a**) SFM, MLO view. It is very difficult to detect the small cluster of microcalcifications in the dense breast parenchyma (*arrow*). The DCIS was missed by both the readers at SFM. (**b**) FFDM, MLO view. The microcalcifications are much better demonstrated at FFDM due to the higher contrast in the image when compared with SFM. Both the FFDM readers gave a true positive score. Histology revealed DCIS of 9 mm

Haygood et al. 2009). Interpretation times for FFDM soft-copy reading has been recorded for one of the European studies. The mean interpretation time for normal examinations in the Oslo I study was 45 s after excluding outliers from analysis (Skaane et al. 2005). A critical analysis of cancers missed during FFDM soft-copy reading indicated that too hasty reading might have caused some false negative interpretations in this study (Skaane et al. 2005). Radiologists' experience in FFDM soft-copy reading and reading environments are important aspects which might have contributed to the conflicting results in the European studies, but these aspects have not been satisfactory addressed.

10.4 Conclusions from the European Studies

In conclusion, the 11 European studies comparing SFM and FFDM in breast cancer screening have shown a higher recall rate and a higher cancer detection rate for FFDM when compared with SFM in most of the studies. Overall, the PPV_1 was comparable for SFM and digital mammography in the 11 studies (Table 10.2). The detection rates of DCIS was higher for FFDM in six of the seven European studies that reported results for both the imaging techniques, and the higher detection rate for DCIS using FFDM was statistically significant in three of the six studies and of borderline significance in one study (Table 10.3). However, it has been pointed out that there is often too much focus on the cancer detection rate and less focus on recall rate. Higher recall rates are usually associated with higher cancer detection rates. The question is therefore whether the higher cancer detection rate for FFDM is the result of a better technique or simply the result of a change in the operating point along the same ROC curve (Gur 2005).

However, it can be concluded from the European studies that digital mammography with soft-copy reading in batch mode is at least as accurate as SFM in population-based screening programs. There are several potential benefits of using FFDM in mammography screening, including simplified archival, retrieval, and transmission of images; dose reduction; reduction of technical failure recalls; higher work-flow (patient throughput); improved diagnostic accuracy due to higher contrast resolution; easier implementation of CAD; potential for telemammography and teleconsultations; the potential for screening program reorganizations; and the potential for implementing new advanced applications including CAD and tomosynthesis in mammography screening (Andersson et al. 2008; Diekmann and Bick 2007; Gilbert et al. 2008). Norway was the first European country to implement FFDM in a population-based mammography screening program in January 2000 (the Oslo I study). The Irish National Breast Screening Program became the first European (and worldwide) mammography screening program in 2008 to become fully digitized (Hambly et al. 2008b). The German breast cancer screening program, the largest mammography screening program in Europe with 94 units and a target population of 10.4 million women in the age group of 50–69 years, has been digitized to about 88% at the end of 2008. About 54% of the German screening units are using CR technology and 34% FFDM with soft-copy reading. The Netherlands are moving fast into digital technology in their mammography screening program. The digitization process in breast cancer screening is rapidly developing in most of the Western European countries.

References

Andersson I, Aspegren K, Janzon L, et al (1988) Mammographic screening and mortality from breast cancer: the Malmo mammographic screening trial. BMJ 297:943–948

Andersson I, Ikeda DM, Zackrisson S, et al (2008) Breast tomosynthesis and digital mammography: a comparison of breast cancer visibility and BIRADS classification in a population of cancers with subtle mammographic findings. Eur Radiol 18:2817–2825

Beam CA, Layde PM, Sullivan DC (1996) Variability in the interpretation of screening mammograms by US radiologists. Arch Intern Med 156:209–213

Berg WA, Campassi C, Langenberg P, et al (2000) Breast imaging reporting and data system: inter- and intraobserver variability in feature analysis and final assessment. Am J Roentgenol 174: 1769–1777

Berns EA, Hendrick RE, Solari M, et al (2006) Digital and screen-film mammography: comparison of image acquisition and interpretation times. Am J Roentgenol 187:38–41

Bick U, Diekmann F (2007) Digital mammography: what do we and what don't we know? Eur Radiol 17:1931–1942

Bick U, Diekmann F, Fallenberg EM (2008) Workflow in digital screening mammography. Radiologe 48:335–344 (in German)

Bjurstam N, Bjorneld L, Warwick J, et al (2003) The Gothenburg breast screening trial. Cancer 97:2387–2396

Bjurstam N, Hofvind S, Pedersen K, et al (2006) Full-field digital mammography screening in the population-based screening program in North-Norway: preliminary results Radiology 241(P):392 (abstr.)

Burnside ES, Park JM, Fine JP, et al (2005) The use of batch reading to improve the performance of screening mammography. Am J Roentgenol 185:790–796

Ciatto S, Ambrogetti D, Risso G, et al (2005) The role of arbitration of discordant reports at double reading of screening mammograms. J Med Screen 12:125–127

Collette HJA, Day NE, Rombach JJ, et al (1984) Evaluation of screening for breast cancer in a non-randomized study (the DOM project) by means of a case-control study. Lancet I:1224–1226

Cornford EJ, Evans AJ, James JJ, et al (2005) The pathological and radiological features of screen-detected breast cancers diagnosed following arbitration of discordant double reading opinions. Clin Radiol 60:1182–1187

Del Turco MR, Mantellini P, Ciatto S, et al (2007) Full-field digital versus screen-film mammography: comparative accuracy in concurrent screening cohorts. Am J Roentgenol 189:860–866

Diekmann F, Bick U (2007) Tomosynthesis and contrast-enhanced digital mammography: recent advances in digital mammography. Eur Radiol 17:3086–3092

Dinnes J, Moss S, Melia J, et al (2001) Effectiveness and cost-effectiveness of double reading of mammograms in breast cancer screening: findings of a systematic review. Breast 10: 455–463

Duffy SW, Tabar L, Chen HH, et al (2002) The impact of organized mammography service screening on breast cancer mortality in seven Swedish counties. Cancer 95:458–469

Elmore JG, Wells CK, Lee CH, et al (1994) Variability in radiologists" interpretations of mammograms. N Engl J Med 331: 1493–1499

Frisell J, Eklund G, Hellstrom L, et al (1991) Randomized study of mammography screening – preliminary report on mortality in the Stockholm trial. Breast Cancer Res Treat 18:49–56

Gabe R, Tryggvadottir L, Sigfusson B, et al (2007) A case-control study to estimate the impact of the Icelandic

population-based mammography screening program on breast cancer death. Acta Radiol 48:948–955
Gershon-Cohen J, Hermel MB, Berger SM (1961) Detection of breast cancer by periodic X-ray examinations: a five-year survey. JAMA 176:1114–1116
Ghate SV, Soo MS, Baker JA, et al (2005) Comparison of recall and cancer detection rates for immediate versus batch interpretations of screening mammograms. Radiology 233:31–35
Gilbert FJ, Astley SM, Gillan MGC, et al (2008) Single reading with computer-aided detection for screening mammography. N Engl J Med 359:1675–1684
Gold RH, Bassett LW, Widoff BE (1990) Highlights from the history of mammography. RadioGraphics 10:1111–1131
Gur D (2005) Technology and practice assessment: in search of a "desirable" statement. Radiology 234:659–660.
Hambly N, Phelan N, Hargaden G, et al (2008a) Impact of digital mammography in breast cancer screening: initial experience in a national breast screening program. In Krupinski EA: IWDM 2008. Springer, Berlin, Heidelberg, pp 55–60
Hambly N, Hargaden GC, Phelan N, et al (2008b) Comparison of full field digital mammography and screen film mammography in breast cancer screening: a retrospective review in the Irish National Breast Screening Program. Radiology 249(P):325 (abstr.)
Haygood TM, Wang J, Atkinson EN, et al (2009) Timed efficiency of interpretation of digital and film-screen screening mammograms. Am J Roentgenol 192:216–220
Heddson B, Roennow K, Olsson M, et al (2007) Digital versus screen-film mammography: a retrospective comparison in a population-based screening program. Eur J Radiol 64: 419–425
Hofvind S, Geller BM, Rosenberg R, Skaane P: Screening-detected breast cancers: Discordant independent double reading in a population-based screening program. Radiology (in press)
IARC Handbooks of Cancer Prevention (2002) Volume 7: breast cancer screening. Chapter 3: use of breast cancer screening. IARC Press, Lyon, pp 47–86
Irwig L, Houssami N, Armstrong B, et al (2006) Evaluating new screening tests for breast cancer. BMJ 332:678–679
Juel I, Hofvind SS, Hoff SR, et al (2008) Screen-film mammography versus full-field digital mammography in a population-based mammography screening program: the Sogn and Fjordane study Radiology 249(P):325–326 (abstr.)
Karssemeijer N, Beijerinck D, Visser R, et al (2008) Effect of introduction of digital mammography with CAD in a population based screening program. Eur Radiol Suppl. 1: 151–152 (abstr.)
Keen JD (2006) Digital and film mammography. N Engl J Med 354:766
Kopans DB (2008) DMIST results: technologic or observer variability? Radiology 248:703–704
Lewin JM, Hendrick RE, D'Orsi CJ, et al (2001) Comparison of full-field digital mammography with screen-film mammography for cancer detection: results of 4,945 paired examinations. Radiology 218:873–880
Lewin JM, D'Orsi CJ, Hendrick RE, et al (2002) Clinical comparison of full-field digital mammography and screen-film mammography for detection of breast cancer. Am J Roentgenol 179:671–677
Lewin J (2006) Clinical trials in full-field digital mammography. Semin Breast Dis 9:87–91.
Lundgren B, Jakobsson S (1976) Single view mammography. A simple and efficient approach to breast cancer screening. Cancer 38:1124–1129
Matcham NJ, Ridley NTF, Taylor SJ, et al (2004) Breast screening: the use of consensus opinion for all recalls. Breast 13: 184–187
Moral Aldaz A, Aupee M, Batal-Steil S, et al (1994) Cancer screening in the European Union. Eur J Cancer 30A:860–872
Nystrom L, Andersson I, Bjurstam N, et al (2002) Long-term effects of mammography screening: updated overview of the Swedish randomised trials. Lancet 359:909–919
Olsen AH, Njor SH, Vejborg I, et al (2005) Breast cancer mortality in Copenhagen after introduction of mammography screening: cohort study. BMJ 330:220–225
Palli D, Del Turco MR, Buiatti E, et al (1986) A case-control study of the efficacy of a non-randomized breast cancer screening program in Florence (Italy). Int J Cancer 38:501–504
Perry N, Broeders M, de Wolf C, Tornberg S, Holland R, von Karsa L (2006) (eds) European guidelines for quality assurance in breast cancer screening and diagnosis, 4th edn. European Communities. European Commission, Luxembourg
Pisano ED, Gatsonis C, Hendrick E, et al (2005) Diagnostic performance of digital versus film mammography for breast-cancer screening. N Engl J Med 353:1773–1783
Pisano ED, Hendrick RE, Yaffe MJ, et al (2008) Diagnostic accuracy of digital versus film mammography: exploratory analysis of selected population subgroups in DMIST. Radiology 246:376–383
Roberts MM, Alexander FE, Anderson TJ, et al (1990) Edinburgh trial of screening for breast cancer: mortality at seven years. Lancet 335:241–246
Sala M, Commas M, Macia F, et al (2009) Implementation of digital mammography in a population-based breast cancer screening program: Effect of screening round on recall rate and cancer detection 252:31–39
Shapiro S, Strax P, Venet L (1971) Periodic breast cancer screening in reducing mortality from breast cancer. JAMA 215: 1777–1785
Shaw CM, Flanagan FL, Fenlon HM, et al (2009) Consensus review of Discordant findings maximizes cancer detection rate in double-reader screening mammography: Irish National Breast Screening Program experience. Radiology 250:354–362.
Skaane P, Young K, Skjennald A (2003) Population-based mammography screening: comparison of screen-film mammography and full-field digital mammography using soft-copy reading: the Oslo I study. Radiology 229:877–884
Skaane P, Skjennald A (2004) Screen-film mammography versus full-field digital mammography with soft-copy reading: randomized trial in a population-based screening program – The Oslo II study. Radiology 232:197–204
Skaane P, Skjennald A, Young K, et al (2005) Follow-up and final results of the Oslo I study comparing screen-film mammography and full-field digital mammography with soft-copy reading. Acta Radiol 46:679–689
Skaane P, Niklason L (2006) Receiver operating characteristic analysis: a proper measurement for performance in breast cancer screening? Am J Roentgenol 186:579–580
Skaane P, Hofvind S, Skjennald A (2007) Randomized trial of screen-film versus full-field digital mammography with soft-copy reading in population-based screening program: follow-up and final results of Oslo II study. Radiology 244:708–717

Skaane P, Diekmann F, Balleyguier C, et al (2008) Observer variability in screen-film mammography versus full-field digital mammography with soft-copy reading. Eur Radiol 18:1134–1143

Skaane P (2009) Studies comparing screen-film mammography and full-field digital mammography in breast cancer screening: updated review. Acta Radiol 50:3–14

Tabar L, Fagerberg CJG, Gad A, et al (1985) Reduction in mortality from breast cancer after mass screening with mammography. Lancet I:829–832

Verbeek ALM, Hendriks JHCL, Holland R, et al (1984) Reduction of breast cancer mortality through mass screening with modern mammography: first results of the Nijmegen project, 1975–1981. Lancet I:1222–1224

Vigeland E, Klaasen H, Klingen TA, et al (2008) Full-field digital mammography compared to screen film mammography in the prevalent round of a population-based screening programme: the Vestfold County study. Eur Radiol 18: 183–191

Vinnicombe S, Pinto Pereira SM, McCormack VA, et al (2009) Full-field digital versus screen-film mammography: comparison within the UK breast screening program and systematic review of published data. Radiology 251: 347–358

de Waard F, Kirkpatrick A, Perry NM, et al (1994) Breast cancer screening in the framework of the Europe against cancer programme. Eur J Cancer Prev 3(Suppl. 1):3–5

Mammographic Signs of Malignancy: Impact of Digital Mammography on Visibility and Appearance

11

Ulrich Bick

CONTENTS

KEY POINTS

The two main aspects affecting lesion visibility in digital mammography are image noise and image processing. A high amount of noise in low-dose mammographic images adversely affects detection of subtle microcalcifications and characterization of mass lesions, but not mass detection in general. Using a higher energy spectrum for digital mammograms while keeping the mean glandular dose constant will lead to a reduction in the image noise and thus to an improved visibility of small microcalcifications, which are in a similar size range as the individual image pixels.

Tailored image processing with improved contrast in areas of dense breast parenchyma is probably the main reason behind the improved sensitivity of digital mammography in women with dense breasts. However, processing algorithms that try to equalize the range of signal intensities throughout the breast may also reduce the contrast and with this the visibility of subtle mass lesions.

11.1 Introduction

Detection of breast cancer in digital mammography is based on the same physical principle of X-ray projection imaging as is conventional film-screen mammography. However, differences in image acquisition, processing, and viewing will result in changes in the image appearance, which will affect the visibility of various mammographic signs in different ways (Cole et al. 2005). For a correct diagnosis, it is crucial to get familiar with these differences in image appearance, which may vary between detector type, acquisition

Ulrich Bick, MD
Department of Radiology, CCM, Charité – Universitätsmedizin Berlin, Charitéplatz 1, 10117 Berlin, Germany

parameters, and processing algorithms used by a digital mammography or workstation vendor. However, with all discussions regarding differences in image quality between film-screen and digital mammography and between different digital mammography systems, one has to bear in mind that the two main determinants for successful detection of breast cancer in mammography are the skills of the technologist obtaining the images and the experience of the radiologist reading the image. These factors far outweigh any differences in image acquisition technology. This is easily demonstrated by the fact that in the paired screening trials comparing film-screen with digital mammography, around a third of cancers were only found by obtaining a second set of mammographic images independently read by one or more additional radiologists, while differences in cancer detection between digital and film-screen mammography on the whole were almost negligible (Bick and Diekmann 2007).

11.2 Mass Lesions

Mass Lesions or "Masses" are defined in the BI-RADS® Lexicon (American College of Radiology 2003) for mammography as space-occupying lesions, which can be seen in two different projections. Along with microcalcifications, masses are one of the main mammographic signs of malignancy, more commonly associated with invasive forms of breast cancer.

11.2.1 Determining the Presence of a True Mass Lesion

Probably, the most difficult task in reading mammograms is determining if an area of density in the breast represents a true mass lesion or just normal breast parenchyma. Oftentimes, this cannot be decided based on the two-view screening mammogram alone (Fig. 11.1), but the woman needs to be recalled for additional imaging such as a spot compression view or a tailored ultrasound. Although the visibility of masses in women with dense breasts is facilitated by digital mammography, the task of distinguishing between normal parenchymal densities and true mass lesions is as difficult with digital mammography as with conventional film-screen mammography. The better visibility of both benign densities and malignant masses (along with the higher conspicuity of microcalcifications) may also explain the slightly higher recall rate observed with digital mammography. Interestingly, the task of detecting mass lesions in digital mammography is largely unaffected by higher image noise (Ruschin et al. 2007; Samei et al. 2007; Saunders et al. 2007). This can be explained by the fact that visibility of masses is mainly limited by anatomical (structural) noise (Fig. 11.2) and not by the quantum noise (Grosjean and Muller 2006). It is important to realize that processing algorithms that try to equalize the range of signal intensities throughout the breast such as contrast-limited adaptive histogram equalization (CLAHE) may reduce the visibility of subtle mass lesions by an absolute or relative reduction in contrast between the mass lesion and the surrounding normal anatomical structures (Sivaramakrishna et al. 2000). Another factor that may influence the apparent density of mass lesions on processed digital mammograms are algorithms increasing (equalizing) the gray values in the periphery of the breast near the skin line (Bick et al. 1995; Byng et al. 1997; Stefanoyiannis et al. 2003). At present, almost no systematic, large-scale (retrospective or prospective) studies are available addressing the impact of different image processing on the visibility of mass lesions in digital mammography (Pisano et al. 2000). This is further hindered by the fact that many institutions – due to storage cost considerations – will not save the "for processing" images in their long-term archive in addition to the mandatory storage of the "for display" version of the mammogram used for making the diagnosis. However, "for processing" images of subtle lesions diagnosed in mammography screening would be necessary for at least a retrospective clinical trial comparing different processing algorithms.

11.2.2 Distinguishing Benign and Malignant Mass Lesions

Differentiation between benign and malignant masses is based on a variety of criteria such as location, size, shape, margin, density, the presence of calcifications or architectural distortions in or around the mass, as well as on so-called associated findings such as skin or nipple changes (American College of Radiology 2003). Of the listed mass descriptors, "margin" and "density" may be affected most by the introduction of digital mammography with routine image processing. The direct relationship between the density of a mass relative to its size and histological parameters such as

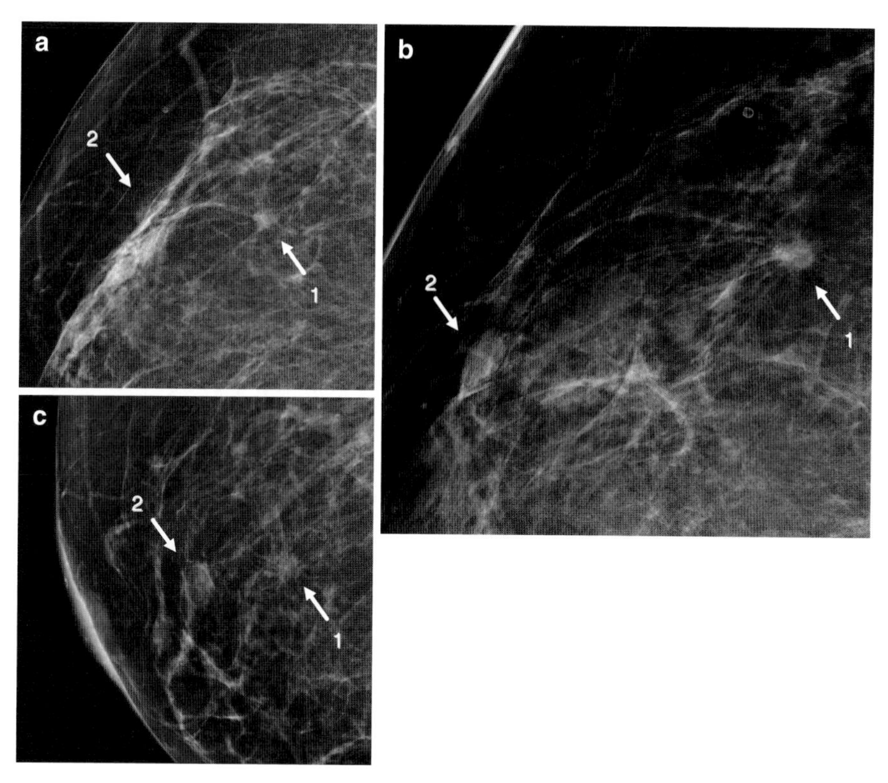

Fig. 11.1. Interdependency of the mass lesion detection and classification task in mammography screening. Fifty-five-year-old asymptomatic female with a 15-mm invasive ductal carcinoma in the right breast detected in organized population-based digital mammography screening. Enlarged regions-of-interest from the right CC (**a**) and MLO (**b**) views show two small densities in the right breast. The more posterior lesion (#1) corresponding to the small invasive carcinoma exhibits a somewhat indistinct margin already on the initial screening views (**a, b**), compared with the slightly larger, but more circumscribed second lesion (#2). The spot compression view obtained with 1.8x geometric magnification (**c**) performed during the second assessment visit after recall from screening confirms a suspicious irregular spiculated mass in position #1 and a second benign-appearing well-circumscribed mass in position #2. Additional fine microcalcifications in lesion #1 are only seen on the magnified spot compression view (**c**). Since small benign circumscribed masses are very common in mammography, the detection of small invasive cancers ≤15 mm is only possible with an acceptable recall rate, if the image quality of the screening mammograms is sufficient to reliably recognize suspicious features in small mass lesions such as indistinct or spiculated margins

cell density is as least partially lost with digital mammography. Depending on the location of the mass lesion (centrally or along the periphery of the breast) and the applied postprocessing algorithm, a lesion may appear denser or less dense relative to their appearance on conventional film-screen mammography (Fig. 11.3). Whereas detection of mass lesions on digital mammography may be less dependent on technical factors than the detection of microcalcifications, characterization of masses, especially identification of fine margin irregularities or spicules benefits significantly from high image quality and low noise (Fig. 11.1). In digital screening mammography, an increasing proportion of subtle mass lesions in dense parenchyma are now detected through subtle associated architectural distortions or spiculations, which may be better visible and enhanced through postprocessing of digital mammograms. This further questions the approach of a low-dose screening mammogram as pursued by some vendors (Fig. 11.4), since not only the detection of microcalcifications but also that of mass lesions will be impaired with this imaging strategy.

11.3 Calcifications

Calcifications are a common finding in mammography and may be related to both benign and malignant findings (Tse et al. 2008). Traditionally,

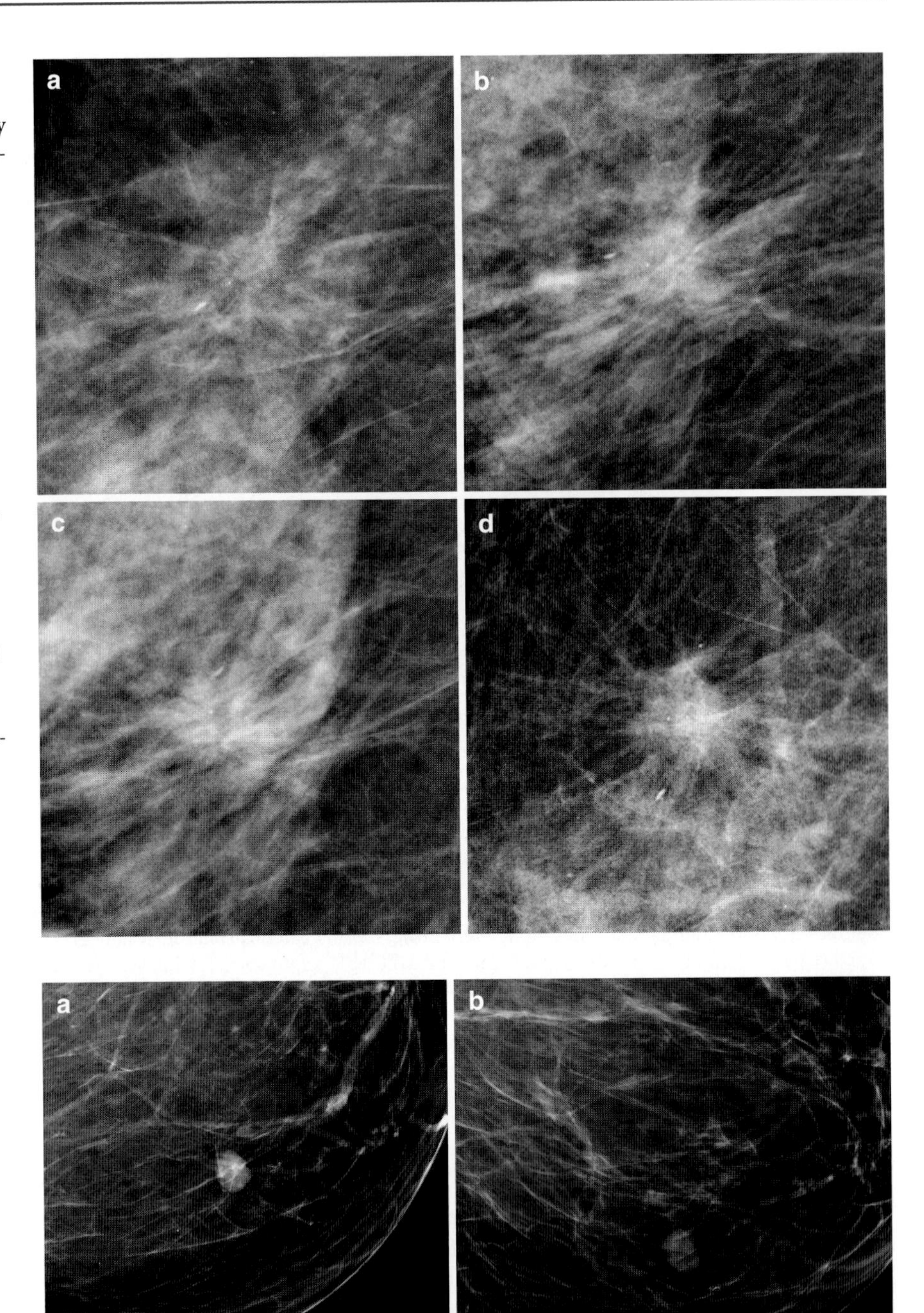

Fig. 11.2. Masking of subtle mass lesions and architectural distortions by normal parenchymal structures. Sixty-six-year-old asymptomatic female with a 6-mm invasive ductal carcinoma in the right breast detected in organized population-based digital mammography screening. Enlarged regions-of-interest from the right CC (**a**), MLO (**b**), and ML (**c**) views show a subtle area of architectural distortion with a few associated microcalcifications. Visibility of the lesion varies between the different views depending on overlying parenchymal background structures. The full mass lesion is only seen on the spot compression view (**d**), which reduces the amount of overlying normal tissue

Fig. 11.3. Variations in mass lesion density as a result of image processing in digital mammography. Fifty-five-year-old asymptomatic female participating in organized population-based digital mammography screening. The enlarged regions-of-interest from the left CC (**a**) and MLO (**b**) views show a small, probably benign circumscribed mass lesion in the periphery of the left breast. Depending on the distance from the skin line and the used algorithm for peripheral density correction, the density of the lesion is significantly higher in the CC view (**a**) than in the MLO view (**b**)

breast calcifications have been divided into most often benign "Macrocalcifications" and potentially malignant smaller "Microcalcifications" (MONSEES 1995). However, the size cutoff – usually set somewhere between 0.5 and 1 mm – to separate macro- from microcalcifications is relatively arbitrary and the BI-RADS® Lexicon (AMERICAN COLLEGE OF RADIOLOGY 2003) uses therefore the neutral term

"Calcifications" (Bassett 1992; Sickles 1986). For a long time, the high image quality requirements necessary for the detection of subtle microcalcifications with a size sometimes as small as around 100 µm was considered the main reason behind the delayed introduction of digital mammography when compared with other digital radiographic imaging procedures, which are now available for more than 20 years. This has changed with the introduction of high-resolution CR mammography systems with a pixel pitch of 50 µm as well as modern solid-state integrated digital detectors for mammography around 10 years ago. At the latest with publication of the favorable DMIST results in the New England Journal in 2005 (Pisano et al. 2005), digital mammography has gained widespread acceptance. Ironically, contrary to earlier believes, detection of microcalcifications appears to be one of the strengths of digital mammography with digital screening-units easily achieving a DCIS detection rate of more than 20% in a general screening population (Del Turco et al. 2007; Vigeland et al. 2008; Weigel et al. 2007).

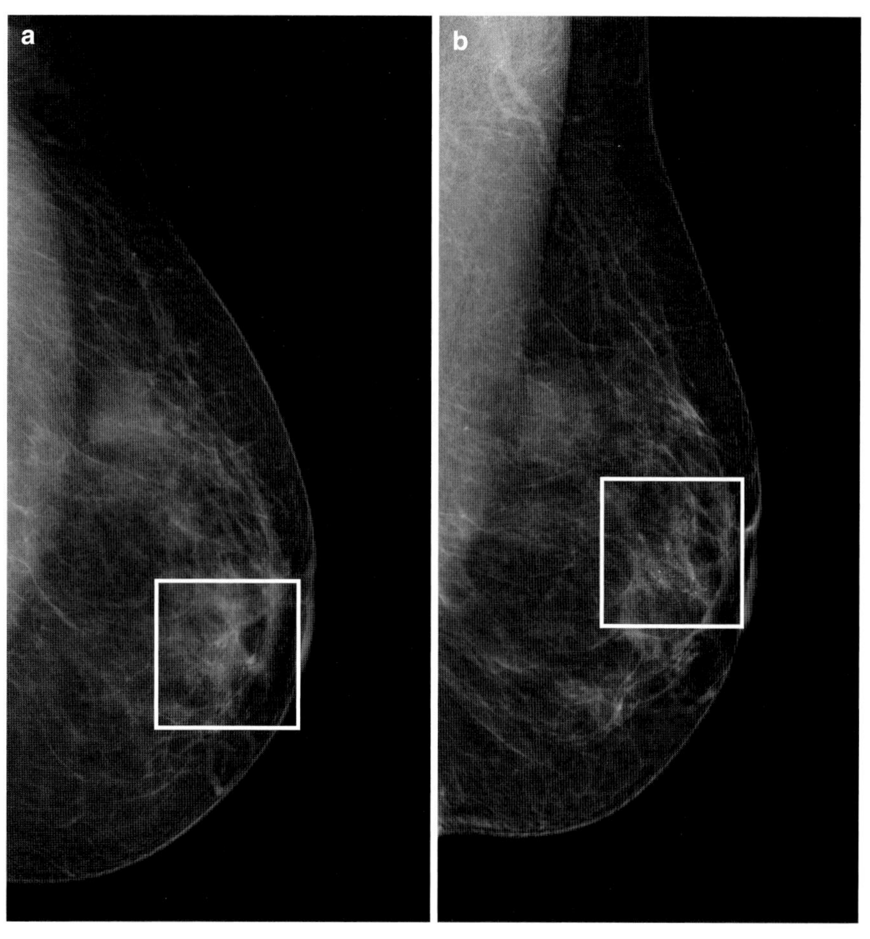

Fig. 11.4. Low-dose screening mammograms in a 40-year-old asymptomatic female with a 40-mm high-grade DCIS in the left breast. Left CC view (**a, c**) at a compressed breast thickness of 5.5 cm and an average glandular dose (AGD) of 0.52 mGy and left MLO view (**b, d**) at a compressed breast thickness of 5.0 cm and an average glandular dose of 0.54 mGy. This corresponds to less than a quarter of the maximum acceptable average glandular dose for this breast thickness as specified in the European guidelines for quality assurance (Perry et al. 2006). For each view, the full image (**a, b**) with the position of the lesion marked and an enlarged region of interest (**c, d**) is shown. The typical highly suspicious pleomorphic calcifications corresponding to the large high-grade DCIS are clearly visualized even on the low-dose images, despite the lower signal-to-noise ratio associated with the lower dose. For reference, the corresponding magnified digital specimen radiography image (**e**) of the final surgical specimen (after histological confirmation of the DCIS through vacuum-assisted stereotactic biopsy) is shown

Fig. 11.4. Continued

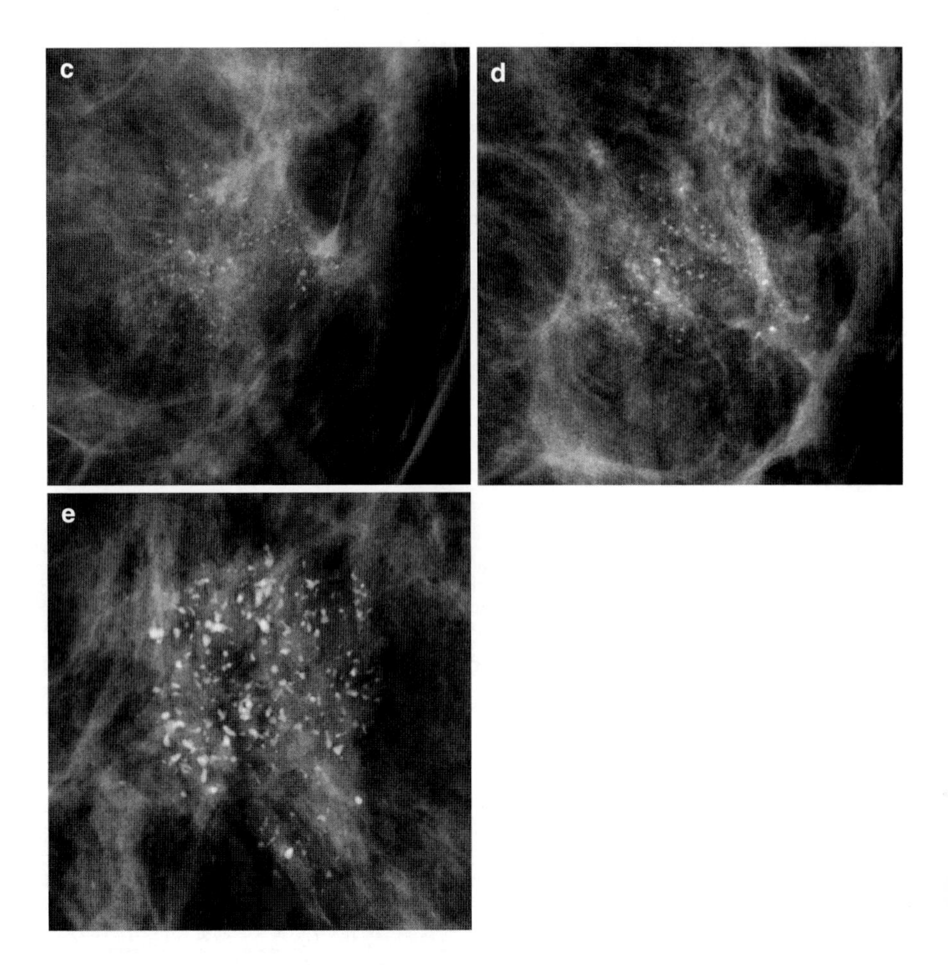

11.3.1 Detection of Subtle Microcalcifications

Small microcalcifications with a size of less than 0.5 mm are a hallmark of breast malignancy. More than 50% of breast cancers will contain at least some of those small microcalcifications, this figure is even higher, when magnification specimen radiography is considered. Most authors will define a suspicious group (also called cluster) of microcalcifications as at least five microcalcifications in a breast volume of 1 cm^3 (Fig. 11.5) (Sickles 1986). The smallest microcalcifications visible on an overview mammogram (without magnification) will be around 130 µm (Cowen et al. 1997; Karssemeijer et al. 1993), although this threshold will vary somewhat with breast density and thickness. Since histologic examination is able to demonstrate even smaller calcifications in breast lesions, many pathologists will differentiate in their report between mammography-relevant calcifications of >100 µm and other smaller calcifications. With digital mammography, two aspects may impair the visibility of microcalcifications with a size at or below the size of the detector elements of the digital mammography system. A microcalcification with a size smaller than the detector element will produce a signal that is larger and of lower contrast than the microcalcification itself. The same is true for a microcalcification with a size similar to the detector element, if it projects between two detector elements. However, since all digital mammography systems currently in clinical use have a detector element size of ≤100 µm, which is smaller than the smallest detectable microcalcifications in mammography, this is not relevant in clinical practice. With the exception of motion blur (Fig. 11.6), the key determinant for visibility of small microcalcifications is the amount of image noise relative to the microcalcification signal, which is dependent on the DQE of the detector and the amount of radiation reaching the detector (Fig. 11.7). Since smaller microcalcifications will produce a signal sometimes as small as a single pixel, the pixel-to-pixel variation related to image noise will impair the

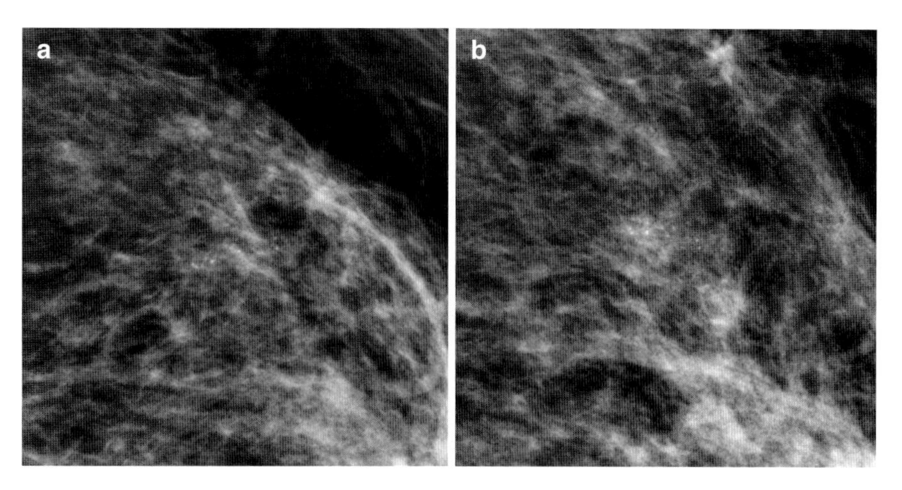

Fig. 11.5. Impact of sufficient visualization of subtle microcalcifications on the detection of small DCIS. Sixty-two-year-old asymptomatic female participating in organized population-based digital mammography screening. The enlarged regions-of-interest from the left CC (**a**) and MLO (**b**) views show a subtle group of fine, predominantly amorphous, or indistinct calcifications, which lead to the detection of a 12-mm intermediate-grade DCIS in the left breast

visibility of small microcalcifications to a much larger degree than the visibility of masses (Fig. 11.8). To assure optimal visualization of small structures down to 0.1 mm in size, the European guidelines for quality assurance in breast cancer screening and diagnosis define achievable and acceptable threshold contrast levels for objects ranging from 0.1 to 2.0 mm in size, which have to be met by digital mammography systems at acceptable glandular dose levels (Perry et al. 2006). For optimal visualization of microcalcifications at a given glandular dose, it is usually beneficial to improve the signal-to-noise ratio by using slightly higher beam energies than traditionally with film-screen mammography (e.g. by switching to Rh/Rh or W/Rh target/filter combinations and by using higher kVp settings). Especially for larger breasts, this will lead to a higher detector dose without increase in average glandular dose, which in turn will result in a lower image noise level relative to the microcalcification signal (Bernhardt et al. 2006; Berns et al. 2003; Huda et al. 2003).

11.3.2 Classification of Benign and Malignant Breast Calcifications

Several criteria such as number, distribution, size, shape, and change over time are used to differentiate between benign and malignant calcifications. Whereas larger typically benign calcifications usually do not pose a diagnostic challenge, classification of smaller, especially indistinct or amorphous microcalcifications may be difficult due to substantial overlap in imaging features of benign and malignant microcalcifications (Fig. 11.5). Of the criteria used to classify microcalcifications, the determination of shape is probably most dependent on the characteristics of the imaging system. In the early days of digital mammography, there was concern that the nominally lower spatial resolution of digital mammography systems compared with film-screen systems may negatively influence the shape information of microcalcifications. And indeed, with a digital detector system with a detector element size of 100 μm, the smallest visible microcalcifications in mammography will be represented by a single pixel. Even with electronic magnification (zoom), these small microcalcifications will always appear as a square or circular, depending on whether pixel replication or interpolation is used for the electronic zoom. However, even conventional film-screen mammography does not contain relevant shape information for microcalcifications that small due to scatter radiation and screen-related unsharpness. In conventional film-screen mammography, additional true geometric (air-gap) magnification views are routinely used for better visualization and characterization of subtle indeterminate microcalcification groups. This technique remains indispensible in digital mammography and cannot be replaced by electronic magnification (zoom) at the viewing workstation (Fig. 11.9) (Kim et al. 2009).

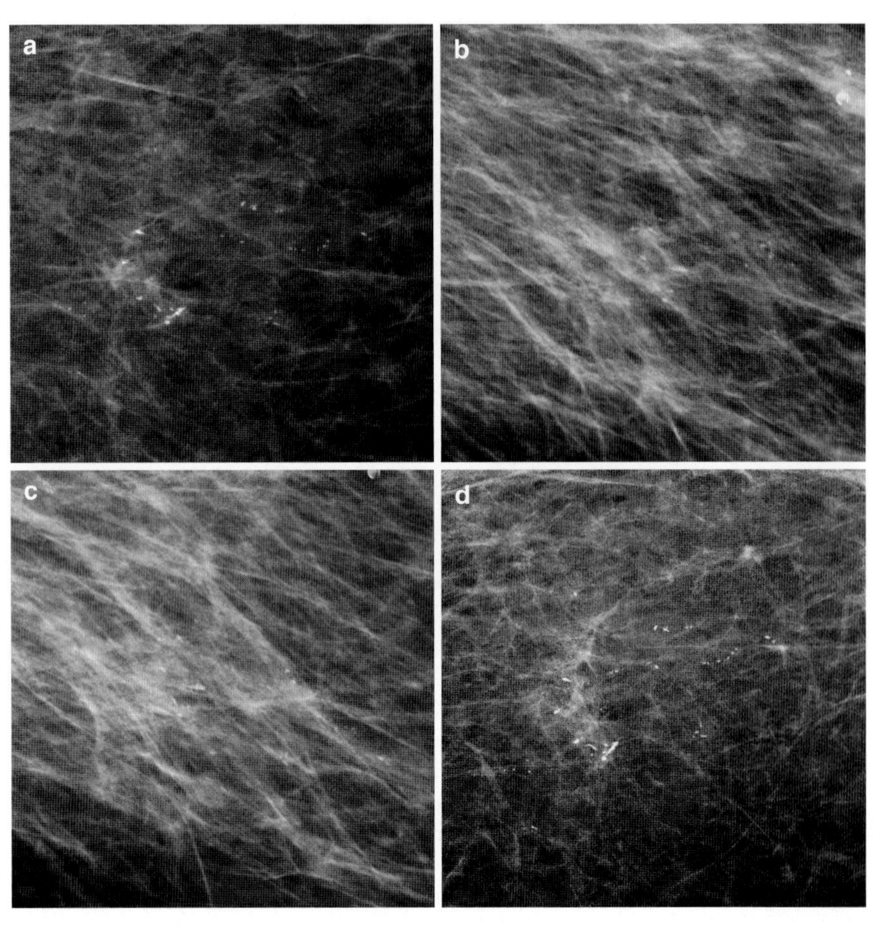

Fig. 11.6. Degradation of visibility of microcalcifications through motion blur. Sixty-seven-year-old asymptomatic female with an extensive 63-mm intermediate-grade DCIS in the left breast detected in organized population-based digital mammography screening. Enlarged regions-of-interest from the right CC (**a**), MLO (**b**), and ML (**c**) views show extensive multifocal microcalcifications, which are confirmed in the magnification view in cranio-caudal projection (**d**). Especially the occurrence of fine-linear calcifications and the in part linear distribution of the calcifications is highly suggestive of malignancy. The microcalcifications are less well seen on the MLO (**b**) and ML (**c**) views, which is predominantly caused by a slight motion blur in these images. In contrast to mass lesions and architectural distortions, visibility of microcalcifications is less affected by overlying parenchymal structures

11.4 Other Mammographic Signs of Malignancy

11.4.1 Architectural Distortions

Architectural distortions are defined as changes in the normal regular architecture of the breast, typically seen in mammography as an accumulation of abnormal straight lines often converging to a single center point. Possible differential diagnosis besides malignancy include complex sclerosing lesions (radial scars) and postsurgical changes. Architectural distortions as the sole finding leading to the detection of malignancy are rare. Oftentimes, they may be found in conjunction with other findings such as masses or microcalcifications and the presence of an accompanying architectural distortion will elevate the level of suspicion. Along with microcalcifications, the fine straight lines representing architectural distortions are the most demanding in terms of image quality. Subtle impairments in image quality due to an elevated image noise level or other factors such as motion blur will make detection of architectural distortions more difficult. However, with all discussion regarding possible differences between the various imaging technologies, it is important to bear in mind that one key determinant

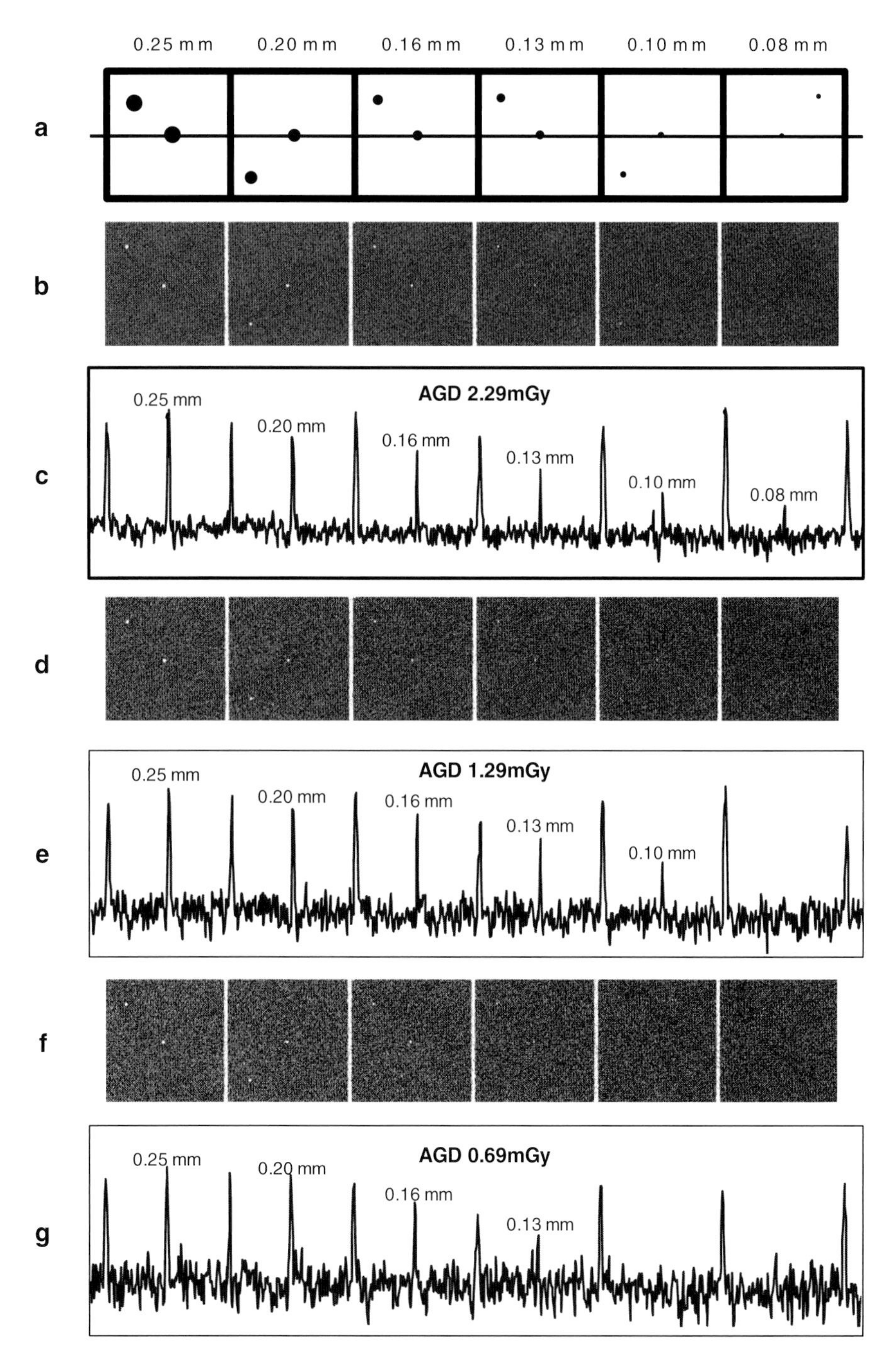

Fig. 11.7. Influence of average glandular dose (AGD) on visibility of small details in mammographic images. CD-MAM phantom (Version 3.4, Artinis Medical Systems, The Netherlands) images were obtained using a selenium-based digital mammography system according to the European guidelines for quality assurance (PERRY et al. 2006) with 2-cm PMMA above and below the phantom resulting in an attenuation equivalent to a compressed breast thickness of 6 cm. For all images, a combination of a Tungsten (W) Target and a Rhodium (Rh) filter was used with a tube voltage of 29 kV. Tube loadings (mAs values) were varied to create images with an average glandular dose of 2.29 mGy (**b, c**), 1.29 mGy (**d, e**), and 0.69 mGy (**f, g**). For better illustration, only the six squares of the CD-MAM phantom corresponding to 2.0 μm thick gold disks ranging in diameter between 0.08 and 0.25 mm are selected. The correct position of the gold disks for these six squares is shown in a schematic drawing (**a**). At each exposure level, the original gray-scale images (**b, d, f**) and the corresponding plot of the gray values in a center line (**c, e, g**) are shown. Window width and level settings were adjusted to normalize for the signal of the largest 0.25 mm disks in the images. With decreasing average glandular dose, the smallest 0.08 and 0.10 mm details are lost in the increasing background noise of the images

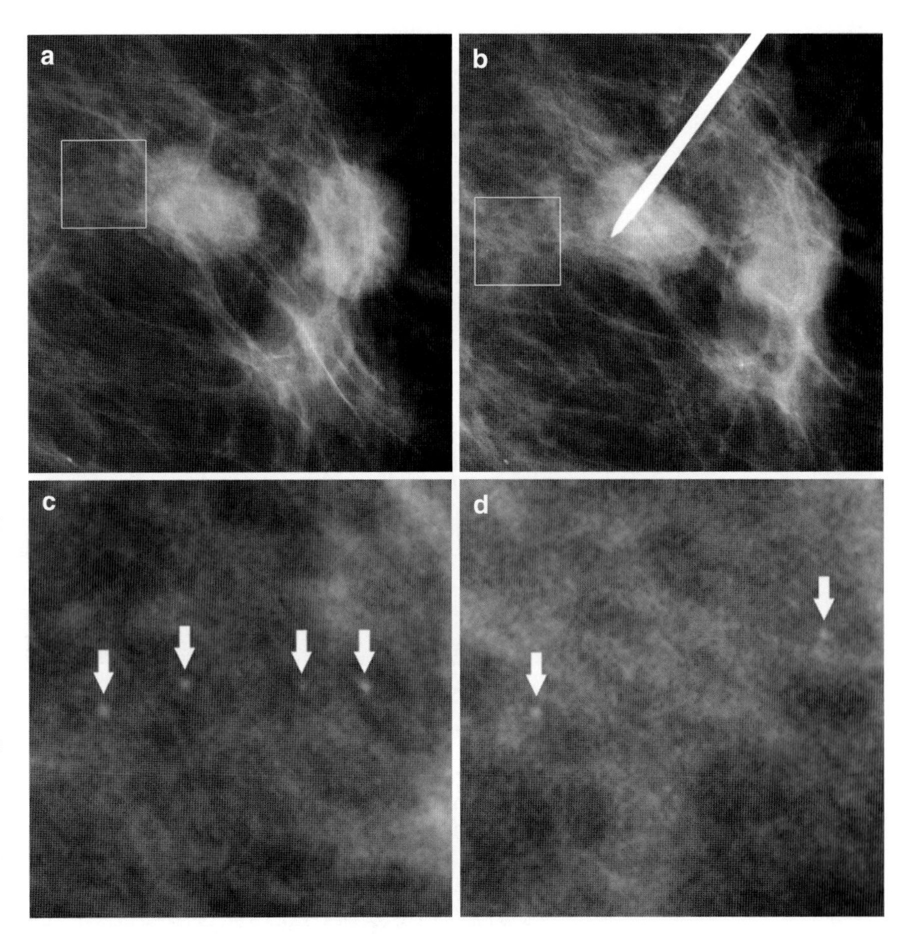

Fig. 11.8. Example of a breast interventional procedure using reduced dose mammographic images. Seventy-six-year-old female with bifocal invasive-ductal carcinoma surrounded by high-grade DCIS. Normal-dose mammographic image (**a**) and needle localization image at 50% reduced dose (**b**) with enlarged area of microcalcifications (**c, d**) in the vicinity of the main tumor. Both masses are equally well seen on the reduced dose image during the localization procedure. However, the individual microcalcifications (*arrows*) are less well depicted on the lower dose image (**d**) than on the normal dose image (**c**) due to a slightly higher amount of image noise in the lower dose image (Reprinted with permission from: Bick and Diekmann 2007)

regarding the detection of architectural distortions (as well as of masses) will be the relationship of the finding relative to the surrounding or overlying normal parenchymal structures. Just slight differences in positioning will determine whether an abnormality becomes visible or is being hidden by anatomical background (Fig. 11.2). This fact is impressively demonstrated by studies involving double exposures (two sets of mammographic images obtained at the same time point), in which up to 30% of cancer were visible only on one set of exposures (Bick and Diekmann 2007).

11.4.2 Asymmetries

Both focal and global asymmetries in the amount and density of the left and right breast parenchyma are common and most often represent normal variations (Sickles 2007). In the few remaining cases of asymmetry actually representing malignancy, the diagnosis is most often made based on accompanying new clinical symptoms or interval changes in the radiological appearance. Preservation of relative parenchymal density during digital image processing is therefore the key for correctly diagnosing malignancies associated with breast parenchymal asymmetries (Fig. 11.10).

11.4.3 Associated Findings

One important advantage of digital mammography is the possibility to correct for regional differences in breast thickness, especially along the periphery of the breast near the skin line through specialized processing algorithms (Bick et al. 1995; Byng et al. 1997; Stefanoyiannis et al. 2003). This enables to view the entire breast from the chest wall to the skin in a similar gray value range. Associated signs such as skin thickening or retraction are therefore much easier to detect with digital mammography than with

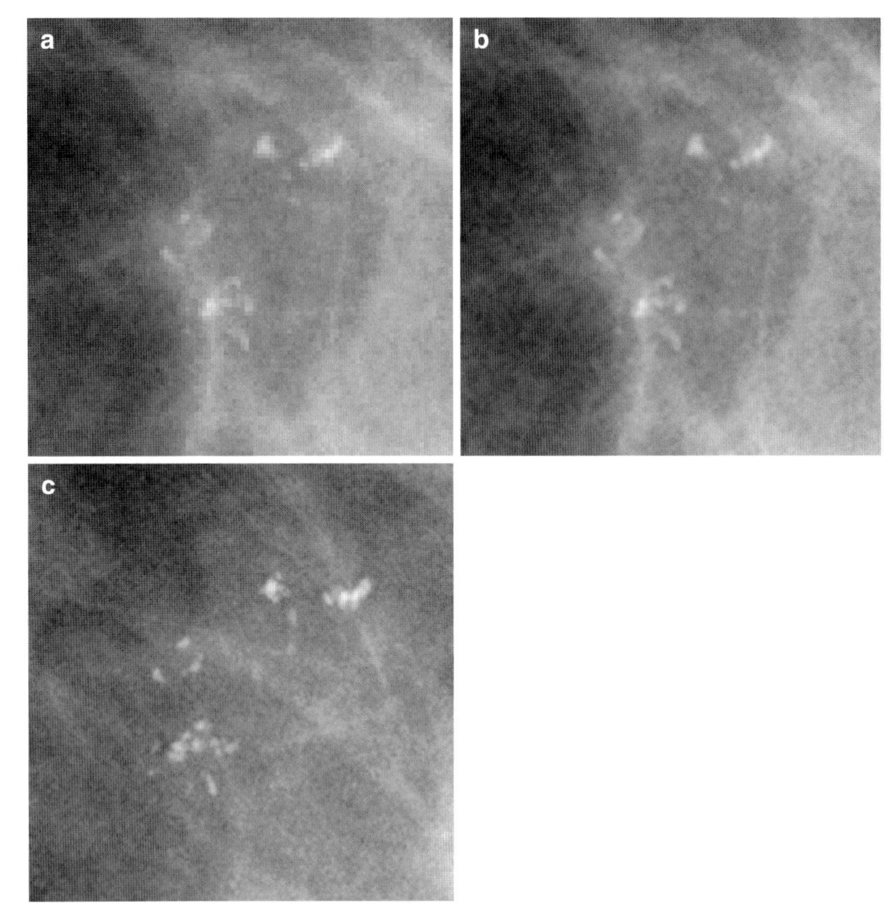

Fig. 11.9. Impact of true geometric magnification on the visibility of microcalcifications. Fifty-two-year-old female with a suspicious cluster of microcalcifications on mammography (Diagnosis: high-grade DCIS). Electronic magnification of the overview digital mammogram using pixel replication (**a**) and bicubic interpolation (**b**) compared with the geometric 1.8× spot magnification view (**c**). The shape of the individual microcalcifications is much better defined on the geometric magnification view, in addition several additional smaller microcalcifications are seen. The images shown correspond to an area of approximately 1 × 1 cm (Reprinted with permission from: Bick and Diekmann 2007)

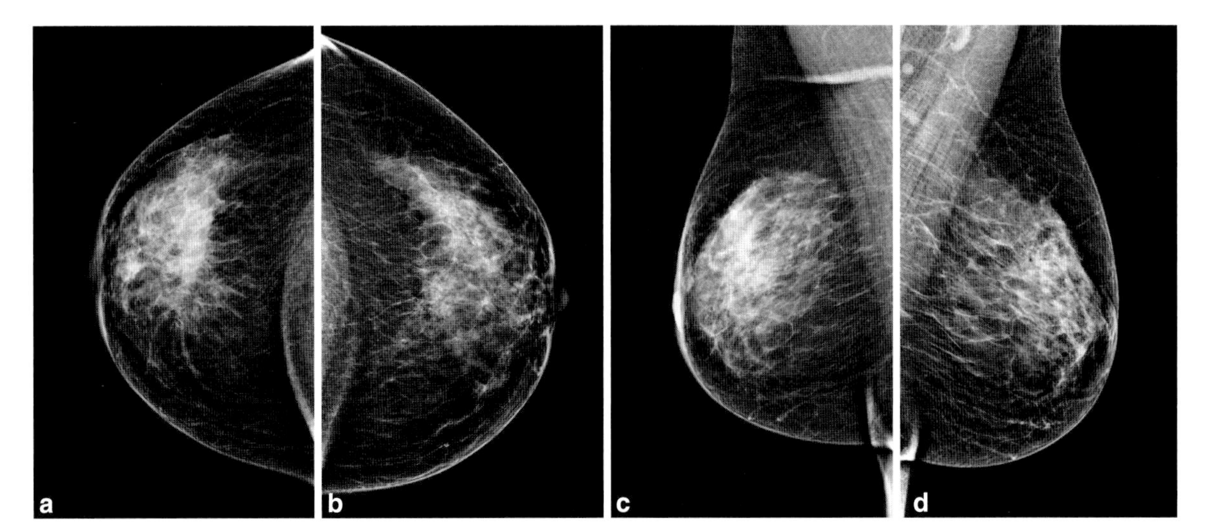

Fig. 11.10. Locally advanced poorly differentiated invasive ductal carcinoma with extensive peritumoral high-grade DCIS and axillary lymph-node involvement in the right breast (pT2 pN2a). Despite the fact that both the primary tumor and the axillary lymph nodes on the right side were palpable in retrospect, the carcinoma was detected in the initial round of an organized population-based digital mammography screening program. In the four-view screening images (**a–d**), the parenchyma in the right CC (**a**) and MLO (**c**) view appears denser and more compact than in the contralateral left breast. Preservation of density differences between the right and left breast during image processing is crucial for the detection of this type of finding. Associated findings such as the subtle architectural distortion of the posterior parenchymal border in the right CC view (**a**) and the periareolar skin thickening in the right MLO view (**c**) are easier to see in the digital processed images when compared with conventional film-screen mammography

conventional film-screen mammography and can be easily correlated in location with abnormalities within the breast in one single view (Fig. 11.10).

References

American College of Radiology A (2003) ACR BI-RADS®. Breast imaging and reporting data system, breast imaging atlas. Mammography, breast ultrasound, magnetic resonance imaging. American College of Radiology, Reston, VA

Bassett LW (1992) Mammographic analysis of calcifications. Radiol Clin North Am 30(1):93–105

Bernhardt P, Mertelmeier T, Hoheisel M (2006) X-ray spectrum optimization of full-field digital mammography: simulation and phantom study. Med Phys 33:4337–4349

Berns EA, Hendrick RE, Cutter GR (2003) Optimization of technique factors for a silicon diode array full-field digital mammography system and comparison to screen-film mammography with matched average glandular dose. Med Phys 30:334–340

Bick U, Diekmann F (2007) Digital mammography: what do we and what don't we know? Eur Radiol 17:1931–1942

Bick U, Giger ML, Schmidt RA, et al (1995) Automated segmentation of digitized mammograms. Acad Radiol 2:1–9

Byng JW, Critten JP, Yaffe MJ (1997) Thickness-equalization processing for mammographic images. Radiology 203: 564–568

Cole EB, Pisano ED, Zeng D, et al (2005) The effects of gray scale image processing on digital mammography interpretation performance. Acad Radiol 12:585–595

Cowen AR, Launders JH, Jadav M, et al (1997) Visibility of microcalcifications in computed and screen-film mammography. Phys Med Biol 42:1533–1548

Del Turco MR, Mantellini P, Ciatto S, et al (2007) Full-field digital versus screen-film mammography: comparative accuracy in concurrent screening cohorts. AJR 189:860–866

Grosjean B, Muller S (2006) Impact of textured background on scoring of simulated CDMAM phantom. In: Astley SM, Brady M, Rose C, Zwiggelaar R (eds) Digital mammography proceedings of the IWDM 2006, LNCS 4046. Springer, Berlin, Heidelberg, pp 460–467

Huda W, Sajewicz AM, Ogden KM, et al (2003) Experimental investigation of the dose and image quality characteristics of a digital mammography imaging system. Med Phys 30: 442–448

Karssemeijer N, Frieling JTM, Hendriks JHCL (1993) Spatial resolution in digital mammography. Invest Radiol 28:413–419

Kim MJ, Kim EK, Kwak JY, et al (2009) Characterization of microcalcification: can digital monitor zooming replace magnification mammography in full-field digital mammography? Eur Radiol 19:310–317

Monsees BS (1995) Evaluation of breast microcalcifications. Radiol Clin North Am 33:1109–1121

Perry N, Broeders M, de Wolf C, et al (2006) European guidelines for quality assurance in breast cancer screening and diagnosis, 4th edn. Office for Official Publications of the European Communities, Luxembourg

Pisano ED, Gatsonis C, Hendrick E, et al (2005) Diagnostic performance of digital versus film mammography for breast-cancer screening. N Engl J Med 353:1773–1783

Pisano ED, Cole EB, Major S, et al (2000) Radiologists' preferences for digital mammographic display. Radiology 216:820–830

Ruschin M, Timberg P, Bath M, et al (2007) Dose dependence of mass and microcalcification detection in digital mammography: free response human observer studies. Med Phys 34: 400–4007

Samei E, Saunders RS, Jr., Baker JA, et al (2007) Digital mammography: effects of reduced radiation dose on diagnostic performance. Radiology 243:396–404

Saunders RS, Jr., Baker JA, Delong DM, et al (2007) Does image quality matter? Impact of resolution and noise on mammographic task performance. Med Phys 34:3971–3981

Sickles EA (1986) Breast calcifications: mammographic evaluation. Radiology 160:289–293

Sickles EA (2007) The spectrum of breast asymmetries: imaging features, work-up, management.Radiol Clin North Am 45: 765–771

Sivaramakrishna R, Obuchowski NA, Chilcote WA, et al (2000) Comparing the performance of mammographic enhancement algorithms: a preference study. AJR 175:45–51

Stefanoyiannis AP, Costaridou L, Skiadopoulos S, et al (2003) A digital equalisation technique improving visualisation of dense mammary gland and breast periphery in mammography. Eur J Radiol 45:139–149

Tse GM, Tan PH, Pang AL, et al (2008) Calcification in breast lesions: pathologists' perspective J Clin Pathol 61:145–151

Vigeland E, Klaasen H, Klingen TA, et al (2008) Full-field digital mammography compared to screen film mammography in the prevalent round of a population-based screening programme: the Vestfold County Study. Eur Radiol 18:183–191

Weigel S, Girnus R, Czwoydzinski J, et al (2007) Digital mammography screening: average glandular dose and first performance parameters. Fortschr Röntgenstr 179:892–825

Contrast-Enhanced Digital Mammography

12

Clarisse Dromain and Corinne Balleyguier

CONTENTS

C. Dromain
C. Balleyguier
Department of Radiology, Institut Gustave-Roussy 39, rue Camille Desmoulins, 94805 Villejuif Cedex, France

KEY POINTS

CEDM is a new advanced application of digital mammography using the intravenous injection of an iodinated contrast agent in conjunction with a mammography examination. Two basic CEDM techniques are under development: temporal subtraction and dual-energy techniques. Temporal subtraction CEDM imaging produces high-energy digital mammography images and subtraction of precontrast from the postcontrast images. Dual-energy subtraction imaging exploits the energy dependence of X-ray attenuation through materials of different compositions in the breast, specifically iodine and soft tissues. A pair of low- and high-energy images is obtained after contrast and then the two images are combined to enhance areas of contrast uptake. This new breast imaging method can be easily implemented clinically using a current digital mammography system with minor adaptations and commercially available iodinated contrast agent. Initial clinical experience has shown the ability of CEDM to map the distribution of neovasculature induced by cancer using mammography. CEDM has been shown to improve the probability of malignancy and the BIRADS assessment when compared with the conventional mammography alone. The potential clinical applications are the determination of the extent of disease, the assessment of recurrent disease, the clarification of mammographically equivocal lesions, the detection of occult lesions on standard mammography, particularly in dense breast, and the monitoring of the response to chemotherapy. CEDM should be a useful adjunct to diagnostic mammography and a promising problem-solving and a staging tool.

12.1 Introduction

Full-field digital mammography developments have been rapid, enabling high-quality breast images with higher contrast resolution, an improved dynamic range, and rapid processing of data and images when compared with screen-film mammography (Skaane et al. 2005; Bick and Diekmann 2007). Its diagnostic accuracy has been shown to be at least equivalent to film-screen mammography (Skaane et al. 2005; Pisano et al. 2005). However, some limitations persist, mainly in dense breast tissue, fibrocystic disease, and during follow-up after breast-conserving therapy. The highly improved technology, now available for full-field digital mammography systems, also offers new advanced applications. Contrast-enhanced digital mammography (CEDM) with the injection of an iodinated contrast agent is one of them. Contrast agent has been used for many years by both CT and MR techniques to explore angiogenesis in breast carcinoma by tracking the uptake and washout of contrast agent in tissue. Iodinated contrast-enhanced CT was shown to be useful for detecting breast carcinoma (Hagay et al. 1996; Chang et al. 1982). However, breast CT results in a high radiation dose to the breast and chest wall. Breast MRI, using gadolinium-based contrast agent, is currently considered the most sensitive imaging technique for the detection of breast carcinoma. So multiple and ever-increasing indications have been established (American College of Radiology 2004) However, MR imaging has a limited specificity, is not widely available, and is expensive.

In this chapter, we review the basic CEDM technique and discuss and illustrate its potential clinical applications.

12.2 Basic Concepts of the Technique

Two basic CEDM techniques are under development: temporal subtraction and dual-energy techniques. Both these techniques use intravenous injection of an iodine-based contrast agent. However, the low-energy exposures used in mammography are not optimal for the visualization of iodine. Indeed, the X-ray beam generated from a molybdenum/rhodium/tungsten anode and a molybdenum /rhodium/silver filter in conventional mammography was developed to maximize the contrast between the glandular parenchyma and fat. To increase the sensitivity of digital mammography in low iodine concentrations, without increasing the dose delivered to the patient, the X-ray spectrum must be shaped so that the energy output of X-rays is above the K-edge of iodine (33.2 keV). One way to obtain these high-energy exposures was to use voltages ranging between 45 and 49 kVp in conjunction with a copper filter instead of the 26–32 kVp used for conventional digital mammography (Skarpathiotakis et al. 2002). Most investigators used the same iodine-based contrast agent as that used for computed tomography imaging at a similar concentration of about 1–1.5 mL/kg body weight. Another approach could be to use a specific noniodinated contrast agent for mammography. Indeed, a recent report suggests that bismuth has a better contrast-to-noise ratio than iodine for the low energy typically used in mammography (Diekmann et al. 2007b).

12.2.1 Dual-Energy Technique

Dual-energy subtraction imaging exploits the energy dependence of X-ray attenuation through materials of different compositions in the breast, specifically iodine and soft tissues. A pair of low- and high-energy images is obtained after the administration of an iodinated contrast medium agent. Then, the two images are combined (Puong et al. 2007) to enhance areas of contrast uptake (Fig. 12.1). In practice, an iodinated contrast agent is first injected, preferably using a power injector at a high flow rate. Then, the breast is compressed and a pair of low- and high-energy images is acquired. The mean time between the acquisition of low-energy and high-energy images is currently about 20 s for the experimental device but this time should be rapidly reduced to a few seconds or a fraction of a second. With a single administration of contrast, it is possible to acquire several pairs of images corresponding to different views (cranio-caudal (CC) and mediolateral oblique (MLO)) or both breasts. The duration of the examination ranges from 5 to 10 min depending on the number of projections. The total X-ray dose delivered to the patient for a pair of low- and high-energy images is estimated between 0.7 and 3.6 mGy depending on breast thickness (30–80 mm) and tissue composition (0–100% glandular tissue). This dose level corresponds to about 1.2 times the dose delivered for a standard mammogram.

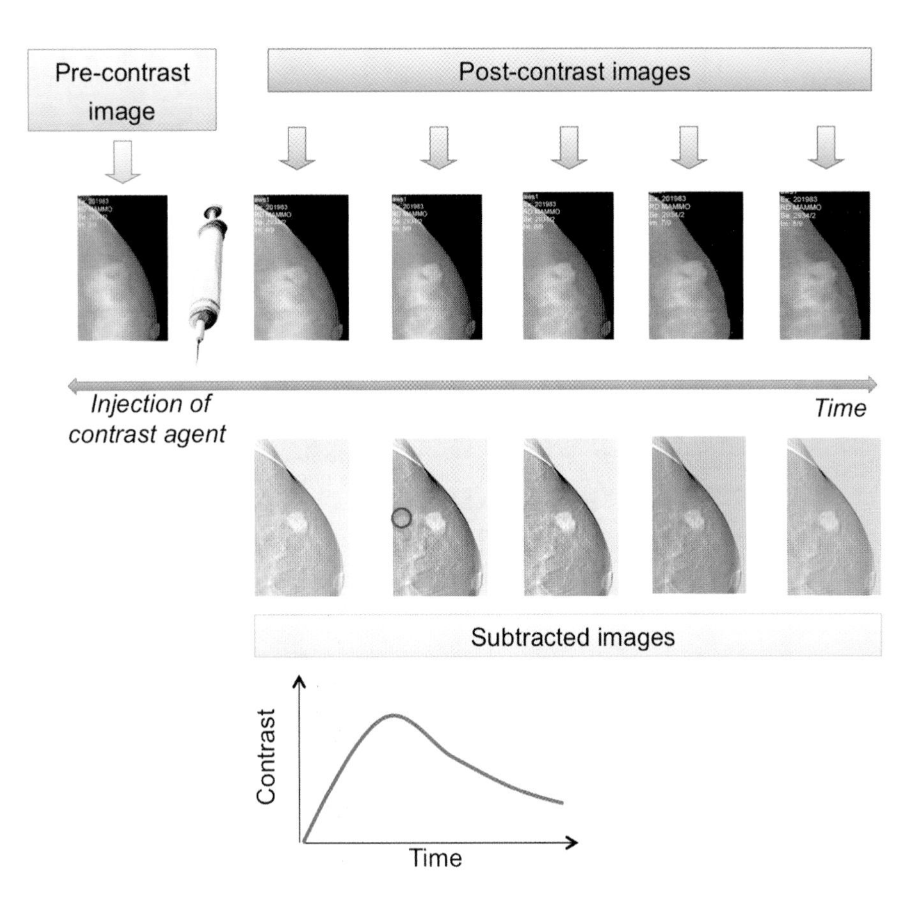

Fig. 12.1. Basic principle of dual-energy CEDM. A pair of low- and high-energy digital mammography images is obtained after administering an iodinated contrast medium agent (**a**). Then, the two images are combined to enhance contrast uptake areas (**b**)

12.2.2 Temporal Subtraction Technique

Temporal subtraction CEDM imaging produces high-energy digital mammography images before and after contrast medium injection. To enhance the visualization of contrast medium in lesions, the pre-contrast image is subtracted from the postcontrast images (Fig. 12.2). In practice, all images must be acquired within a single breast compression. That is why the patient must be comfortably settled to minimize motions during the image sequence acquisition. The CC view, allowing the patient to sit, was preferred in most of the clinical studies. Light breast compression is used for all mammography images, strong enough to limit motion but limited to avoid reducing blood flow. Once the breast has been compressed, the examination begins with the acquisition of a single mask mammogram. A monophasic intravenous injection is then performed, preferably using a power injector at a high flow rate, and several postinjection images are captured.

The total temporal CEDM examination duration is approximately 15 min and the total X-ray dose delivered to the patient depends on breast thickness, tissue composition, and the number of images acquired in the CEDM sequence. About one-fifth of the dose delivered for a standard mammogram is obtained using a high-energy spectrum for the acquisition of CEDM images.

12.3 Image Analysis and Interpretation

Image analysis of dual-energy CEDM images involves combining low-energy and high-energy images through appropriate image processing to generate an image with contrast uptake information. To achieve the temporal analysis of CEDM images, the precontrast images must be subtracted from the postcontrast images. Image registration software is also required to compensate for frequent patient motions that could generate artifacts in the subtracted images. On subtracted temporal CEDM images, regions of interest can be placed in areas of early enhancement and in adjacent breast tissue to analyze the uptake and the

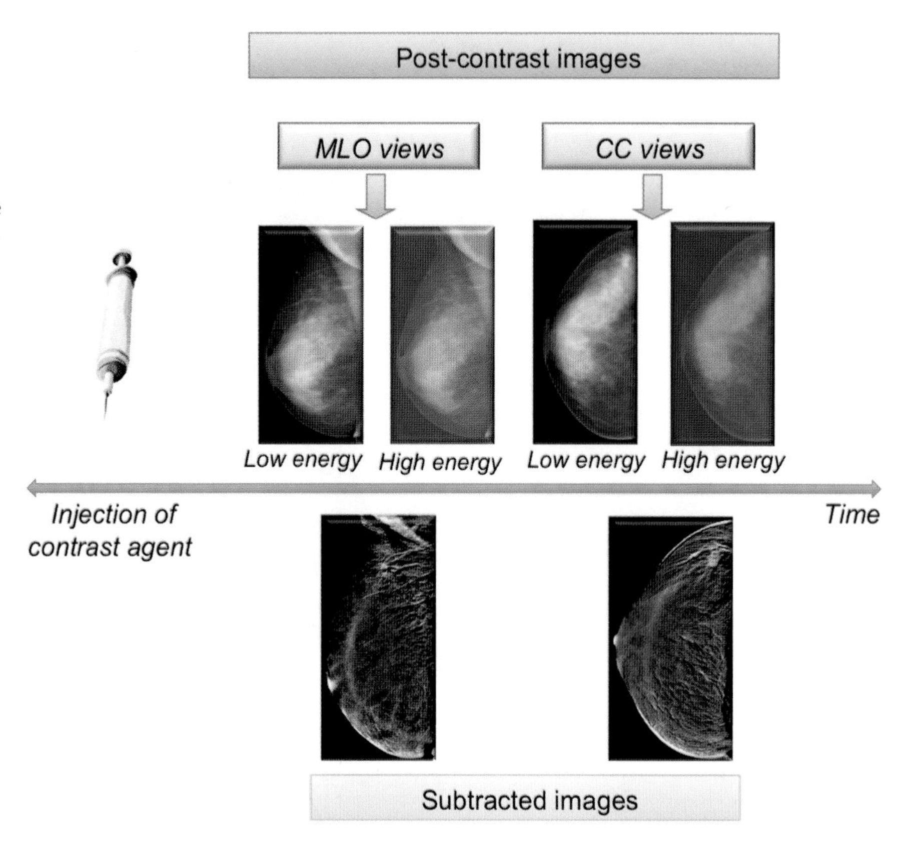

Fig. 12.2. Basic principle of temporal subtraction CEDM. High-energy digital mammography images are obtained before and after contrast medium injection. Then, the precontrast image is subtracted from the post-contrast images to enhance contrast uptake areas

washout of the contrast agent. Values of differential enhancement between the lesion and normal breast tissue can then be plotted versus time (Fig. 12.3).

Interpretation of CEDM images requires simultaneous evaluation with mammography. A dual-energy CEDM examination provides two native images of each view, one acquired with low energy similar to a conventional mammogram with a high contrast-to-noise ratio of microcalcifications and one acquired with high energy with a high contrast-to-noise ratio of iodine. It is therefore useful to analyze the two native images in addition to the subtracted image. Furthermore, the low-energy native images are highly similar to conventional mammograms and can be used to interpret the images instead of conventional mammography. However, as the low-energy image is acquired after contrast medium injection, it could be more informative than a conventional mammogram acquired without contrast agent. The resulting subtracted images are then reviewed using reading criteria based on contrast enhancement intensity and morphology. Spiculated and irregular margins, an irregular shape, and heterogeneous enhancement are the features most predictive of malignancy. Diffuse regional enhancement also has a high positive predictive value for malignancy.

As several images are acquired in the same position after contrast injection, the temporal subtraction CEDM technique allows the analysis of lesion enhancement kinetics. On breast MRI, a combined model using both morphologic characteristics and a kinetic assessment has been shown to improve diagnostic accuracy (Kuhl et al. 1999). A kinetic assessment is essentially useful to increase specificity in indeterminate lesions depicted using morphologic criteria. In a similar manner to that seen on breast MRI, a washout pattern on postcontrast CEDM images has a high positive predictive value for malignancy. However, its sensitivity is low, and most breast carcinomas depicted with temporal subtraction CEDM showed progressive enhancement (Jong et al. 2003; Dromain et al. 2006). One speculated hypothesis to explain this lack of washout in most cancers depicted with CEDM is the fact that contrary to 3D MR imaging, CEDM is a 2D projection imaging. So, measurements are made in a column of breast tissue including the lesion but also surrounding normal breast parenchyma. Since both the surrounding breast parenchyma, as well as certain portions of the lesion, will have a contiguous enhancement, the summation of signals in 2D is more often continuous and could mask portions of the lesion with a typical wash-out enhancement.

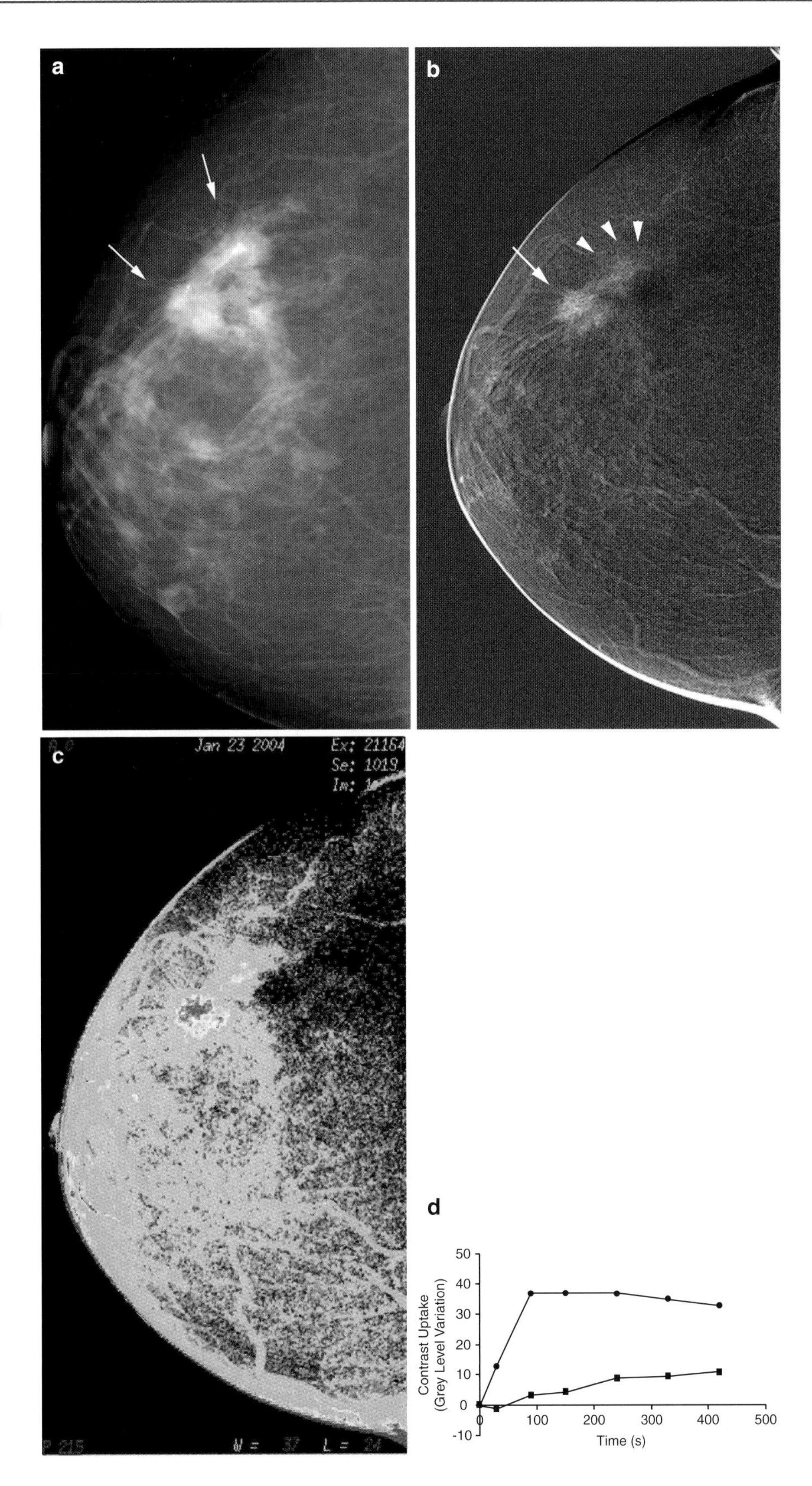

Fig. 12.3. Temporal subtraction CEDM examination. Cranio-caudal mammogram (**a**) shows two opacities in the upper outer quadrant (*arrows*). Subtraction image derived from the 1'30 temporal CEDM images (**b**) and corresponding parametric image of the maximum slope of enhancement (**c**) show two adjacent areas of enhancement, one with strong and spiculated enhancement (*arrow*) and the other with moderate and less circumscribed enhancement (*arrowheads*). Enhancement kinetics curves derived from ROIs drawn in the strong area of enhancement (d) show early enhancement followed by a washout whereas the moderate area of enhancement depicts gradual increasing enhancement. At histopathological examination, the lesion with strong and early enhancement proved to be a ductal carcinoma in situ, whereas the lesion with gradual increasing enhancement proved to be an invasive ductal carcinoma (IDC)

12.4 Comparison of Dual-Energy vs. Temporal Technique

Motion and misregistration artifacts, specific to subtraction imaging used during interpretation, are more likely to occur with the temporal technique than with the dual-energy technique because of longer compression time and because contrast medium is administered under breast compression. These artifacts may degrade image quality and confound interpretation. To decrease patient motion, the technologist must ensure that the patient is comfortable and must counsel the patient regarding the importance of remaining perfectly still during the examination, and especially during the administration of the contrast agent. This is why most clinical images of temporal CEDM were obtained with CC views, which are more comfortable than the MLO views.

The dual-energy CEDM technique allows a shorter acquisition duration than the temporal technique with better acceptance from patients, better reproducibility of the quality of subtracted images. Another main advantage of the dual-energy technique is that contrast is injected without breast compression, thus avoiding heat effect motions due to the arrival of contrast and better acceptance by patients and by technologists. Moreover, it has less impact on breast perfusion and contrast medium diffusion in the breast. Finally, the dual-energy technique is capable of imaging two different views of a breast or both with a single injection (Fig. 12.4), allows more accurate localization of

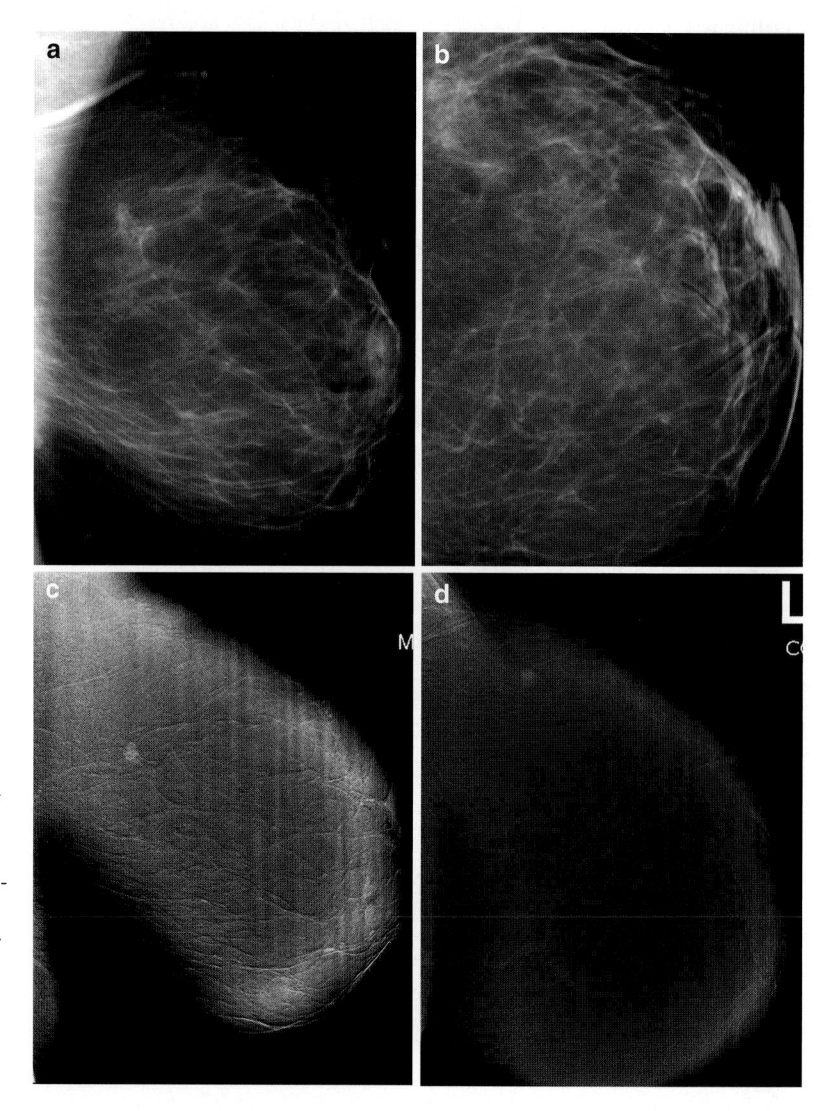

Fig. 12.4. Small IDC detected with dual-energy CEDM examination. Mammograms show a small opacity in the upper quadrant of the breast best seen on mediolateral view (**a**) than on cranio-caudal view (**b**). Subtracted dual-energy CEDM MLO (**c**) and CC (**d**) images, acquired respectively 2 and 4 min after injection of an iodinated contrast agent, readily depict a small enhancing nodule corresponding to a 6-mm invasive ductal carcinoma

the lesion, better assessment of disease extent, and provides better guidance for additional or second-look breast ultrasonography.

12.5 Advantages and Disadvantages of CEDM

CEDM is a modification of digital mammography and can be performed using a current digital mammography system with minor adaptations. Its implementation in a mammography facility is easy with immediate availability in the mammography suite. No special training of the technologist is needed for patient positioning nor for image acquisition, which are similar to that of conventional mammography.

In our experience, the CEDM examination is well accepted by patients who are pleased to have a complete assessment without any remaining questionable findings. Indeed, when screening mammograms demonstrate questionable or additional findings, additional MRI or follow-up examinations may be required, which are a source of unnecessary anxiety for patients. The CEDM examination can be performed immediately following mammography, in the same mammography suite without a new appointment and without loss of time. Moreover, it provides faster imaging when compared with MRI. Another exclusive advantage of CEDM relative to breast US and MRI is direct correlation with conventional mammograms. Suspicious contrast uptake detected on subtracted CEDM images can be easily analyzed retrospectively on conventional mammograms. The "subtracted" CEDM images can also be interpreted very easily and rapidly by radiologists and understood by oncologists and surgeons. We believe that the CEDM examination could be a useful imaging method to guide biopsy or wire localization in the future because it is easier, faster, and cheaper than MRI.

The main potential disadvantage of CEDM is the need for intravenous administration of an iodinated contrast agent. CEDM is therefore not indicated in patients with a contraindication to iodinated contrast agents. Another disadvantage compared with MRI is that CEDM is an irradiating technique. However, a controlled glandular dose is delivered to the patient, which is approximately equivalent to two conventional mammography views.

12.6 First Clinical Experience

Technical and clinical experience has been acquired and encouraging results have been published during the last few years on CEDM as an adjunct to mammography. All these studies used a prototype developed from a commercialized full-field digital silicon flat-panel digital breast (Senographe 2000D and Senographe DS, General Electric Healthcare) modified for the acquisition of high-energy exposures. The temporal subtraction technique was first tested with an approach similar to that of breast MRI (Jong et al. 2003; Dromain et al. 2006; Diekmann 2005). These first studies, in a limited sample of patients showed that digital mammography was capable of depicting tumor angiogenesis in invasive breast cancer. Jong et al. (2003) examined 22 women with 10 cancers and 12 benign lesions with temporal CEDM. Enhancement was observed in 89% of the invasive cancers. The two false negatives corresponded to one case of ductal carcinoma in situ and one case of invasive ductal carcinoma. Among the 12 patients with a benign breast lesion, initially considered worrisome at conventional mammography or US, 7 had no enhancement and 5 had nodular enhancement corresponding to 3 fibroadenomas and 2 cases of a fibrocystic change with focal intraductal hyperplasia. The kinetic curves did not consistently demonstrate a distinctly different pattern for benign and malignant lesions. In our study (Dromain et al. 2006), temporal CEDM examinations were performed in 20 patients with 22 malignant tumors. The sensitivity of CEDM for the detection of breast carcinoma was 80%. There was an excellent correlation between the histopathology size and the size of the enhancement measured on CEDM subtracted images. As in the study by Jong et al., most of the cancers showed progressive enhancement without a washout pattern. A poor correlation between the intratumor mean vascular density evaluated on CD34-immunostained histological sections and quantitative characteristics of enhancement kinetics curves was observed, probably because it was difficult to perform quantitative assessment on projection images acquired from breasts under compression. Indeed, CEDM images are projections of the entire breast and enhancement depends on the size of the tumor. Moreover, enhancement is not exclusively due to the number of vessels but is also most likely related to functional parameters such as vessel permeability, especially when using contrast agent that is migrating to the extracellular fluid space. A multicentric retrospective reader study aimed at comparing the diagnostic accuracy of temporal CEDM as an adjunct to mammography (MX) with the diagnostic accuracy of mammography alone was performed (Diekmann et al. 2007a, b). Data were collected at four centers in Europe and North America in 75 women with 85 lesions (17 benign and 68 malignant). The average sensitivity and specificity across the five readers for the probability of cancer increased with mammography + CEDM from 0.81 to 0.86 and 0.62 to 0.66, respectively. The area under the ROC curve for mammography + CEDM using the probability of cancer scale, was higher than the area under the ROC curve for MX alone for all readers (with statistical significance for two readers).

The study by Lewin et al. (2003) is the only published preliminary clinical experience using the dual-energy technique. They studied 26 women with 13 invasive carcinomas. Eight (85%) of the invasive carcinoma exhibited strong enhancement, 3 showed moderate enhancement, and 1 showed weak enhancement. Among these lesions, 5 were not detected on conventional mammograms. More recently, we tested the dual-energy CEDM technique in 120 patients referred to our institution for a screening mammograms recall with uncertain mammographic findings (Dromain et al. 2008). For each CEDM examination, a pair of low- and high-energy images was acquired using a modified full-field digital mammography system (Senographe DS GE healthcare) with an MLO view at 2 min and a CC view at 4 min after injection of 1.5 mL/kg of an iodinated contrast agent. Compared to mammography alone for the BIRADS assessment, CEDM yielded significantly higher sensitivity (93% vs. 78%) and a negative predictive value (87% vs. 67%) and had a clearly greater area under the ROC curve. When we compared CEDM with the standard of care (mammography + US), this difference was less evident but still present. In particular, dual-energy had significantly higher specificity than mammography + US.

12.7 Potential Clinical Applications

CEDM is of interest in the nonscreening setting. It would not replace mammography but could be helpful in selected cases where there are unanswered questions after clinical and conventional mammography evaluation (Fig. 12.5). Although the Food and Drug Administration has not yet approved this technology, CEDM has the potential to increase the cancer detection rate, to improve staging, and to improve the selection of patients for biopsy. Indeed, if it is negative, it could help avoid unnecessary biopsies. If positive, it could heighten confidence in the presence and the suspicion of cancer, improve cancer localization and size assessment, and help US focus on anomalies detected by CEDM.

Basically, the potential indications are closely similar to those of breast MRI. The potential indications for CEDM include the evaluation of newly diagnosed breast cancer, to assess the extent of disease (Figs. 12.6 and 12.7) and to evaluate possible contralateral disease, a problem-solving method for better characterization when conventional mammography assessment is equivocal (Fig. 12.5), the evaluation of residual disease after a lumpectomy with positive margins, the evaluation of breast cancer recurrence, the detection of breast cancer in women with axillary metastasis.

Only two classic indications for breast MRI do not appear, *a priori*, to be good indications for the CEDM. They are screening of high-risk women (BRCA1/2) because of their sensitivity to radiation exposure and the assessment of response to neoadjuvant chemotherapy because several imaging examinations are needed and will best be performed using a nonradiating imaging method.

12.8 Future Improvements

One of the future promising improvements of CEDM is its combination with digital breast tomosynthesis (DBT) into a single technique. Indeed, one of the most important limitations of mammography is the combination of structures in two-dimensional views than can obscure a cancer or create a false-positive lesion. DBT is another modification of the digital mammography unit to enable the acquisition of multiple projected views with different angles that are processed into a series of 1-mm slices that depict the tissue structure in three dimensions. Early experience indicates that DBT has the potential to ameliorate the visibility of the lesion by reducing or eliminating tissue overlap. It provides a better delineation of the lesion border that could result in easier differentiation between benign and malignant lesions (Niklason et al. 1997; Poplack et al. 2007; Andersson et al. 2008). It also provides

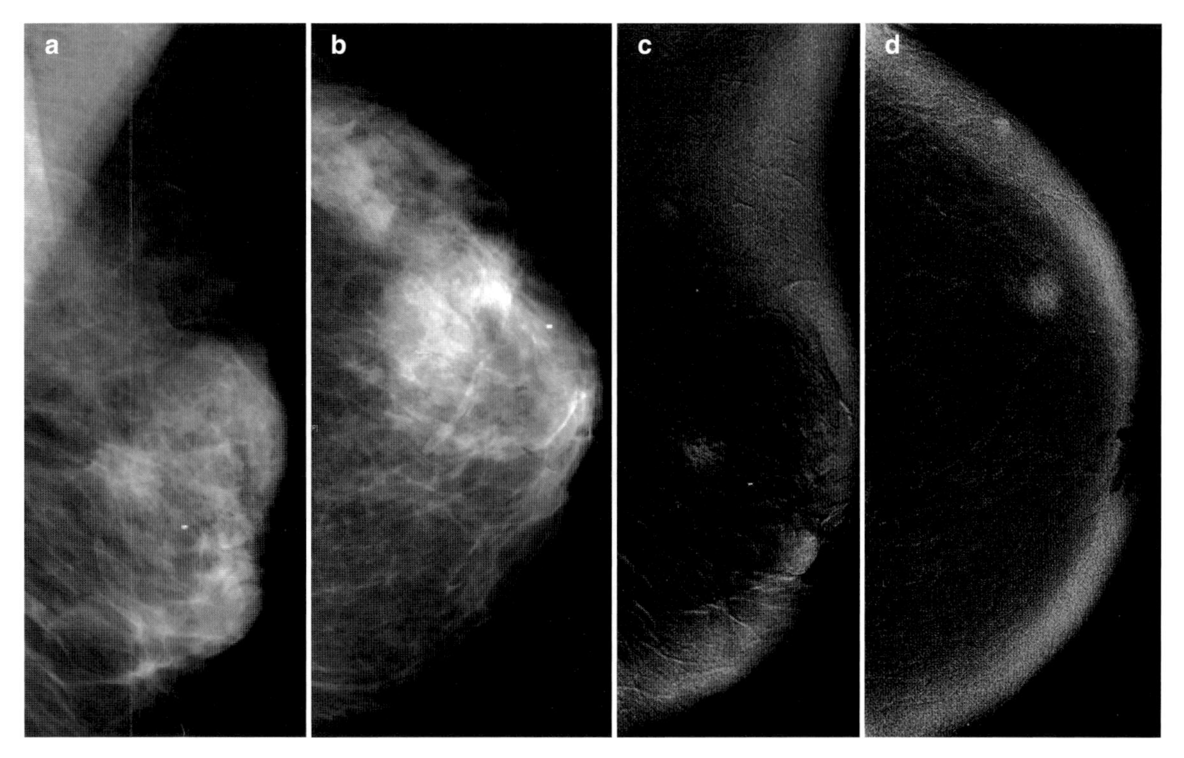

Fig. 12.5. Clarification of equivocal lesion using dual-energy CEDM examination. Mediolateral (**a**) and cranio-caudal (**b**) mammograms show asymmetry of density in the upper and outer quadrant. Subtracted dual-energy CEDM MLO (**c**) and CC (**d**) images, acquired, respectively, 2 and 4 min after the injection of an iodinated contrast agent, readily depict an enhancing nodule with irregular margins highly suggestive of malignancy. The CEDM examination has greater sensitivity and specificity, especially in dense breasts

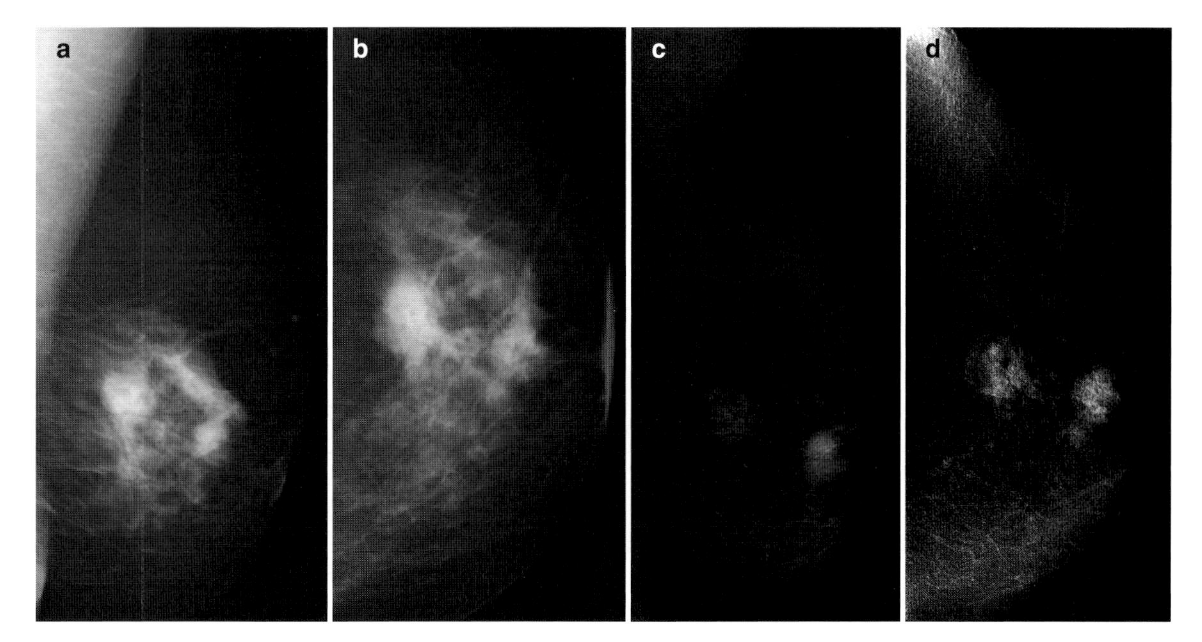

Fig. 12.6. Determining the extent of disease using dual-energy CEDM. Mediolateral (**a**) and cranio-caudal (**b**) mammograms show areas of density, poorly circumscribed, in the central area of the breast suggestive of a multifocal tumor. Mammographic findings are not sufficient for precise preoperative staging. Subtracted dual-energy CEDM MLO (**c**) and CC (**d**) images provide additional information and enable accurate determination of the size and the extent of disease. The size and the extent of enhancing lesions demonstrated on subtracted CEDM images were closely correlated with pathological findings

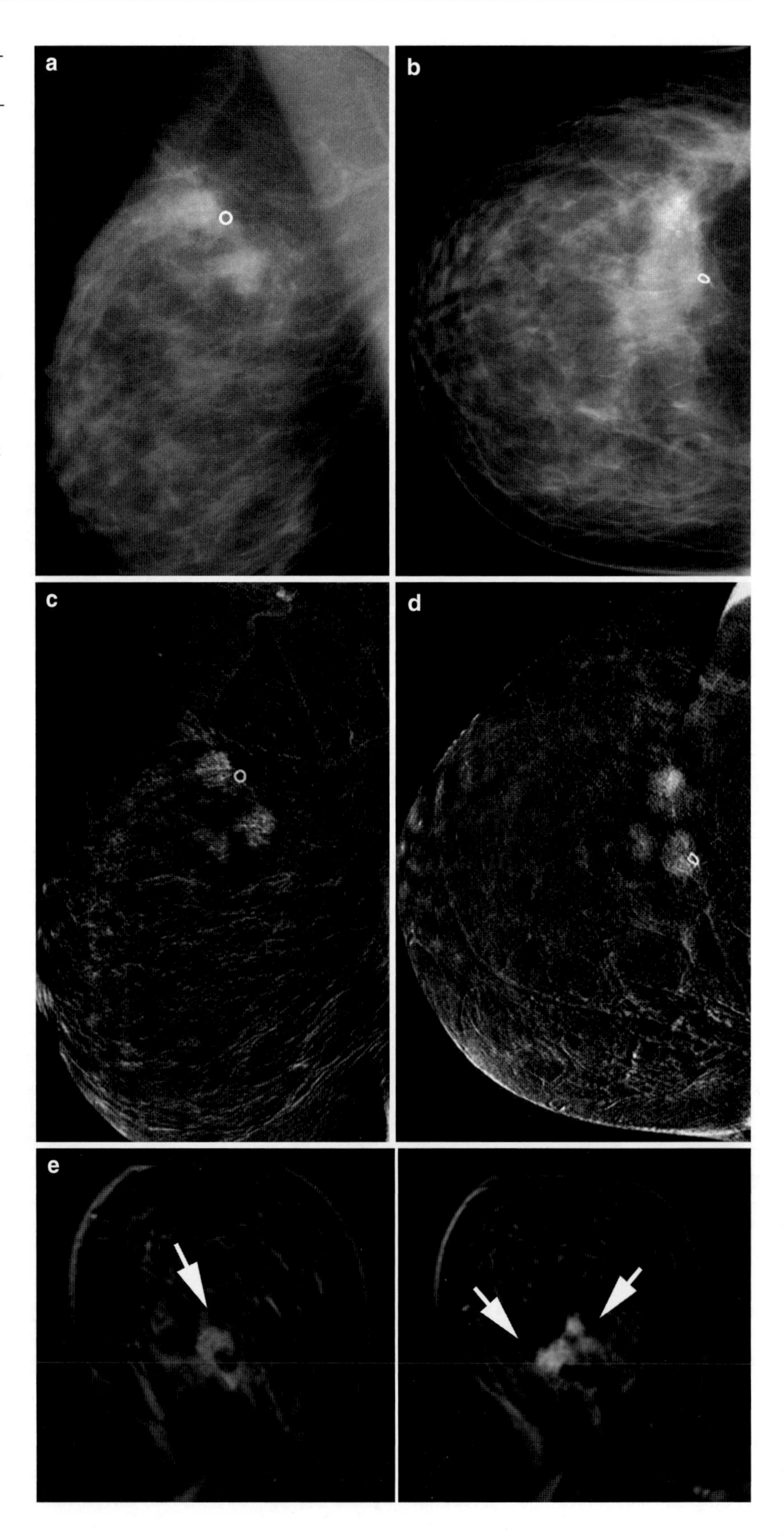

Fig. 12.7. Invasive ductal carcinoma in a 44-year-old woman with a palpable mass. Mediolateral (**a**) and cranio-caudal (**b**) mammograms show a poorly circumscribed opacity in the upper and outer quadrant. Subtracted dual-energy CEDM MLO (**c**) and CC (**d**) images, acquired respectively 2 and 4 min after the injection of an iodinated contrast agent, show three adjacent enhancing nodules and enable accurate determination of the size and the extent of disease. CEDM findings are closely correlated with MRI findings (*arrows*) (**e**)

more accurate 3D tumor localization for surgical planning. Combining CEDM with DBT may offer the benefits of both approaches. Only a few studies have investigated the use of contrast-enhanced DBT (Chen et al. 2007; Carton 2007). In a pilot study, Chen et al. demonstrated the feasibility of contrast-enhanced digital breast tomography (CE-DBT) in 13 patients. CE-DBT was performed at 49 kVp with a rhodium target and a copper filter. A temporal subtraction technique was used with acquisitions of pre- and postcontrast DBT image sets in the medial lateral oblique projection with slight compression. Each image set consisted of nine images acquired over a 50° arc and was obtained with a mean glandular X-ray dose comparable to two conventional mammographic views.

Another potential improvement is the use of the slot-scan system with a photo-counting detector. In this detector, individual X-ray quanta, which interact in the detector, are counted. Indeed, these mammography systems enable dual-energy visualization and distinction of high-energy from low-energy photons with a single image acquisition and thus, without resort to subtraction. Experimental results suggest that they may yield images free of motion artifacts (Diekmann et al. 2008).

12.9 Conclusion

Contrast enhanced digital mammography is a new advanced application of full-field digital mammography enabling the detection of angiogenesis in breast carcinoma. This new breast imaging method can be easily implemented clinically using a current digital mammography system with minor adaptations and commercially available iodinated contrast agent. CEDM should be a useful adjunct to diagnostic mammography and a promising problem-solving and staging tool.

Acknowledgments The authors thank General Electric Medical Systems and Serge Muller for technical assistance.

References

American College of Radiology (2007) ACR practice guidelines for the performance of magnetic resonance imaging (MRI) of the breast. Arch Oncol 15(1-2):37–41

Andersson I, Ikeda DM, Zackrisson S, et al (2008) Breast tomosynthesis and digital mammography: a comparison of breast cancer visibility and BIRADS classification in a population of cancers with subtle mammographic findings. Eur Radiol 18 (12):2817–2825

Bick U, Diekmann F (2007) Digital mammography: what do we and what don't we know? Eur Radiol 17(8):1931–1942

Carton AK (2007) Technical development of contrast-enhanced digital breast tomosynthesis. Radiology 238 (Suppl):318

Chang CH, Nesbit DE, Fisher DR, et al (1982) Computed tomographic mammography using a conventional body scanner. Am J Roentgenol 138:553–558

Chen SC, Carton AK, Albert M, et al (2007) Initial Clinical experience with Contrast-Enhanced Digital Breast Tomosynthesis. Acad Radio 14:229–238

Diekmann F (2005) Digital mammography using iodine-based contrast media: initial clinical experience with dynamic contrast medium enhancement. Invest radiol 40(7):397–404

Diekmann, F, Fredenberg E, Lundqvist M, et al (2008) Contrast agents in digital mammography: a new diagnostic option resulting from the use of photon-counting detectors. RSNA abstract book, p 121

Diekmann F, Marx C, Jong R, Dromain C, et al (2007a) Diagnostic accuracy of contrast-enhanced digital mammography as an adjunct to mammography. Abstract. Europ Radiol 17:174

Diekmann F, Sommer A, Lawaczeck R, et al (2007b) Contrast-to-noise ratios of different elements in digital mammography: evaluation of their potential as new contrast agents. Invest Radiol 42:319–325

Dromain C, Balleyguier C, Muller S, et al (2006) Evaluation of tumor angiogenesis of breast carcinoma using contrast enhanced digital mammography. Am J Roentgenol Nov 187(5):W528–W537

Dromain C, Balleyguier C, Tardivon A, et al (2008) Dual energy contrast enhanced digital mammography: preliminary clinical results. RSNA abstract book:121

Hagay C, Cherel PJ, de Maulmont CE, et al (1996) Contrast-enhanced CT: value for diagnosing local breast cancer recurrence after conservative treatment. Radiology 200: 631–638

Jong RA, Yaffe MJ, Skarpathiotakis M, et al (2003) Contrast-enhanced digital mammography: initial clinical experience. Radiology 228:842–850

Kuhl CK, Mielcareck P, Klaschik S, Leutner C, et al (1999) Dynamic breast MR imaging: are signal intensity time course data useful for differential diagnosis of enhancing lesions? Radiology Apr 211(1):101–110

Lewin JM, Isaacs PK, Vance V, et al (2003) Dual-energy contrast-enhanced digital subtraction mammography: feasibility. Radiology 229:261–268

Niklason LT, Christian BT, Niklason LE, et al (1997) Digital tomosynthesis in breast imaging. Radiology 205(2):399–406

Pisano ED, Gatsonis C, Hendrick E, et al (2005) Digital Mammographic Imaging Screening Trial (DMIST) Investigators Group. Diagnostic performance of digital mammography versus film mammography for breast-cancer screening. N Engl J Med Oct 27 353(17):1773–1783

Poplack SP, Tosteson TD, Kogel CA, et al (2007) Digital breast tomosynthesis: initial experience in 98 women with abnormal digital screening mammography. Am J Roentgen 616–623

Puong S, Bouchevreau X, Patoureaux F, et al (2007) Dual-Energy Contrast-Enhanced Digital Mammography using a new approach for breast tissue cancelling. Proceeding of SPIE, Medical Imaging

Skaane P, Balleyguier C, Diekmann F, et al (2005) Breast lesion detection and classification: comparison of screen-film mammography and full-field digital mammography with soft-copy reading-observer performance study. Radiology Oct 237(1):37–44

Skaane P, Skjennald A (2004) Screen-film mammography versus full-field digital mammography with soft-copy reading: randomized trial in a population-based screening program–the Oslo II Study. Radiology July 232(1): 197–204

Skarpathiotakis M, Yaffe MJ, Bloomquist AK, et al (2002) Development of contrast digital mammography. Med Phys 29:2419–2426

Digital Breast Tomosynthesis and Breast CT

13

Felix Diekmann

CONTENTS

Felix Diekmann
Department of Radiology, Charité,
Klinik für Strahlenheilkunde, Augustenburger Platz 1,
13353 Berlin, Germany

KEY POINTS

The new techniques of three-dimensional visualization of the breast that have been developed in recent years have the potential to markedly improve diagnostic breast imaging. Many experts even expect the new 3D techniques such as digital breast tomosynthesis (DBT) or breast CT to completely replace "conventional" digital mammography in the intermediate term as they have definitive advantages over 2D mammography. Various approaches are theoretically conceivable and have to be evaluated in the clinical setting, including the use of different reconstruction algorithms, the use of different DBT angles, slit-scan DBT vs. flat-panel DBT vs. breast CT, different dose settings, and DBT in one plane vs. two planes. Finally, DBT might also have advantages in combination with other tools for improving the diagnostic accuracy of breast imaging such as computer-aided diagnosis and contrast mammography.

13.1 Introduction

Simple screening mammography has only moderate sensitivity. In the so-called DMIST study, sensitivity was only about 40% in women with very dense breast tissue (Pisano et al. 2005). There are several reasons for the poor sensitivity of mammography when used as a stand-alone technique. First, a focal lesion depicted on a mammogram is often difficult to characterize. Since it is not possible to definitely characterize all inconclusive focal breast lesions by additional diagnostic procedures, the radiologist has to use a set of criteria for differentiating benign vs. malignant breast lesions (e.g., BIRADS lexicon). Such criteria have limited validity and are affected by many factors (Chlebowski and Khalkhali 2005). It has been shown for instance that morphologic criteria for the characterization of breast lesions vary according to the risk for breast cancer of the patient population examined and the imaging modality used (Schrading and Kuhl 2008). Experience indicates that correct lesion characterization is also more difficult in dense glandular tissue, one reason being that it is more difficult to classify a lesion as "smoothly marginated" if the margins are obscured by superimposed glandular tissue. Another problem is that dense glandular tissue often obscures a lesion altogether for two reasons: first, lesion conspicuity may be poor because a lesion lacks contrast with surrounding tissue. This is the case for tumors that differ only little from normal glandular tissue and often pose a diagnostic dilemma. Invasive lobular carcinoma is a case in point (Diekmann et al. 2004). Second, a focal lesion may be obscured by massive overlying dense glandular tissue, rendering a lesion very difficult to detect by two-dimensional mammography (Carney et al. 2003). The limited sensitivity of mammography has inspired numerous new concepts to overcome this problem. For instance, it is theoretically conceivable to visualize neoangiogenesis using laser wavelengths tuned to blood products. Preliminary results suggest that this concept is basically feasible using a combination of molecular imaging (visualization of microvessels/neoangiogenesis) and 3D mammography (CT-Scan) (Poellinger et al. 2008); however, large prospective clinical studies are not yet available.

Impedance measurement is another approach that has been proposed to specifically improve the detection of breast cancer in dense glandular tissue. This approach exploits the differences in conductivity between normal glandular tissue and cancer (Diebold et al. 2005). The use of contrast agents can improve contrast between some tumors and breast tissue in digital mammography; this approach is presented in one of the preceding chapters by Dr. Dromain (Chen et al. 2007; Diekmann and Bick 2007; Diekmann et al. 2003, 2005, 2007b; Jong et al. 2003; Lawaczeck et al. 2003; Lewin et al. 2003). Great hope is placed on new 3D approaches such as DBT or dedicated breast CT, which are expected to improve the visualization of tumors otherwise obscured by dense glandular tissue (Karellas et al. 2008; Poplack et al. 2006). These new 3D mammography techniques are described in more detail in this chapter.

13.2 Hardware

Both new 3D techniques – breast CT and DBT – are based on the well-established concept of tomography and the principles used in computed tomography. Breast CT using dedicated equipment is the 3D technique for the breast that most closely resembles conventional CT scanners (Chen and Ning 2002; Shaw et al. 2005; Tang et al. 2005; Yang et al. 2007). A dedicated CT scanner images the breast hanging through a hole in the table into the gantry. The patient is positioned prone on the table and the breast is scanned, as in conventional computed tomography, by a device rotating around the breast under the table. Using this approach, the breast can be scanned as in incremental CT (such a scanner was used for obtaining the breast CT presented in Fig. 13.1) or with devices performing a spiral movement (and thereby imitating spiral CT). Both types of scanners allow 360° rotation around the breast.

In contrast, DBT uses a smaller scan angle. The X-ray source moves around the breast within the range of the angle. In flat-panel DBT, the detector is stationary and, unlike in conventional tomography, does not move opposite to the X-ray source. A diagram of the setup for flat-panel DBT from three different scan directions is shown in Fig. 13.2. The figure also illustrates the principle of slit-scan DBT. As in flat-panel DBT, the X-ray source moves around the breast at a predefined angle. The detector also moves around the breast (as known from digital mammography using the slot-scan technique). The setup for slit-scan DBT is also shown in a diagram presented in Fig. 13.2.

Many of the parameters used in DBT can be varied including the scan angle and the radiation dose (see Sect. 13.5). Since slit-scan tomography uses multiple detectors for generating a scan rather than a single slit for one detector and these detectors scan a certain angle, the term "DBT angle" has two meanings. The angle in both meanings of the term can be varied. One might assume that radiologists should prefer the angle for DBT to be as large as possible. A maximum scan angle of 360° would then represent a dedicated breast CT scanner. However, images obtained with a smaller scan angle might be easier/faster to interpret. Radiation exposure is another issue in 3D imaging that is still under discussion. It is usually attempted to perform DBT with a radiation exposure that is below the total dose of two-plane mammography. Because of the anisotropic voxel size

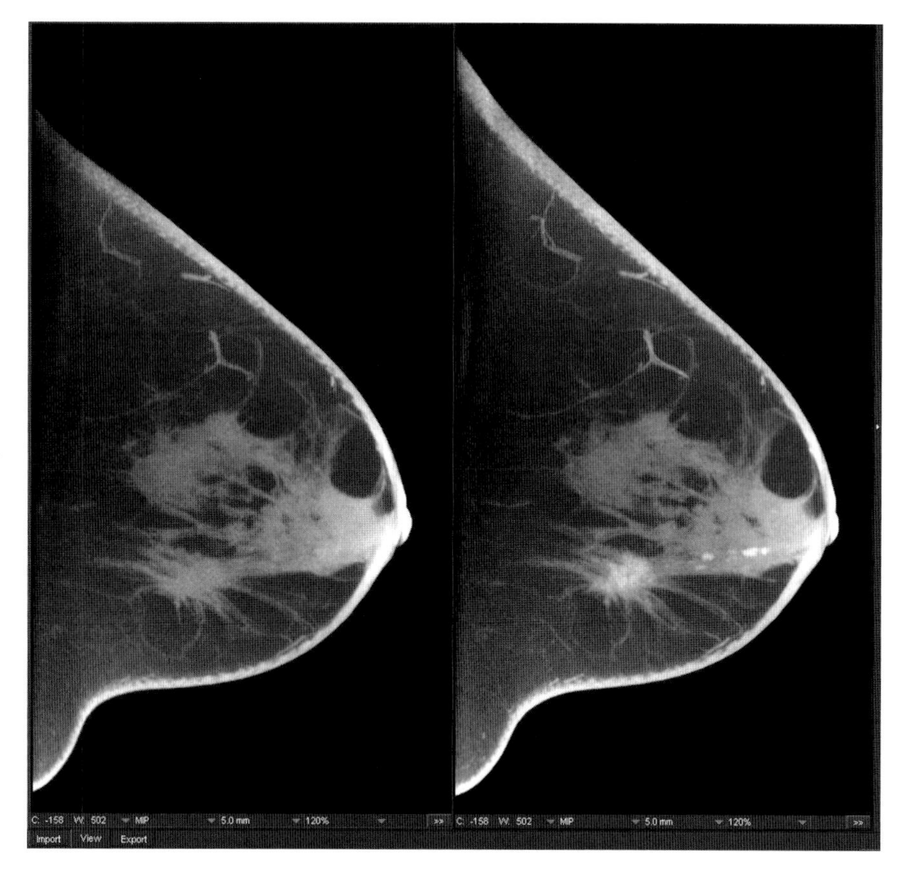

Fig. 13.1. Forty-five-year-old female presents for evaluation of left lump. Precontrast (*left*) and postcontrast (*right*) sagittal cone beam CT (Koning Corporation, West Henrietta, New York) images of the left breast. Spiculated mass (Invasive Ductal Carcinoma) in the lower part of the left breast showing strong enhancement after administration of nonionic contrast material. Additional small areas of enhancement extending in a linear distribution toward the nipple representing additional Invasive Ductal Carcinoma. Images courtesy of Stamatia Destounis, Elizabeth Wende Breast Care, LLC

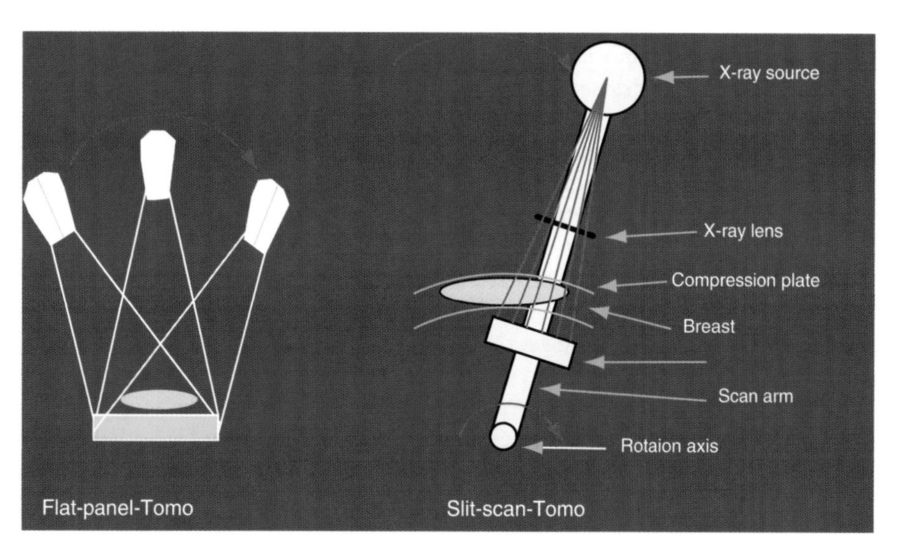

Fig. 13.2. Principles of flat-panel DBT and slit-scan DBTs (with reprint permission from Diekmann and Bick 2007)

in the reconstructed slices, however, it is currently still open whether DBT should be performed in two planes, just like conventional mammography. This question must be answered especially if DBT is to be used instead of rather than as an adjunct to conventional mammography.

While numerous quality controls have been established for the hardware used in mammography, quality criteria are still lacking for DBT. Since one of the central concerns in developing DBT was to remove interfering overlying structures for sharp visualization of the structures of interest, it is conceivable to use the accuracy with which the separation of interfering structures from adjacent slices can be accomplished as a quality criterion for DBT. Radiologists are used to virtually 100% selectivity from computed tomography, where the impact of so-called partial volume effects is minimal and decreases with the slice thickness used. In DBT, on the other hand, structures above and below the target slice massively interfere with the image generated of the target slice. This phenomenon is known as smearing and the degree of smearing varies with the scan angle, the reconstruction algorithm used, the contrast of the smearing structures, and other factors. Smearing may not always be undesirable but can be exploited to advantage. It is possible that the detection of microcalcifications by a radiologist viewing a stack of images can be improved if they are visible in multiple slices (though blurred) above and below the slice in which they are highlighted. Figure 13.3 shows three slices from a DBT of a phantom containing simulated microcalcifications. The first (uppermost) DBT slice shows marginal calcifications in the center. However, the slice also contains blurred additional microcalcifications at the bottom of the image. Smearing was investigated by Ruben van Engen (Nijmegen) for slit-scan DBT. The effect is illustrated in Fig. 13.4. Looking at slices away from the target slice, in which the microcalcifications are sharply depicted, the calcifications appear larger, less conspicuous, and more blurred as the distance from the target slice increases. This effect may impair interpretation of the images (e.g., in evaluating distribution/morphology) or, on the contrary, may improve detection in that a radiologist "leafing" through a stack of images "is being prepared" for the calcifications, which are no longer just highlighted in a single slice. Clinical studies are needed to determine whether this effect is actually beneficial or should be minimized because it is distracting rather than helpful. Alternatively, it is conceivable to use the smearing effect or degree of separation as a quality assurance criterion for DBT. While a very refined system of quality controls has been established for conventional digital mammography (see Chap. 3), binding guidelines are still lacking for DBT. Further investigations are needed to identify suitable quality control mechanisms for this new technique.

The role of the smearing effect might be pivotal for the decision whether DBT or dedicated breast CT is a more suitable technique for 3D mammography. DBT is not CT – the very high resolution of DBT in the xy-plane – which is expected to develop in direct proportion to the higher resolution of new detector generations in digital mammography comes at the expense of resolution in the z-direction, which gives rise to the smearing effect. Computed tomography, on the other hand, is characterized by an excellent separation of individual voxels, which also applies in the z-plane. While imaging using isotropic voxels is becoming the dominant approach in whole-body imaging, it is still open whether isotropic voxels are also desirable for mammography (see Sect. 13.7).

13.3 Software: Image Viewing Systems and Reconstruction Algorithms

Numerous algorithms are available for processing DBT source data sets. The most widely known algorithm is the shift-and-add algorithm. This algorithm is very similar to a simple backprojection algorithm. A diagram of the shift-and-add algorithm is presented in Fig. 13.5. Images from different angles are acquired by tilting the tube. These are subsequently added together; only one plane appears sharp because the images are shifted against each other. This reconstruction algorithm closely imitates conventional tomography, in which the X-ray source moves in opposite direction to the detector. Detector movement is electronically simulated using shift and add.

Backprojection algorithms are also used in computed tomography. Today, filtered backprojection is the most widely used algorithm for the reconstruction of CT data sets. However, the reconstruction technique originally proposed by Hounsfield in his patent specification for CT in 1972 is iterative reconstruction of 3D data sets. This reconstruction technique uses a set of linear equations. Since the number of unknowns in the equation system increases exorbitantly with the resolution of the system, this approach was impractical for a long time due to the demands on computing capacity. Nevertheless, this approach was more accurate and more suitable, at least for the originally used matrix of 64 pixels, compared with the filtered backprojection approach used today. There is a tendency in DBT to return to algebraic reconstruction. Algebraic algorithms such as SART (simultaneous algebraic reconstruction technique) or ML-EM (maximum likelihood expectation maximization) are used experimentally for the reconstruction of 15 or 21 different scan angles (Wu et al. 2004; Zhang et al. 2006).

The software used in DBT is not only required for primary single-slice reconstruction but also for

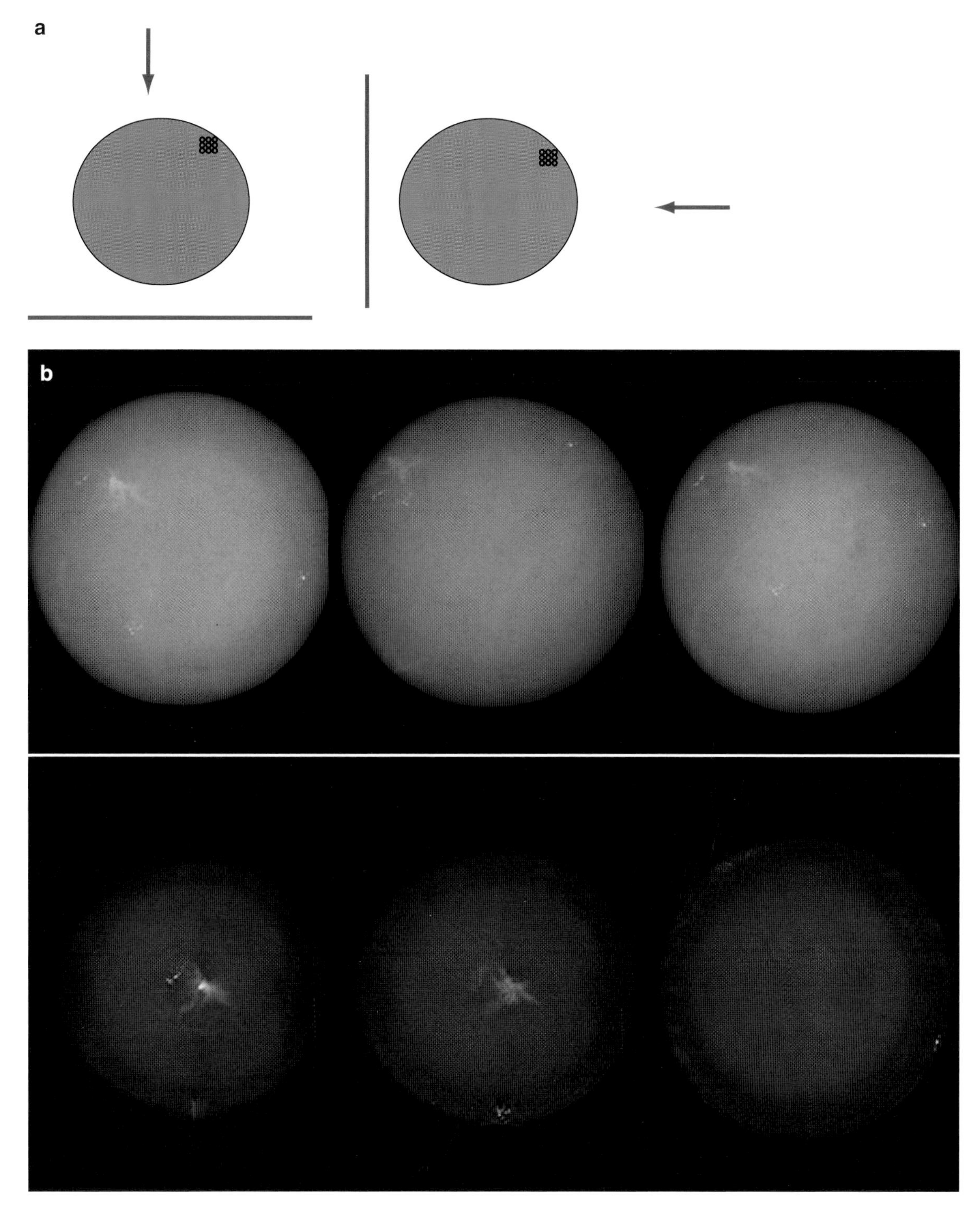

Fig. 13.3. (**a**) Diagram of specimen radiography in craniocaudal and mediolateral projections (radiography in 2 planes). Although the microcalcification cluster is contiguous with the outer margin, both images will show a clear margin around the luster. Performed intraoperatively, specimen radiography in 2 planes would result in R1 resection.(**b**) Rubber ball with simulated marginal microcalcifications. Neither of the three projections (craniocaudal, mediolateral, and oblique; *upper panel*) shows that the microcalcifications are contiguous with the margin. Selected DBT slices (*lower panel*) clearly show the microcalcifications to be contiguous with the margin: calcifications in the uppermost image (*first image*), marginal calcification at the bottom (*second image*), cluster contiguous with the right margin (*third image*). Images courtesy of Sylvie Puong and Serge Muller, GE Healthcare

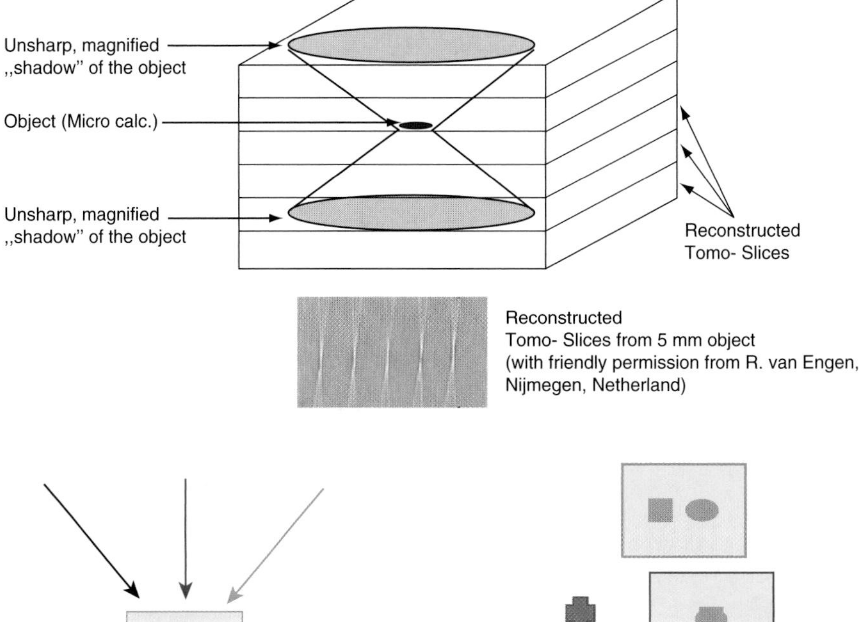

Fig. 13.4. Smearing effect in DBT. The simulated microcalcifications appear larger/more blurred with increasing distance of the slice from the actual site of the calcifications

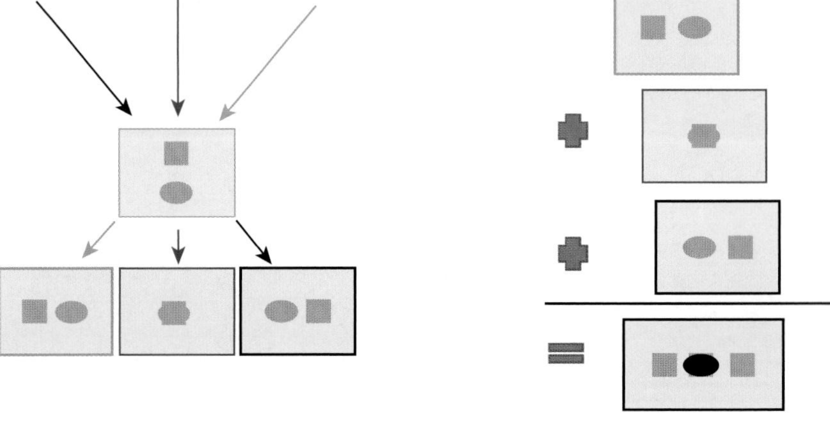

Fig. 13.5. Diagram of the shift-and-add algorithm. The left drawing represents three images obtained from different directions. The right drawing illustrates how the shifted images are added together, resulting in sharp visualization of one object (*oval*)

secondary reconstruction, which serves to present the large image data sets in a manner that allows efficient interpretation. The question as to how presentation of image data affects interpretation by the radiologist has been extensively investigated for conventional mammography (see the Chap. 7). The problem of adequate image presentation is potentiated for 3D imaging techniques, where the time required for interpretation and reporting and the radiologist's diagnostic quality are even more dependent on effective display of the image data. Manufacturers therefore have developed software tools to first present the image data sets in a different format and not as a set of thin slices (e.g., 3D structure or thicker slices reconstructed from the primary thin slices). The problem has been known from computed tomography since the advent of multislice CT scanners. The primary thin slices are degraded by a high noise level and are often more difficult and time-consuming to interpret than secondary reconstructions. Different algorithms are available for secondary reconstruction of thin slices (e.g., the known maximum intensity projection [MIP], minimum intensity projection, averaging, or complex functions, see (Diekmann et al. 2007a, b)). The choice of reconstruction algorithm determines whether (flat) low-contrast objects or high-contrast objects such as microcalcifications tend to be highlighted.

13.4 Clinical Applications

As already mentioned at the beginning of this chapter, the major limitations to be overcome to improve the sensitivity of mammography are the at times poor lesion conspicuity and the fact that tumors may be obscured by overlying structures. This is illustrated in Fig. 13.6, which shows a tumor that is hard to see on the original digital mammogram due to overlying dense glandular tissue. The same tumor is

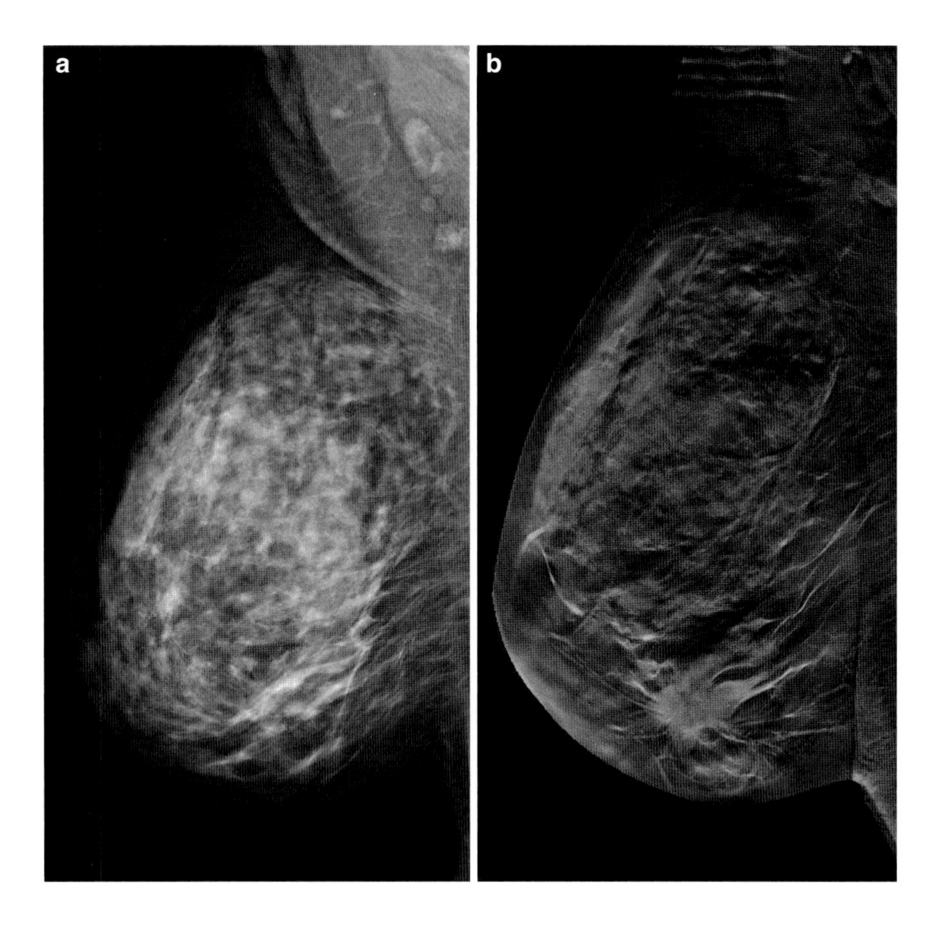

Fig. 13.6. Original mammogram (**a**) of a tumor in the lower portion of the oblique view. The tumor is difficult to see. A selected 1-mm flat panel DBT slice (**b**) from the same patient shows a tumor that still has low contrast but is much more conspicuous when compared with the mammogram.
Images courtesy of Dr. Andersson, Malmö University

clearly visualized by (flat-panel) DBT. The problem of poor tumor conspicuity cannot be overcome by a technique such as DBT if tumor contrast persists to be poor even after superimposed structures have been largely removed by this technique. Figure 13.7 shows an example of invasive lobular breast cancer, which is poorly delineated from surrounding breast tissue due to poor contrast and because the X-ray morphology of the tumor is similar to that of normal breast tissue in both digital mammography and (slit-scan) DBT.

Available data is still inadequate for an accurate appraisal of the future clinical role of DBT. However, one may speculate about the possible benefit of this new technique for a range of different applications based on theoretical considerations. DBT has the potential to improve specimen radiography. Two-dimensional mammography will never allow absolutely accurate evaluation of the site of an abnormality in a surgical specimen. Specimen radiography in two planes provides much more accurate information; nevertheless, it may also fail to demonstrate contiguity of a focal lesion with the resection margin. The schematic drawing in Fig. 13.3a illustrates how specimen radiography in two or three planes may fail to demonstrate the marginal location of microcalcifications in surgical specimens. The same principle is illustrated in Fig. 13.3b using a phantom (rubber ball with simulated microcalcifications). All three projection images (craniocaudal, mediolateral, and oblique views) show microcalcification clusters that do not extend to the outer margin of the specimen. In contrast, selected DBT slices show the calcifications to be contiguous with the margin. Used intraoperatively, this technique might thus help radiologists make a more valid recommendation for extending the resection without subjecting the patient to a second operation. Based on the same principle, DBT can also directly visualize skin calcifications without the need for obtaining additional tangential images as proposed for digital 2D mammography.

Some studies have also investigated the routine use of DBT for the detection of breast cancer. However, larger studies, e.g., in screening populations, for definition of the clinical significance of 3D mammography are still lacking. The results obtained by Poplack et al.

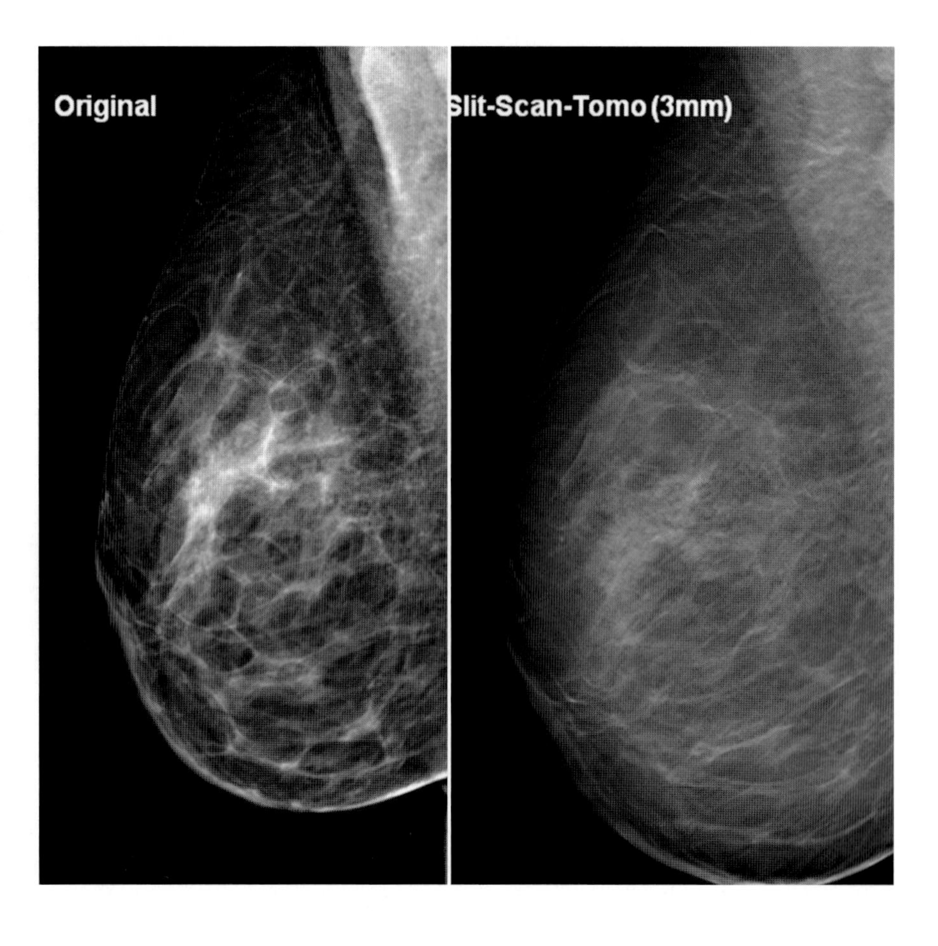

Fig. 13.7. Example of a tumor that is difficult to see on the original digital mammogram (*right*) and in slit-scan DBT. Images courtesy of K. Leiflandt, St. Göran Hospital, Stockholm

in a population of about 100 women suggest that a markedly lower recall rate might result if DBT is used as the primary screening modality (Poplack et al. 2006). Andersson et al. (2008) investigated 36 patients and showed DBT to have a higher sensitivity than conventional digital mammography in this small patient population. In this study, a higher BIRADS category was assigned in 12 of the 36 patients. Good et al. also showed that DBT tended to improve the diagnosis; however, the authors state that a larger population needs to be examined for a more valid comparison. The authors also indicate that the fact that most differences between DBT and conventional digital mammography were not significant might be attributable to the small number of cases investigated.

When looking at the available studies, one must clearly differentiate between different uses of DBT. When used as an add-on modality for conventional digital mammography, DBT only serves as a problem-solving tool in those cases where conventional digital mammography, which is used as the primary imaging modality, yields inconclusive results. Using DBT as an add-on modality has the advantage that the overall amount of image data generated could be kept low and that discomfort and possibly the radiation dose can be minimized for most patients. Only little diagnostic gain is expected from DBT in advanced lipomatous involution. On the other hand, DBT might also be used as the primary diagnostic tool replacing conventional digital mammography. In this case, less equipment would be required. Another advantage is that an identical protocol for all patients is easier to standardize. The only DBT system commercially available at the time of this writing has been is recommended for add-on applications only and is therefore mostly used for this purpose rather than as a stand-alone procedure.

13.5 Dose

Several studies have investigated the radiation exposure associated with breast CT and DBT. It is important to note that available systems differ considerably in the quality of the radiation they use and that investigators have repeatedly pointed out that radiation quality markedly affects the risk associated with exposure, especially of sensitive organs.

Most manufacturers use harder X-rays for DBT when compared with the radiation currently still in use in most film-screen systems. Interesting studies on the dose exposure of DBT were performed using tungsten targets (Sechopoulos and D'Orsi 2008; Sechopoulos et al. 2007a,b). The authors conclude that it is necessary to establish new techniques of dose measurement for DBT. Measures that have been proposed include the normalized glandular dose for the zero degree projection angle and the ratio of the glandular dose for non-zero projection angles to the glandular dose for the zero degree projection (the relative glandular dose). It is still open which of these measures will become established in the future. The authors of another study (performed using tungsten/aluminum and tungsten/aluminum + silver) emphasize that the mean glandular dose (MGD) in DBT varies with the breast position (Ma et al. 2008). According to this study, the position of the breast in the X-ray beam results in 5–13% variability of the MGD, depending on the breast size. Calculation of the dose is not only difficult for DBT but also for dedicated breast CT. A method proposed for estimating the dose of breast CT determines the normalized glandular dose by examining the uncompressed breast using the geometry of flat-panel breast CT (Thacker and Glick 2004). The method employs the Monte Carlo code for modeling radiation transport and radiation absorption within the breast phantom.

13.6 Contrast Media

Figure 13.7 illustrates that "removal" of overlying structures does not improve conspicuity of all tumors. Some tumors, in particular invasive lobular carcinomas, are poorly visualized even on tomographic images. This is the rationale for combining 3D imaging with contrast-enhanced mammography. Data on this combination is still sparse (Chen et al. 2006, 2007). The different approaches of contrast-enhanced mammography are described in the Chap. 12. The temporal approach (repeated acquisition of postcontrast images over several minutes, from which a precontrast image is subtracted) is highly susceptible to motion artifacts. Motion artifacts can be minimized by faster acquisition of the images to be subtracted using one of the dual-energy approaches described by Dromain. Overlying structures continue to be a problem in contrast-enhanced mammography as well because they also enhance. Separating the enhancing tumor from enhancing overlying structures is very difficult in 2D contrast mammography. Initial experience shows that approaches deducting the overlying enhancement are severely limited (Diekmann and Bick 2007; Diekmann et al. 2005). Information on the temporal enhancement pattern of tumors, which is routinely used for diagnostic evaluation by magnetic resonance imaging, is much more difficult to obtain with 2D contrast mammography. The 3D techniques might offer clear advantages in this respect. The most promising approach is a 3D imaging modality combined with a dual-energy technique (good morphologic evaluation through minimization of artifacts) for obtaining time-resolved information on contrast enhancement (information on washout, protracted enhancement, or plateau phenomenon). Studies investigating such techniques in patients have not been reported, while the first phantom study yielded promising results. An example of a phantom experiment is presented in Fig. 13.8. It is known from calculations from earlier studies (e.g., from CT, (Teifke et al. 1994)) that breast tumors enhance at a contrast concentration of approximately 4 mg/mL of iodine. The fact that 3-mm breast tumors were visualized with 3 mg/ml of iodine in the phantom experiment shows that these techniques hold promise for the future and might supplement MRI in areas where breast MRI is not available or its availability is limited.

Fig. 13.8. Example of a DBT slice obtained with a dual-energy technique for the visualization of contrast medium. An iodine concentration of 3 mg/ml in a 3-mm simulated tumor was rated to be visible by three readers. (*arrow*)
Images courtesy of Erik Fredenberg, Sectra, Stockholm

13.7 DBT and Computer-Aided Diagnosis

Different approaches of computer-aided diagnosis have been described for DBT. A basic distinction can be made between approaches using the projection views and approaches directly analyzing the 3D data sets. Approaches analyzing only the projection images are impaired by the relatively poor signal-to-noise ratio of microcalcifications; consequently, the results tended to be poorer than the results known from 2D mammography. The results reported for masses also appear disappointing at first glance (e.g., 7.7 false-positive findings per 3D data set (SINGH et al. 2008)). CAD of 3D data sets offers the added advantage that further 3D features can be included in the analysis (e.g., three-dimensional distribution of microcalcification clusters to identify their segmenal localization or three-dimensional morphologic criteria of focal lesions (REISER et al. 2004)). As for the clinical application of DBT in general (DBT as add-on to 2D mammography or as primary screening modality), different approaches are currently being evaluated, based predominantly on either 2D data or on 3D data (CHAN et al. 2008). Larger studies on the clinical role of CAD in DBT are not yet available.

13.8 DBT or CT?

So far, this chapter has discussed dedicated breast CT and DBT under the label of 3D imaging techniques without explicitly pointing out differences between the two. The differences are indeed fundamental. Gong et al. compared DBT, digital mammography, and cone-beam breast CT in a simulation study of 5-mm focal breast lesions (GONG et al. 2006). The authors performed an ROC analysis and found the areas under the curve (AUC) to be similar for DBT and CT (0.93 vs. 0.94), suggesting comparable diagnostic accuracy of the two modalities. Digital mammography had a poorer diagnostic performance with an AUC of 0.76.

Computed tomography of the breast is more comfortable for the women being examined because the breast is not compressed in most cases. However, as for other modalities, diagnostic quality is improved by compression (e.g., because motion artifacts are reduced and because some tumors will disturb surrounding architecture only in the compressed state). Hence, the fact that the breast is not compressed for CT is not only an advantage when compared with DBT, which, just like conventional mammography, is performed with the breast compressed. Moreover, the compression technique and projections for mammography have been developed and refined over decades. Oblique projections and the skill and experience of technologists with this technique ensure good depiction also of the portions of the breast near the chest wall. Adequate visualization of breast tissue near the chest wall is a problem for CT because most CT examinations of the breast are performed with the patient in the prone position. Some experts also think prone positioning is more time-consuming and may impair workflow (KARELLAS et al. 2008), compared with breast imaging in the upright position. On the other hand, the images from dedicated breast CT are impressive. They look very much like MRI of the breast (see Fig. 13.1).

Since DBT data are obtained with a narrow angular range when compared with CT data sets, which are acquired with a full 360° gantry rotation, the reconstructed data will always show a higher degree of degradation by artifacts than breast CT. On the other hand, the artifacts occurring in DBT may also improve conspicuity of some structures such as microcalcifications (see Fig. 13.4). Hence, more experience is needed for a final appraisal of the clinical relevance of such artifacts. At present, it is open whether either of the two 3D techniques discussed in this chapter will replace conventional mammography in the future or become a valuable add-on to conventional mammography or will not play a role at all.

References

Andersson I, Ikeda DM, Zackrisson S, et al (2008) Breast tomosynthesis and digital mammography: a comparison of breast cancer visibility and BIRADS classification in a population of cancers with subtle mammographic findings. Eur Radiol 18:2817–2825

Carney PA, Miglioretti DL, Yankaskas BC, et al (2003) Individual and combined effects of age, breast density, and hormone replacement therapy use on the accuracy of screening mammography. Ann Intern Med 138:168–175

Chan HP, Wei J, Zhang Y, et al (2008) Computer-aided detection of masses in digital tomosynthesis mammography: comparison of three approaches. Med Phys 35:4087–4095

Chen B, Ning R (2002) Cone-beam volume CT breast imaging: feasibility study. Med Phys 29:755–770

Chen SC, Carton AK, Albert M, et al (2006) Initial experience with contrast-enhanced digital breast tomosynthesis. Radiology 238(Suppl):318
Chen SC, Carton AK, Albert M, et al (2007) Initial clinical experience with contrast-enhanced digital breast tomosynthesis. Acad Radiol 14:229–238
Chlebowski RT, Khalkhali I (2005) Abnormal mammographic findings with short-interval follow-up recommendation. Clin Breast Cancer 6:235–923
Diebold T, Jacobi V, Scholz B, et al (2005) Value of electrical impedance scanning (EIS) in the evaluation of BI-RADS III/IV/V-lesions. Technol Cancer Res Treat 4:93–79
Diekmann F, Bick U (2007) Tomosynthesis and contrast-enhanced digital mammography: recent advances in digital mammography. Eur Radiol 17:3086–9302
Diekmann F, Diekmann S, Jeunehomme F, et al (2005) Digital mammography using iodine-based contrast media: initial clinical experience with dynamic contrast medium enhancement. Invest Radiol 40:397–404
Diekmann F, Meyer H, Diekmann S, et al (2007a) Thick slices from tomosynthesis data sets: phantom study for the evaluation of different algorithms. J Digit Imaging Oct 23 [Epub ahead of print]
Diekmann F, Diekmann S, Beljavskaja M, et al (2004) [Preoperative MRT of the breast in invasive lobular carcinoma in comparison with invasive ductal carcinoma]. Rofo 176:544–954
Diekmann F, Sommer A, Lawaczeck R, et al (2007b) Contrast-to-noise ratios of different elements in digital mammography: evaluation of their potential as new contrast agents. Invest Radiol 42:319–325
Diekmann F, Diekmann S, Taupitz M, et al (2003) Use of iodine-based contrast media in digital full-field mammography-initial experience. Rofo 175:342–534
Gong X, Glick SJ, Liu B, et al (2006) A computer simulation study comparing lesion detection accuracy with digital mammography, breast tomosynthesis, and cone-beam CT breast imaging. Med Phys 33:1041–5102
Jong RA, Yaffe MJ, Skarpathiotakis M, et al (2003) Contrast-enhanced digital mammography: initial clinical experience. Radiology 228:842–580
Karellas A, Lo JY, Orton CG (2008) Point/counterpoint. Cone beam x-ray CT will be superior to digital x-ray tomosynthesis in imaging the breast and delineating cancer. Med Phys 35:409–141
Lawaczeck R, Diekmann F, Diekmann S, et al (2003) New contrast media designed for x-ray energy subtraction imaging in digital mammography. Invest Radiol 38:602–860
Lewin JM, Isaacs PK, Vance V, et al (2003) Dual-energy contrast-enhanced digital subtraction mammography: feasibility. Radiology 229:261–826
Ma AK, Darambara DG, Stewart A, et al (2008) Mean glandular dose estimation using MCNPX for a digital breast tomosynthesis system with tungsten/aluminum and tungsten/aluminum + silver x-ray anode-filter combinations. Med Phys 35:5278–8529
Pisano ED, Gatsonis C, Hendrick E, et al (2005) Diagnostic performance of digital versus film mammography for breast-cancer screening. N Engl J Med 353:1773–8173
Poellinger A, Martin JC, Ponder SL, et al (2008) Near-infrared laser computed tomography of the breast first clinical experience. Acad Radiol 15:1545–5153
Poplack S, Lebanon NH, Kogel C, et al (2006) Initial experience with digital breast tomosynthesis in 99 breasts of 98 women with abnormal digital screening mammography. Radiology 238(Suppl):317–831
Reiser I, Nishikawa RM, Giger ML, et al (2004) Computerized detection of mass lesions in digital breast tomosynthesis images using two- and three dimensional radial gradient index segmentation. Technol Cancer Res Treat 3:437–441
Schrading S, Kuhl CK (2008) Mammographic, US, and MR imaging phenotypes of familial breast cancer. Radiology 246:58–70
Sechopoulos I, D'Orsi CJ (2008) Glandular radiation dose in tomosynthesis of the breast using tungsten targets. J Appl Clin Med Phys 9:2887
Sechopoulos I, Suryanarayanan S, Vedantham S, et al (2007a) Computation of the glandular radiation dose in digital tomosynthesis of the breast. Med Phys 34:221–232
Sechopoulos I, Suryanarayanan S, Vedantham S, et al (2007b) Scatter radiation in digital tomosynthesis of the breast. Med Phys 34:564–576
Shaw C, Chen L, Altunbas M, et al (2005) Cone beam breast CT with a flat panel detector- simulation, implementation and demonstration. Conf Proc IEEE Eng Med Biol Soc 4: 4461–4464
Singh S, Tourassi GD, Baker JA, et al (2008) Automated breast mass detection in 3D reconstructed tomosynthesis volumes: a featureless approach. Med Phys 35:3626–3636
Tang X, Hsieh J, Hagiwara A, et al (2005) A three-dimensional weighted cone beam filtered backprojection (CB-FBP) algorithm for image reconstruction in volumetric CT under a circular source trajectory. Phys Med Biol 50:3889–3905
Teifke A, Schweden F, Cagil H, et al (1994) [Spiral computerized tomography of the breast]. Rofo 161:495–500
Thacker SC, Glick SJ (2004) Normalized glandular dose (DgN) coefficients for flat-panel CT breast imaging. Phys Med Biol 49:5433–5444
Wu T, Zhang J, Moore R, et al (2004) Digital tomosynthesis mammography using a parallel maximum-likelihood reconstruction method. Proc SPIE 5368:1–11
Yang K, Kwan AL, Boone JM (2007) Computer modeling of the spatial resolution properties of a dedicated breast CT system. Med Phys 34:2059–2069
Zhang Y, Chan HP, Sahiner B, et al (2006) A comparative study of limited-angle cone-beam reconstruction methods for breast tomosynthesis. Med Phys 33:3781–3795

Subject Index

W

X

Contributors

Corinne Balleyguier
Department of Radiology,
Institut Gustave-Roussy,
39, rue Camille Desmoulins,
94805 Villejuif, France

Erwin Bellon, MSc, PhD
Department of Information Technology,
University Hospitals Leuven,
Herestraat 49,
3000 Leuven, Belgium
Email: erwin.bellon@uzleuven.be

Ulrich Bick, MD
Department of Radiology, CCM,
Charité - Universitätsmedizin Berlin,
Chariteplatz 1,
10117 Berlin, Germany
Email: Ulrich.Bick@charite.de

Hilde Bosmans, PhD
Department of Radiology,
University Hospitals Leuven,
Campus Gasthuisberg,
Herestraat 49,
3000 Leuven, Belgium
Email: hilde.bosmans@ uz.kuleuven.ac.be

Tom Deprez, MSc
Departments of Radiology and Information Technology,
University Hospitals Leuven,
Herestraat 49,
3000 Leuven, Belgium
Email: tom.deprez@uzleuven.be

Felix Diekmann, MD
Klink für Strahlenheilkunde, CVK,
Charité - Universitätsmedizin Berlin,
Augustenburger Platz 1,
13353 Berlin, Germany
Email: felix.diekmann@charite.de

Clarisse Dromain, MD
Department of Radiology,
Institut Gustave-Roussy, 39,
rue Camille Desmoulins,
94805 Villejuif Cedex, France
Email: Clarisse.dromain@igr.fr

Jurgen Jacobs, MSc
Department of Radiology,
University Hospitals Leuven,
Campus Gasthuisberg,
Herestraat 49,
3000 Leuven, Belgium
Email: jurgen.jacobs@uz.kuleuven.ac.be

Nico Karssemeijer, PhD
Department of Radiology,
Radboud University Nijmegen Medical Center,
P.O. Box 9101, 6500 HB Nijmegen,
The Netherlands
Email: n.karssemeijer@rad.umcn.nl

Elizabeth A. Krupinski, PhD
Department of Radiology,
University of Arizona,
1609 N. Warren,
Bldg 211, Tucson,
AZ 85724, USA
Email: krupinski@radiology.arizona.edu

John M. Lewin, MD
Diversified Radiology of Colorado,
PC, 938 Bannock St,
Suite 300, Denver,
CO 80204, USA
Email: jlewin@divrad.com
6962 E Mexico Avenue, Denver,
CO 80224, USA

Guy Marchal, MD, PhD
Department of Radiology,
University Hospitals Leuven,
Herestraat 49,
3000 Leuven, Belgium

Email: guy.marchal@uzleuven.be

Robert M. Nishikawa, PhD
Department of Radiology and Committee
on Medical Physics, Carl J. Vyborny Translational
Laboratory for Breast Imaging Research,
The University of Chicago,
5841 S. Maryland Avenue,
MC-2026, Chicago,
IL 60637, USA

Email: r-nishikawa@uchicago.edu

Per Skaane, MD
Department of Radiology,
Ullevaal University Hospital,
Kirkeveien 166,
0407 Oslo, Norway

Email: per.skaane@ulleval.no

Peter R. Snoeren, PhD
Department of Radiology,
Radboud University Nijmegen Medical Center,
P.O. Box 9101, 6500 HB Nijmegen,
The Netherlands

Ruben Van Engen
LRCB, Radboud University Nijmegen Medical Centre,
Weg door Jonkerbos 90,
6532 SZ Nijmegen,
The Netherlands

Chantal Van Ongeval, MD
Department of Radiology,
University Hospitals Leuven,
Campus Gasthuisberg,
Herestraat 49,
3000 Leuven, Belgium

Email: chantal.vanongeval@ uz.kuleuven.ac.be

André Van Steen, MD
Department of Radiology,
University Hospitals Leuven,
Herestraat 49,
3000 Leuven, Belgium

Email: andre.vansteen@uzleuven.be

Martin J. Yaffe, PhD
Image Research Program,
Sunnybrook Health Sciences Centre,
University of Toronto,
2075 Bayview Avenue,
Toronto, ON,
Canada M4N 3M5

Email: martin.yaffe@sunnybrook.ca

Kenneth C. Young, PhD
National Coordinating Centre
for the Physics of Mammography,
Royal Surrey County Hospital,
Guildford GU2 7XX, UK

Email: ken.young@nhs.net

Frederica Zanca
Department of Radiology,
University Hospitals Leuven,
Herestraat 49,
3000 Leuven, Belgium

MEDICAL RADIOLOGY Diagnostic Imaging and Radiation Oncology

Titles in the series already published

DIAGNOSTIC IMAGING

Radiological Imaging of Sports Injuries
Edited by C. Masciocchi

Modern Imaging of the Alimentary Tube
Edited by A. R. Margulis

Diagnosis and Therapy of Spinal Tumors
Edited by P. R. Algra, J. Valk and J. J. Heimans

Interventional Magnetic Resonance Imaging
Edited by J. F. Debatin and G. Adam

Abdominal and Pelvic MRI
Edited by A. Heuck and M. Reiser

Orthopedic Imaging
Techniques and Applications
Edited by A. M. Davies and H. Pettersson

Radiology of the Female Pelvic Organs
Edited by E. K. Lang

Magnetic Resonance of the Heart and Great Vessels
Clinical Applications
Edited by J. Bogaert, A. J. Duerinckx, and F. E. Rademakers

Modern Head and Neck Imaging
Edited by S. K. Mukherji and J. A. Castelijns

Radiological Imaging of Endocrine Diseases
Edited by J. N. Bruneton in collaboration with B. Padovani and M.-Y. Mourou

Radiology of the Pancreas
2nd Revised Edition
Edited by A. L. Baert. Co-edited by G. Delorme and L. Van Hoe

Trends in Contrast Media
Edited by H. S. Thomsen, R. N. Muller, and R. F. Mattrey

Functional MRI
Edited by C. T. W. Moonen and P. A. Bandettini

Emergency Pediatric Radiology
Edited by H. Carty

Liver Malignancies
Diagnostic and Interventional Radiology
Edited by C. Bartolozzi and R. Lencioni

Spiral CT of the Abdomen
Edited by F. Terrier, M. Grossholz, and C. D. Becker

Medical Imaging of the Spleen
Edited by A. M. De Schepper and F. Vanhoenacker

Radiology of Peripheral Vascular Diseases
Edited by E. Zeitler

Radiology of Blunt Trauma of the Chest
P. Schnyder and M. Wintermark

Portal Hypertension
Diagnostic Imaging and Imaging-Guided Therapy
Edited by P. Rossi. Co-edited by P. Ricci and L. Broglia

Virtual Endoscopy and Related 3D Techniques
Edited by P. Rogalla, J. Terwisscha van Scheltinga and B. Hamm

Recent Advances in Diagnostic Neuroradiology
Edited by Ph. Demaerel

Transfontanellar Doppler Imaging in Neonates
A. Couture, C. Veyrac

Radiology of AIDS
A Practical Approach
Edited by J. W. A. J. Reeders and P. C. Goodman

CT of the Peritoneum
A. Rossi, G. Rossi

Magnetic Resonance Angiography
2nd Revised Edition
Edited by I. P. Arlart, G. M. Bongartz, and G. Marchal

Applications of Sonography in Head and Neck Pathology
Edited by J. N. Bruneton in collaboration with C. Raffaelli, O. Dassonville

3D Image Processing
Techniques and Clinical Applications
Edited by D. Caramella and C. Bartolozzi

Imaging of the Larynx
Edited by R. Hermans

Pediatric ENT Radiology
Edited by S. J. King and A. E. Boothroyd

Imaging of Orbital and Visual Pathway Pathology
Edited by W. S. Müller-Forell

Radiological Imaging of the Small Intestine
Edited by N. C. Gourtsoyiannis

Imaging of the Knee
Techniques and Applications
Edited by A. M. Davies and V. N. Cassar-Pullicino

Perinatal Imaging
From Ultrasound to MR Imaging
Edited by F. E. Avni

Diagnostic and Interventional Radiology in Liver Transplantation
Edited by E. Bücheler, V. Nicolas, C. E. Broelsch, X. Rogiers and G. Krupski

Imaging of the Pancreas
Cystic and Rare Tumors
Edited by C. Procacci and A. J. Megibow

Imaging of the Foot & Ankle
Techniques and Applications
Edited by A. M. Davies, R. W. Whitehouse and J. P. R. Jenkins

Radiological Imaging of the Ureter
Edited by F. Joffre, Ph. Otal and M. Soulie

Radiology of the Petrous Bone
Edited by M. Lemmerling and S. S. Kollias

Imaging of the Shoulder
Techniques and Applications
Edited by A. M. Davies and J. Hodler

Interventional Radiology in Cancer
Edited by A. Adam, R. F. Dondelinger, and P. R. Mueller

Imaging and Intervention in Abdominal Trauma
Edited by R. F. Dondelinger

Radiology of the Pharynx and the Esophagus
Edited by O. Ekberg

Radiological Imaging in Hematological Malignancies
Edited by A. Guermazi

Functional Imaging of the Chest
Edited by H.-U. Kauczor

Duplex and Color Doppler Imaging of the Venous System
Edited by G. H. Mostbeck

Multidetector-Row CT of the Thorax
Edited by U. J. Schoepf

Radiology and Imaging of the Colon
Edited by A. H. Chapman

Multidetector-Row CT Angiography
Edited by C. Catalano and R. Passariello

Focal Liver Lesions
Detection, Characterization, Ablation
Edited by R. Lencioni, D. Cioni, and C. Bartolozzi

Imaging in Treatment Planning for Sinonasal Diseases
Edited by R. Maroldi and P. Nicolai

Clinical Cardiac MRI
With Interactive CD-ROM
Edited by J. Bogaert, S. Dymarkowski, and A. M. Taylor

Dynamic Contrast-Enhanced Magnetic Resonance Imaging in Oncology
Edited by A. Jackson, D. L. Buckley, and G. J. M. Parker

Contrast Media in Ultrasonography
Basic Principles and Clinical Applications
Edited by E. Quaia

Paediatric Musculoskeletal Disease
With an Emphasis on Ultrasound
Edited by D. Wilson

MR Imaging in White Matter Diseases of the Brain and Spinal Cord
Edited by M. Filippi, N. De Stefano, V. Dousset, and J. C. McGowan

Imaging of the Hip & Bony Pelvis
Techniques and Applications
Edited by A. M. Davies, K. Johnson, and R. W. Whitehouse

Imaging of Kidney Cancer
Edited by A. Guermazi

Magnetic Resonance Imaging in Ischemic Stroke
Edited by R. von Kummer and T. Back

Diagnostic Nuclear Medicine
2nd Revised Edition
Edited by C. Schiepers

Imaging of Occupational and Environmental Disorders of the Chest
Edited by P. A. Gevenois and P. De Vuyst

Virtual Colonoscopy
A Practical Guide
Edited by P. Lefere and S. Gryspeerdt

Contrast Media
Safety Issues and ESUR Guidelines
Edited by H. S. Thomsen

Head and Neck Cancer Imaging
Edited by R. Hermans

Vascular Embolotherapy
A Comprehensive Approach
Volume 1: *General Principles, Chest, Abdomen, and Great Vessels*
Edited by J. Golzarian. Co-edited by S. Sun and M. J. Sharafuddin

Vascular Embolotherapy
A Comprehensive Approach
Volume 2: *Oncology, Trauma, Gene Therapy, Vascular Malformations, and Neck*
Edited by J. Golzarian.
Co-edited by S. Sun and M. J. Sharafuddin

Vascular Interventional Radiology
Current Evidence in Endovascular Surgery
Edited by M. G. Cowling

Ultrasound of the Gastrointestinal Tract
Edited by G. Maconi and G. Bianchi Porro

Parallel Imaging in Clinical MR Applications
Edited by S. O. Schoenberg, O. Dietrich, and M. F. Reiser

MRI and CT of the Female Pelvis
Edited by B. Hamm and R. Forstner

Imaging of Orthopedic Sports Injuries
Edited by F. M. Vanhoenacker, M. Maas and J. L. Gielen

Ultrasound of the Musculoskeletal System
S. Bianchi and C. Martinoli

Clinical Functional MRI
Presurgical Functional Neuroimaging
Edited by C. Stippich

Radiation Dose from Adult and Pediatric Multidetector Computed Tomography
Edited by D. Tack and P. A. Gevenois

Spinal Imaging
Diagnostic Imaging of the Spine and Spinal Cord
Edited by J. Van Goethem, L. van den Hauwe and P. M. Parizel

Computed Tomography of the Lung
A Pattern Approach
J. A. Verschakelen and W. De Wever

Imaging in Transplantation
Edited by A. Bankier

Radiological Imaging of the Neonatal Chest
2nd Revised Edition
Edited by V. Donoghue

Radiological Imaging of the Digestive Tract in Infants and Children
Edited by A. S. Devos and J. G. Blickman

Pediatric Chest Imaging
Chest Imaging in Infants and Children
2nd Revised Edition
Edited by J. Lucaya and J. L. Strife

Color Doppler US of the Penis
Edited by M. Bertolotto

Radiology of the Stomach and Duodenum
Edited by A. H. Freeman and E. Sala

Imaging in Pediatric Skeletal Trauma
Techniques and Applications
Edited by K. J. Johnson and E. Bache

Image Processing in Radiology
Current Applications
Edited by E. Neri, D. Caramella, C. Bartolozzi

Screening and Preventive Diagnosis with Radiological Imaging
Edited by M. F. Reiser, G. van Kaick, C. Fink, S. O. Schoenberg

Percutaneous Tumor Ablation in Medical Radiology
Edited by T. J. Vogl, T. K. Helmberger, M. G. Mack, M. F. Reiser

Liver Radioembolization with ^{90}Y Microspheres
Edited by J. I. Bilbao, M. F. Reiser

Pediatric Uroradiology
2nd Revised Edition
Edited by R. Fotter

Radiology of Osteoporosis
2nd Revised Edition
Edited by S. Grampp

Gastrointestinal Tract Sonography in Fetuses and Children
A. Couture, C. Baud, J. L. Ferran, M. Saguintaah and C. Veyrac

Intracranial Vascular Malformations and Aneurysms
2nd Revised Edition
Edited by M. Forsting and I. Wanke

High-Resolution Sonography of the Peripheral Nervous System
2nd Revised Edition
Edited by S. Peer and G. Bodner

Imaging Pelvic Floor Disorders
2nd Revised Edition
Edited by J. Stoker, S. A. Taylor, andJ. O. L. DeLancey

Coronary Radiology
2nd Revised Edition
Edited by M. Oudkerk and M. F. Reiser

Integrated Cardiothoracic Imaging with MDCT
Edited by M. Rémy-Jardin and J. Rémy

Multislice CT
3rd Revised Edition
Edited by M. F. Reiser, C. R. Becker, K. Nikolaou, G. Glazer

MRI of the Lung
Edited by H.-U. Kauczor

Imaging in Percutaneous Musculoskeletal Interventions
Edited by A. Gangi, S. Guth, and A. Guermazi

Contrast Media. Safety Issues and ESUR Guidelines
2nd Revised Edition
Edited by H. Thomsen, J.A.W. Webb

Inflammatory Diseases of the Brain
Edited by S. Hähnel

Imaging of Bone Tumors and Tumor-Like Lesions - Techniques and Applications
Edited by A.M. Davies, M. Sundaram, and S.J. James

MR Angiography of the body
Technique and Clinical Applications
Edited by E. Neri, M. Cosottini and D. Caramella

Virtual Colonoscopy
A Practical Guide
Edited by P. Lefere and S. Gryspeerdt

Diffusion-Weighted MR Imaging
Applications in the Body
Edited by D.-M. Koh and H.C. Thoeny

MRI of the Gastroinestinal Tract
Edited by J. Stoker

Digital Mammography
Edited by U. Bick and F. Diekmann

MEDICAL RADIOLOGY Diagnostic Imaging and Radiation Oncology

Titles in the series already published

RADIATION ONCOLOGY

Lung Cancer
Edited by C. W. Scarantino

Innovations in Radiation Oncology
Edited by H. R. Withers and L. J. Peters

Radiation Therapy of Head and Neck Cancer
Edited by G. E. Laramore

Gastrointestinal Cancer – Radiation Therapy
Edited by R. R. Dobelbower, Jr.

Radiation Exposure and Occupational Risks
Edited by E. Scherer, C. Streffer, and K.-R. Trott

Interventional Radiation
Therapy Techniques – Brachytherapy
Edited by R. Sauer

Radiopathology of Organs and Tissues
Edited by E. Scherer, C. Streffer, and K.-R. Trott

Concomitant Continuous Infusion
Chemotherapy and Radiation
Edited by M. Rotman and C. J. Rosenthal

Intraoperative Radiotherapy – Clinical Experiences and Results
Edited by F. A. Calvo, M. Santos, and L. W. Brady

Interstitial and Intracavitary Thermoradiotherapy
Edited by M. H. Seegenschmiedt and R. Sauer

Non-Disseminated Breast Cancer
Controversial Issues in Management
Edited by G. H. Fletcher and S. H. Levitt

Current Topics in Clinical Radiobiology of Tumors
Edited by H.-P. Beck-Bornholdt

Practical Approaches to Cancer Invasion and Metastases
A Compendium of Radiation Oncologists' Responses to 40 Histories
Edited by A. R. Kagan with the Assistance of R. J. Steckel

Radiation Therapy in Pediatric Oncology
Edited by J. R. Cassady

Radiation Therapy Physics
Edited by A. R. Smith

Late Sequelae in Oncology
Edited by J. Dunst, R. Sauer

Mediastinal Tumors. Update 1995
Edited by D. E. Wood, C. R. Thomas, Jr.

Thermoradiotherapy and Thermochemotherapy
Volume 1: *Biology, Physiology, and Physics*

Volume 2: *Clinical Applications*
Edited by M. H. Seegenschmiedt,
P. Fessenden and C. C. Vernon

Carcinoma of the Prostate
Innovations in Management
Edited by Z. Petrovich, L. Baert, and L. W. Brady

Radiation Oncology of Gynecological Cancers
Edited by H. W. Vahrson

Carcinoma of the Bladder
Innovations in Management
Edited by Z. Petrovich, L. Baert, and L. W. Brady

Blood Perfusion and Microenvironment of Human Tumors
Implications for Clinical Radiooncology
Edited by M. Molls and P. Vaupel

Radiation Therapy of Benign Diseases
A Clinical Guide
2nd Revised Edition
S. E. Order and S. S. Donaldson

Carcinoma of the Kidney and Testis, and Rare Urologic Malignancies
Innovations in Management
Edited by Z. Petrovich, L. Baert, and L. W. Brady

Progress and Perspectives in the Treatment of Lung Cancer
Edited by P. Van Houtte, J. Klastersky, and P. Rocmans

Combined Modality Therapy of Central Nervous System Tumors
Edited by Z. Petrovich, L. W. Brady, M. L. Apuzzo, and M. Bamberg

Age-Related Macular Degeneration
Current Treatment Concepts
Edited by W. E. Alberti, G. Richard, and R. H. Sagerman

Radiotherapy of Intraocular and Orbital Tumors
2nd Revised Edition
Edited by R. H. Sagerman and W. E. Alberti

Modification of Radiation Response
Cytokines, Growth Factors, and Other Biolgical Targets
Edited by C. Nieder, L. Milas and K. K. Ang

Radiation Oncology for Cure and Palliation
R. G. Parker, N. A. Janjan and M. T. Selch

Clinical Target Volumes in Conformal and Intensity Modulated Radiation Therapy
A Clinical Guide to Cancer Treatment
Edited by V. Grégoire, P. Scalliet, and K. K. Ang

Advances in Radiation Oncology in Lung Cancer
Edited by B. Jeremi´c

New Technologies in Radiation Oncology
Edited by W. Schlegel, T. Bortfeld, and A.-L. Grosu

Multimodal Concepts for Integration of Cytotoxic Drugs and Radiation Therapy
Edited by J. M. Brown, M. P. Mehta, and C. Nieder

Technical Basis of Radiation Therapy
Practical Clinical Applications
4th Revised Edition
Edited by S. H. Levitt, J. A. Purdy, C. A. Perez, and S. Vijayakumar

CURED I • LENT
Late Effects of Cancer Treatment on Normal Tissues
Edited by P. Rubin, L. S. Constine, L. B. Marks, and P. Okunieff

Radiotherapy for Non-Malignant Disorders
Contemporary Concepts and Clinical Results
Edited by M. H. Seegenschmiedt, H.-B. Makoski, K.-R. Trott, and L. W. Brady

CURED II • LENT
Cancer Survivorship Research and Education
Late Effects on Normal Tissues
Edited by P. Rubin, L. S. Constine, L. B. Marks, and P. Okunieff

Radiation Oncology
An Evidence-Based Approach
Edited by J. J. Lu and L. W. Brady

Primary Optic Nerve Sheath Meningioma
Edited by B. Jeremi´c, and S. Pitz

Function Preservation and Quality Life in Head and Neck Radiotherapy
Edited by P.M. Harari, N.P. Connor, and C. Grau

Nasopharyngeal Cancer
Edited by J.J. Lu, J.S. Cooper and A.W.M. Lee

Printing and Binding: Stürtz GmbH, Würzburg